Nonlinear and Hybrid Systems
in Automotive Control

Rolf Johansson and Anders Rantzer (Eds.)

Nonlinear and Hybrid Systems in Automotive Control

With 160 Figures

Rolf Johansson, MD, PhD
Anders Rantzen, PhD
Department of Automatic Control, Lund University, PO Box 118, SE-221 00, Lund, Sweden

Library of Congress Cataloging-in-Publication Data
Nonlinear and hybrid systems in automotive control / Rolf Johansson and Anders Rantzer (eds.).
p.cm.
Includes bibliographical references and index.
1.Automotive computers. 2. Automobiles—Automatic control. 3. Adaptive control systems. I. Johansson, Rolf. II. Rantzer, Anders, 1963
TL272.53.N65 2002
629.2'7—dc21

First published 2003 by Springer-Verlag London Limited

This edition published by:	SAE International
	400 Commonwealth Drive
	Warrendale, PA 15096-0001,USA.
	Phone: (724) 776-4841
	Fax: (724) 776-5760
	E-mail: publications@sae.org
	http://www.sae.org

© Springer-Verlag London 2003

ISBN 0-7680-1137-X

All rights reserved.

Permission to photocopy for internal or personal use, or the internal or personal use of specific clients, is granted by SAE for libraries and other users registered with the Copyright Clearance Centre (CCC), provided that the base fee of $0.50 per page is paid directly to CCC, 222 Rosewood Dr., Danvers, MA 01923. Special requests should be addressed to the SAE Publications Group.

0-7680-1137-X
SAE Order No R-348

The publishers are not responsible for any statement made in this publication. Data, discussion, and conclusions developed by the Authors are for information only and are not intended for use without independent substantiating investigation on the part of potential users. Opinions expressed are those of the Authors and are not necessarily those of the publishers.

Printed and bound in the United States of America.

Preface

In the past decade, there has been an enormous surge of activity in automotive control, both in terms of research and development and in terms of capturing the imagination of the general public as to its seemingly endless and diverse opportunities.

This period has been accompanied by a technological maturation as well, for example, in the three-way catalytic conversion, in anti-lock brake system (ABS) devices and in navigational aids. Several areas of automotive control have now become well established. As of the writing of this book, the field is on the verge of a new explosion in areas of growth, involving active suspension systems, new engine designs for improved fuel efficiency and emission standards.

Given the state of maturity of the subject and the diversity of students who study this material, we felt the need for a book that presents an overview of the research area and collects contributions from a number of leading scientists in the field.

This book on automotive control contains contributions on:

- vehicle dynamics and active suspension;

- anti-lock braking systems (ABS), brake dynamics, friction and wheel-slip modeling;

- combustion-engine control;

- system theory and hybrid system analysis for application in automotive control

The contributions represent a panorama of systems and control research issues in the application context of automotive control. Moreover, the contributions present important new efforts to solve theoretical problems arising from problems in automotive control. Important common themes are nonlinear observer and system theory for hybrid systems—*i.e.*, systems involving continuous-time dynamics and logic switching. In addition to their theoretical capacity, the authors represent important cooperation projects involving academic teams and automotive companies.

Authors Mark D. Donahue and J. Karl Hedrick, University of California at Berkeley—Prof. Hedrick also being Director of the PATH program, Partners For Advanced Transit And Highways—present an implementation of active suspension, preview controller for improved ride comfort. A fully active suspension and preview control is utilized to improve ride comfort, which allows increased travelling speed over rough terrain. Previous research is extended and the relevant implementation issues are addressed. Specifically, the methodology of model predictive control has been applied to explicitly address suspension saturation constraints, suspension rate limits, and other system non-linearities. For comparison, the following non-preview controllers were implemented: a skyhook damping controller, a linear quadratic regulator (LQR), and a mock passive suspension controller. Particular attention is given to the hydraulic actuator force controller that tracks commands generated by higher-level controllers. The complete system has been successfully realized on a US military high-mobility multipurpose wheeled vehicle (HMMWV) using a commercially available microprocessor platform. Experimental results show that the power absorbed by the driver is decreased by more than half, significantly improving ride comfort.

Fu-Cheng Wang and Malcolm C. Smith, Cambridge University, present work on active and passive suspension control for vehicle dive and squat. Performance capabilities of passive and active vehicle suspension systems are examined from a mechanical networks point of view. It is known that the reduction of effects of road disturbances is a conflicting requirement with the reduction of effects of inertial loads in a quarter-car model when passive control is used, but not with active control of suitable structure. The extension of these ideas to a half-car trailing-arm model is considered. It is shown that the choice of suspension geometry does not remove the basic trade-offs for passive suspensions. An active control structure to allow the road and load transmission paths to be optimised independently will be presented. The design approach is to be applied to a non-linear trailing-arm vehicle model to demonstrate good anti-dive and anti-squat behavior together with a soft ride in response to road disturbances. The performance of the controller is demonstrated using the multi-body simulation code *AutoSim*.

Johan Bengtsson and Rolf Johansson of Lund University, and Agneta Sjögren, Volvo Technical Development, present their work on adaptive cruise control (ACC) and modeling of drivers' longitudinal behavior. In the last few years, many vehicle manufacturers have introduced advanced driver support in some of their automobiles. One of those new features for driver support is adaptive cruise control (ACC), which extends the conventional cruise control system to control of relative speed and distance to other vehicles. In order to design an ACC controller, it is suitable to have a model on drivers' behavior. The approach to find dynamical models of the drivers' behavior is to use system identification. Basic data analysis was made by means of system identification methodology, and several models of drivers' longitudinal behavior are proposed, including both linear regression models and subspace-based models. In various situations, detection for when a driver's behavior changes or deviates from the normal is useful. To that purpose, a GARCH (generalized autoregressive conditional heteroskedasticity) model was used to model the driver in situations such as arousal.

The next four chapters all originate from the study of ABS control within the EU funded project *Heterogenous Hybrid Control (H2C)* involving DaimlerChrysler and SINTEF together with university partners from Glasgow and Lund. Authors Jens Kalkkuhl, Jens Lüdemann of DaimlerChrysler and Tor Arne Johansen, Norwegian University of Science and Technology, provide a contribution on nonlinear adaptive backstepping with estimator resetting using multiple observers A multiple model-based observer/estimator for the estimation of parameters is used to reset the parameter estimation in a conventional Lyapunov-based nonlinear adaptive controller. Transient performance can be improved without increasing the gain of the controller or estimator. This allows performance to be tuned without compromising robustness and sensitivity to noise and disturbances. The advantages of the scheme are demonstrated in an automotive wheel slip controller.

Another contribution involving the DaimlerChrysler sphere and Lund University is provided by Stefan Solyom and Anders Rantzer. The ABS is an important component of a complex steering system for the modern car. Most of ABS controllers available on the market are table-based on-off controllers. In the latest generation of "brake by wire" systems, the performance requirements on the ABS are much higher. The control objective shifts to maintain a specified tire slip for each wheel during braking. The authors propose a design model and, based on that, a gain-scheduled controller that regulates the tire-slip. Simulation and test results are presented.

Kenneth J. Hunt, Yongji Wang, Michael Schinkel and Tilmann Schmitt-Hartmann, University of Glasgow, present their work on controller design for hybrid systems using simultaneous D-stabilisation and its application to ABS. In recent years, hybrid systems have been widely studied. Many controller design approaches are based on a state space plant model and use full state feedback to satisfy certain stability conditions, for example, the existence of a common Lyapunov function. Pole assignment and LMI-based controller design techniques have also been used. Recent progress is reported on a new controller design method called simultaneous D stabilisation and strong simultaneous D-stabilisation, which can deal with the multiple plant requirement resulting from hybrid systems. Simulation results with a DaimlerChrysler test vehicle for ABS are also presented.

Another report on ABS brakes considers wheel slip control using gain-scheduled constrained LQR. It is contributed by by Idar Petersen from SINTEF, Tor A. Johansen, Norwegian University of Science and Technology, and Jens Kalkkuhl and Jens Lüdemann of DaimlerChrysler. A wheel slip controller for ABS brakes is formulated using an explicit constrained LQR design. The controller gain matrices are designed and scheduled on the vehicle speed based on local linearizations. A Lyapunov function for the nonlinear control system is derived using the Riccati equation solution in order to prove stability and robustness with respect to uncertainty in the road/tire friction characteristic. Experimental results from a test vehicle with electromechanical brake actuators and brake-by-wire show that high performance and robustness are achieved.

Carlos Canudas-de-Wit of Laboratoire d'Automatique de Grenoble P Tsiotras, Georgia Institute of Technology, J. Yi, and R. Horowitz, University of California, Berkeley, present a comprehensive contribution on friction tire/road modeling, estimation, and optimal braking control. A series of results concerning the problem of modeling and estimation of contact road/tire friction are presented. Also, some aspects related to the problem of optimal and emergency braking control are discussed. The modeling part discusses a new dynamic friction force model for the longitudinal road/tire interaction for wheeled ground vehicles. The model is based on a dynamic friction model developed previously for contact-point friction problems, called the *LuGre model*. Next, the problem of tire-road friction estimation using only angular wheel velocity which cannot always been computed from actual sensors is treated. Tire forces information is relevant to problems like: optimization of ABS, traction system, diagnostic of the road friction conditions, etc. These results may suggest alternative traction control methodologies, other than the current ones based on the use of tracking of the "optimal" slip coefficient using, for example, sliding mode control. A control scheme for emergency braking of vehicles. The controller utilizes estimated state feedback control to achieve a near maximum deceleration.

Elbert Hendricks, Technical University of Denmark, presents work on nonlinear observer control of internal combustion engines with EGR. The increased requirements of engine control systems with respect to accuracy, functionality, and emission levels have led to a new generation of control strategies. In contrast to earlier systems, these control systems are based on dynamic physical engine models (Mean Value Engine Models, MVEMs) and nonlinear estimation. In fact, the new second generation Engine Control Units (ECUs) are just going into production and represent the first mass market application of nonlinear observers. One purpose of this paper is to review critically the design principles behind some of the newest ECUs. An attempt will also be made to indicate the general direction of development of the newest systems and possible new applications for this methodology.

Authors Andrea Balluchi from PARADES, Luca Benvenuti, Università di Roma "La Sapienza", Marika Domenica Di Benedetto, Università dell'Aquila, and Alberto L. Sangiovanni-Vincentelli from the University of California at Berkeley provide a contribution on idle-speed controller synthesis using an assume-guarantee approach. The goal of idle control for automotive engines is to maintain the engine speed within a given range, robustly with respect to load torque disturbances acting on the crankshaft. Mean value models have been used in the past to design idle control algorithms. However, the behavior of the torque generation process and the dynamics of the power train are not captured with enough accuracy to guarantee that the idle control specifications as given by car makers are met. A cycle-accurate hybrid model can be used to overcome these obstacles. To tackle the complexity of the controller design, the system is decomposed in three parts. For each part in isolation, a control law is derived for a simplified model, assuming that the other parts can be controlled to yield appropriate inputs. The overall control strategy is then applied to the system. Hence, the correct interaction of the feedback loops is formally verified using an assume-guarantee approach, to ensure that the behavior of the controlled system meets the given specifications.

Dirk Förstner and Jan Lunze, Technical University Hamburg-Harburg, present work on fault diagnosis of switched nonlinear dynamical systems with application to a diesel injection system. For the on-line monitoring of safety- and emission-relevant parts of automotive systems diagnostic methods are needed that take into account the parameter uncertainties and nonlinearities of automotive systems and the real-time restrictions of on-board diagnosis. This paper concerns the diagnosis of switched nonlinear dynamical systems, which is applied to a diesel injection system. A model–based diagnostic method is presented that processes the events generated by the output signals when passing predefined thresholds. For nominal and faulty behavior, a model in form of a nondeterministic automaton is set up. The diagnostic algorithm uses this model to compare the measured event sequences with the nominal or faulty behavior in order to determine faults from inconsistencies. The trade–off between the complexity of the algorithm and the accuracy of the diagnostic result can be found by varying the model depth, which is the number of recent events stored in the model state. The successful application of this method is demonstrated for the power stage of a common-rail Diesel injection system.

Authors Giovanni Fiengo, Luigi Glielmo, and Stefania Santini review modeling of three-way catalytic converter (TWC). New regulations for emission control require the improvement of the system composed of a spark ignition internal combustion engine and TWC. In particular an important problem is to minimize harmful emissions during the transient warm-up phase where the TWC is not working yet and, hence, a large amount of pollutants are emitted in the air. Towards this goal a dynamical thermo-chemical TWC model simple enough for the design and test of warm-up control strategies is presented. The model is obtained through an asymptotic approximation of a more detailed model—i.e., by letting the adsorption coefficient between gas and substrate tend to infinity. The model has been identified and validated with experimental data.

Magnus Gäfvert, Karl-Erik Årzén, Bo Bernhardsson, and Lars Malcolm Pedersen, Lund University, present their work on the control of gasoline direct injection (GDI) engines using torque feedback. A novel approach to the control of a GDI engine is presented. The controller consists of a combination of subcontrollers, where torque feedback is a central part. The subcontrollers are, with a few exceptions, designed using simple linear feedback and feedforward control-design methods, in contrast to traditional table-based engine control. A silent extremum-controller is presented. It is used to minimize the fuel consumption in stratified mode. The controller has been evaluated with good results on the European driving cycle using a dynamic simulation model.

Authors Per Tunestål, Jan-Ola Olsson, and Bengt Johansson, Lund University, present work on a novel combustion engine principle known as homogeneous charge compression ignition (HCCI). The HCCI engine, with its excellent potential for high efficiency and low NO_X emissions, is investigated from a controls perspective. Combustion timing, *i.e.*, where in the thermodynamic cycle combustion takes place, is identified as the most challenging problem with HCCI engine control. A number of different means for controlling combustion timing are suggested, and results using a dual-fuel solution are presented. This solution uses two fuels with different ignition characteristics to control the time of auto-ignition. Cylinder pressure measurement is suggested for feedback of combustion timing. A simple net-heat release algorithm is applied to the measurements, and the crank angle of 50 % burnt is extracted. Open-loop instability is detected in some high-load regions of the operating range. This phenomenon is explained by positive feedback between the cylinder wall heating and ignition timing processes. Closed-loop performance is hampered by time delays and model uncertainties. This problem is particularly pronounced at operating points that are open-loop unstable.

Authors L. Berardi, E. De Santis, M. D. Di Benedetto, G. Pola, Università dell'Aquila, and University of California, Berkeley, treat the problem of approximations of maximal controlled safe sets for hybrid systems. In the determination of the *"maximal safe set"* for a hybrid system, the core problem lies in the computation of a maximal controlled invariant set contained in a constraint set for a continuous time dynamical system. In the case of a linear system, we propose a procedure that, on the basis of a controlled invariant set for the exponential discretization of the continuous-time system, leads to an arbitrarily good approximation of the maximal controlled invariant set for the continuous-time system. The approximating set has the interesting property that the constraints can be satisfied by means of a piecewise constant control. An example of an application of the proposed procedure to idle control is illustrated.

A Hamiltonian formulation of bond graphs is provided by Goran Golo, Arjan van der Schaft, Peter C. Breedveld, University of Twente, and Bernhard M. Maschke, Université Claude Bernard Lyon-1. This paper deals with the mathematical formulation of bond graphs. It is proven that the power continuous part of bond graphs, the junction structure, can be associated with a Dirac structure and that the equations describing a bond graph model correspond to a port Hamiltonian system. The conditions for well-posedness of the modelled system are given, and representations suitable for numerical simulation are derived. The index of the representations is analysed and sufficient conditions for computational efficiency are given. The results are applied to some models arising in automotive applications.

Authors S. Pettersson and B. Lennartson, Chalmers University of Technology, consider stability analysis of hybrid systems and its application to gearbox control. This paper includes an application consisting of an automatic gearbox and cruise controller, which naturally is modelled as a hybrid system including state jumps in the continuous state of the controller. Motivated by this application, we extend existing stability results to include state jumps as well. The proposed stability results are based on Lyapunov techniques. The search for the (piecewise quadratic) Lyapunov functions is formulated as a linear matrix inequality (LMI) problem. It is shown how the proposed stability analysis is applied to the automatic gearbox and cruise controller.

Authors W. P. M. H. Heemels (Eindhoven University of Technology), M. K. Çamlıbel (University of Groningen), A. J. van der Schaft (University of Twente), and J. M. Schumacher (Tilburg University) provide a contribution on the existence and uniqueness of solution trajectories to hybrid dynamical systems. This paper studies the fundamental system-theoretic property of well-posedness for several classes of hybrid dynamical systems. Hybrid systems are characterized by the presence and interaction of continuous dynamics and discrete actions. Many different description formats have been proposed in recent years for such systems; some proposed forms are quite direct, others lead to rather indirect descriptions. The more indirect a description form is, the harder it becomes to show that solutions are well-defined. This contribution intends to provide a survey on the available results on existence and uniqueness of solutions for given initial conditions in the context of various description formats for hybrid systems.

Several of the contributions have been presented in the meetings *Nonlinear and Adaptive Control Network (NACO 2) Workshop on Automotive Control*, Lund, May 18–19, 2001 or *Hybrid Control and Automotive Applications*, Lund, May 5–6, 2000 and Berlin, June 7–8, 2001. The editors are grateful to Ms. Eva Schildt for skilful administration of the Lund meetings.

On behalf of the research project networks, we would like thank the EU Commission for financial support granted to the research projects NACO2 and H2C. Finally, the editors would like to thank Mr. Leif Andersson and the Springer-Verlag editors for valuable editorial advice and help.

In reading this book, we hope that the reader will feel the same excitement that we do about the technological and social prospects for the field of automotive control and the elegance and challenges of the underlying theory.

Rolf Johansson   Anders Rantzer

Contents

1. **Implementation of an Active Suspension, Preview Controller for Improved Ride Comfort** 1
   M. D. Donahue   J. K. Hedrick
   1.1 Introduction 2
   1.2 Controller Structure 3
   1.3 Force Tracking Controller 3
   1.4 Higher-level Controllers 8
   1.5 Preview Information 11
   1.6 Experimental Results 13
   1.7 Conclusions 15
   1.8 References 18
   1.A HMMWV Equipment 19
   1.B Test Track 21
   1.C Nomenclature 21

2. **Active and Passive Suspension Control for Vehicle Dive and Squat** 23
   Fu-Cheng Wang   M. C. Smith
   2.1 Introduction 24
   2.2 Limitations Imposed by Passivity in Vehicle Suspension Design 24
   2.3 Suspension Geometry in the Half-car Trailing-arm Model   27
   2.4 Active Suspension Design for Independence of Disturbance Responses 33
   2.5 Active Suspension Design for the Trailing-arm Model .. 36
   2.6 References 38

3. **Modeling of Drivers' Longitudinal Behavior** 41
   J. Bengtsson   R. Johansson   A. Sjögren
   3.1 Introduction 42
   3.2 Material and Methods 42
   3.3 Results and Validation 48
   3.4 Conclusion 52
   3.5 References 53

4. **Nonlinear Adaptive Backstepping with Estimator Resetting using Multiple Observers** 59
   J. Kalkkuhl   T. A. Johansen   J. Lüdemann
   4.1 Introduction 60
   4.2 Nonlinear Adaptive Backstepping 62
   4.3 Stability Analysis of Parameter Resetting 65
   4.4 Multiple Model Observer (MMO) 71

	4.5	A Second Order Benchmark System	76
	4.6	Wheel Slip Control	77
	4.7	Conclusions	82
	4.8	References	82
5.	**ABS Control—A Design Model and Control Structure**		85
	S. Solyom A. Rantzer		
	5.1	Introduction	86
	5.2	The Design Problem	87
	5.3	The Control Structure	90
	5.4	Simulation and Experimental Results	93
	5.5	Conclusion	95
	5.6	References	95
6.	**Controller Design for Hybrid Systems using Simultaneous D-stabilisation and its Application to Anti-lock Braking Systems (ABS)**		97
	K. J. Hunt Yongji Wang M. Schinkel T. Schmitt-Hartmann		
	6.1	Introduction	98
	6.2	Constraints for SSP with D-stable Regions	101
	6.3	Constraints for SSSP with D-stable Regions	109
	6.4	Numerical Solution Techniques for the SSP and SSSP	111
	6.5	Design Example and Application to ABS Control	114
	6.6	Conclusions	119
	6.7	References	121
7.	**Wheel Slip Control in ABS Brakes using Gain-scheduled Constrained LQR**		125
	I. Petersen T. A. Johansen J. Kalkkuhl J. Lüdemann		
	7.1	Introduction	126
	7.2	Modelling	126
	7.3	Control Problem	130
	7.4	Gain-scheduled LQRC Controller Design and Analysis	130
	7.5	Implementation	138
	7.6	Experimental Results	140
	7.7	Discussion and Conclusions	140
	7.8	References	144
	7.A	Appendix—Details of Proof	145
8.	**Friction Tire/Road Modeling, Estimation and Optimal Braking Control**		147
	C. Canudas-de-Wit P. Tsiotras X. Claeys J. Yi R. Horowitz		
	8.1	Introduction	148
	8.2	Road/Tire Contact Friction Models	150
	8.3	Higher-dimensional Models	165
	8.4	Road/Tire Friction Observers	175
	8.5	General Observer Design	177
	8.6	Optimal Braking	184
	8.7	Observed-based Emergency Braking Control	195

	8.8	Conclusions	205
	8.9	References	206
9.	**Nonlinear Observer Control of Internal Combustion Engines with EGR**		211
	E. Hendricks		
	9.1	Introduction	212
	9.2	Torque Control Feedforward Observer	212
	9.3	Closed-loop Observer	217
	9.4	Possible Improvements	220
	9.5	Conclusions	224
	9.6	Nomenclature	225
	9.7	References	225
10.	**Idle Speed Control Synthesis using an Assume-guarantee Approach**		229
	A. Balluchi L. Benvenuti M.D. Di Benedetto		
	A.L. Sangiovanni-Vincentelli		
	10.1	Introduction	230
	10.2	Plant Hybrid Model	232
	10.3	Idle Speed Control Design	234
	10.4	Closed-loop System Behavior Verification	237
	10.5	Conclusions	242
	10.6	References	243
11.	**Fault Diagnosis of Switched Nonlinear Dynamical Systems with Application to a Diesel Injection System**		245
	D. Förstner J. Lunze		
	11.1	Introduction	246
	11.2	Discrete-event Behaviour of Switched Nonlinear Systems	249
	11.3	Requirements on Models Used for Diagnosis	251
	11.4	Consistency-based Diagnosis	252
	11.5	Representation of Quantised Systems by means of Automata	254
	11.6	A Diagnostic Algorithm for Quantised Systems	256
	11.7	Automotive Application: Fault Diagnosis of a Power Stage	257
	11.8	Conclusions	260
	11.9	References	260
12.	**Modelling the Dynamic Behaviour of Three-way Catalytic Converters during the Warm-up Phase**		263
	G. Fiengo L. Glielmo S. Santini		
	12.1	Motivations	264
	12.2	Basics of the TWC	265
	12.3	A Two-time-scale Infinite-adsorption Model of TWC	268
	12.4	Machine Learning for Reaction Kinetics	275
	12.5	A Phenomenological Model of TWC	278
	12.6	Conclusions	283
	12.A	Appendix—Mathematical Reduction Procedure	283

13. **Control of Gasoline Direct Injection Engines using Torque Feedback: A Simulation Study** 289
   M. Gäfvert   K.-E. Årzén   B. Bernhardsson   L. M. Pedersen
   13.1 Introduction 290
   13.2 GDI Engines 291
   13.3 The GDI Benchmark 292
   13.4 The GDI Engine Model 293
   13.5 Core Control Strategies 296
   13.6 Controller Designs 300
   13.7 Core Controller Results 307
   13.8 A Complete Engine Management System 309
   13.9 Full Benchmark Results and Comparisons 312
   13.10 Torque Estimation and Sensing 314
   13.11 Conclusions 316
   13.12 References 317

14. **Closed-loop Combustion Control of HCCI Engines** 321
   P. Tunestål   J.-O. Olsson   B. Johansson
   14.1 Homogeneous Charge Compression Ignition (HCCI) ... 322
   14.2 Closed-loop Control of Ignition Timing 324
   14.3 Closed-Loop Combustion Control of HCCI Engines 326
   14.4 Conclusion and Discussion 332
   14.5 References 332

15. **Approximations of Maximal Controlled Safe Sets for Hybrid Systems** 335
   L. Berardi   E. De Santis   M. D. Di Benedetto   G. Pola
   15.1 Introduction 336
   15.2 Definition and Properties of Controlled Safe Sets 336
   15.3 Inner Approximations of the Maximal Controlled Invariant Set 339
   15.4 An Example of Application 345
   15.5 Conclusions 348
   15.6 References 348

16. **Hamiltonian Formulation of Bond Graphs** 351
   G. Golo   A. van der Schaft   P. C. Breedveld   B. M. Maschke
   16.1 Introduction 352
   16.2 Bond Graph Models 352
   16.3 Dirac Structures 354
   16.4 Geometric Formulation of a Bond Graphs 355
   16.5 Well-posedness and Equation Suitable for Numerical Simulation 358
   16.6 Index of System 364
   16.7 Example 366
   16.8 Conclusion 371
   16.9 References 371

17. **Stability Analysis of Hybrid Systems —A Gearbox Application** 373
    S. Pettersson   B. Lennartson
    17.1 Introduction 374
    17.2 Application and Hybrid Model 375
    17.3 Exponential Stability 378
    17.4 Linear Matrix Inequalities 380
    17.5 Stability of the Gearbox Application 385
    17.6 Conclusions 387
    17.7 References 387
18. **On the Existence and Uniqueness of Solution Trajectories to Hybrid Dynamical Systems** 391
    W. P. M. H. Heemels   M. K. Çamlıbel   A. J. van der Schaft   J. M. Schumacher
    18.1 Introduction 392
    18.2 Model Classes 393
    18.3 Solution Concepts 396
    18.4 Well-posedness Notions 399
    18.5 Well-posedness of Hybrid Automata 400
    18.6 Well-posedness of Multi-modal Linear Systems 403
    18.7 Complementarity Systems 405
    18.8 Differential Equations with Discontinuous Right Hand Sides 413
    18.9 Summary 419
    18.10 References 419

Author List 423

Author Index 429

Subject Index 435

1

Implementation of an Active Suspension, Preview Controller for Improved Ride Comfort

M. D. Donahue   J. K. Hedrick

Abstract

A fully active suspension and preview control is utilized to improve ride comfort, which allows increased travelling speed over rough terrain. Previous research is extended and the relevant implementation issues are addressed. Specifically, the methodology of model predictive control has been applied to explicitly address suspension saturation constraints, suspension rate limits, and other system non-linearities. For comparison, the following non-preview controllers were implemented: a skyhook damping controller, a linear quadratic regulator, and a mock passive suspension controller. Particular attention is given to the hydraulic actuator force controller that tracks commands generated by higher-level controllers. The complete system has been successfully realized on a US military high-mobility multi-purpose wheeled vehicle (HMMWV) using a commercially available microprocessor platform. Experimental results show that the power absorbed by the driver is decreased by more than half, significantly improving ride comfort.

1.1 Introduction

Significant attention has been paid to the design of active and semi-active vehicle suspensions. This paper focuses on implementation of an active suspension; where, the standard shock absorbers of the passive suspension are replaced with rectilinear hydraulic actuators governed by electrohydraulic servovalves. The overall objective is to improve ride comfort and maintain crisp handling. Active suspensions can modulate the flow of energy to and from the system [13]. Furthermore, the dynamic characteristics of an active suspension can be continuously adjusted in response to driving conditions as measured by sensors mounted on the vehicle, allowing for better resolution of the trade-offs between ride comfort and road holding [7].

Prevalent techniques used for the design of active suspension controllers [9] require that vehicle actuators track a desired input force trajectory. In this paradigm, there are two distinct, interesting research topics: how to control the actuator to obtain the desired force and how to generate the desired force trajectory.

Several authors who have considered the force generation process of the electrohydraulic servosystem for active suspension control [5], [16] warn that ignoring the nonlinear effects of the actuator dynamics could lead to system instability. Prior attempts at classical control solutions to the force tracking problem have proved incapable of producing adequate results [2]. This has made the design and implementation of more complex, nonlinear control algorithms, such as sliding mode control methods [3], [12], a necessity to achieve acceptable performance.

As explained by Alleyne and Liu [4], the inadequacy of simple methods to solve this problem stems from fundamental limitations in the basic force tracking formulation. The absence of a pure damping element in the system, *i.e.* a shock absorber, and the inherent feedback of the piston velocity to the actuator chamber pressure, (1.3), result in a pair of lightly damped zeros on the open-loop force transfer function. These zeros cannot be modified by simple feedback and produce severe bandwidth restrictions on the force tracking controller. Osorio *et al* [10] present an output redefinition solution to this problem. An artificial damping term is added to the system dynamics, making it possible to better damp the zerodynamics of the system.

Approaches to determine the desired force trajectory include skyhook damping and linear quadratic regulators. More interesting, is the potential to use information of the upcoming road and theoretical system models to compute an optimal force trajectory. Tomizuka [17], Hac [8], and Thomson *et al* [15] present possible solutions. However, the method of model predicative control, as discussed by Gopalasamy *et al* [6], will be used here as it explicitly considers system constraints and nonlinearities.

This work combines the findings of previous researchers and implements an active suspension, preview controller on a commercially available microprocessor platform. Experiments to evaluate the performance of the controllers are conducted on a specially equipped, US military HMMWV. The

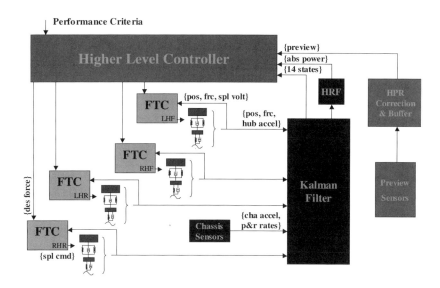

Figure 1.1 Controller system structure, 22 sensors required

vehicle is outfitted with hydraulic actuators, a sensor suite to measure system states, and two preview sensors. More information is provided in Appendix 1.A.

Section 1.2 provides an overview of the control structure. Section 1.3 presents the force tracking controller models and control law derivation. Section 1.4 outlines the different higher-level controller paradigms. Section 1.5 explains the preview correction algorithms. Section 1.6 describes the experimental set-up and presents the experimental results.

1.2 Controller Structure

A hierarchical control structure is used. There are four inner control loops, one outer control loop, and several supporting subsystems, Figure 1.1. Four independent force-tracking controllers (FTCs) operate on quarter-car systems using raw sensor data for feedback signals. The Kalman filter combines sensor information into one consistent set of state information. A higher-level controller operates on this information and generates the desired force for the FTCs. Only one type of higher-level control is active at a given time. The higher-level controllers are updated at a slower sampling rate than the FTCs. Preview information is gathered, corrected and buffered until requested by a higher-level controller.

1.3 Force Tracking Controller

The force tracking controller, Figure 1.2, regulates the force of an individual actuator to the desired force prescribed by a higher-level controller.

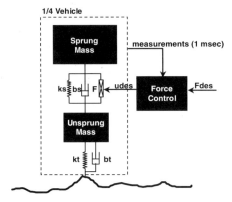

Figure 1.2 Diagram of force tracking controller system and quarter-car model

Although the higher-level controller may make decisions based upon the full car model, it is sufficient to only consider the quarter car dynamics when designing a controller for a single actuator. For the actual system, the added dynamics due to full car motion may be considered as model errors. This and other implementation issues are addressed following the controller derivation.

Plant Models

Quarter-car. A standard quarter-car model is used (1.1), see Figure 1.2 and Appendix 1.C for the respective schematic and nomenclature. One item to note is the existence of a pure damping element in parallel with the hydraulic actuator. In a typical application, the shock is removed. However, the model behaves closer to the actual system when a pure, low damping coefficient, damping element is used.

$$\begin{aligned} \Sigma F_{ms} &= c_s(\dot{x}_u - \dot{x}_s) + k_s(x_u - x_s) + F_a - F_f \\ &= m_s \ddot{x}_s \\ \Sigma F_{mu} &= c_t(\dot{r} - \dot{x}_u) + k_t(r - x_u) + k_s(x_s - x_u) \\ &\quad + c_s(\dot{x}_s - \dot{x}_u) - F_a + F_f \\ &= m_u \ddot{x}_u \end{aligned} \qquad (1.1)$$

It is convenient to define the state space, state vector as follows:

$$\beta = [r - x_u, \dot{x}_u, x_u - x_s, \dot{x}_s] \qquad (1.2)$$

Hydraulic actuator. The hydraulic actuators are governed by electrohydraulic servovalves and are mounted in parallel to the suspension springs, allowing for the generation of forces between the sprung and unsprung masses.

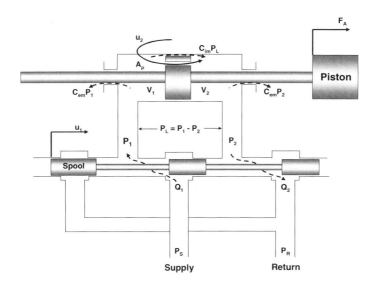

Figure 1.3 Physical schematic and variables for the hydraulic actuator

The electrohydraulic system consists of an actuator, a primary power, spool valve, and a secondary bypass valve. As seen in Figure 1.3, the hydraulic actuator cylinder lies in a follower configuration to a critically centered electrohydraulic power spool valve with matched and symmetric orifices. Positioning of the spool u_1 directs high-pressure fluid flow to either one of the cylinder chambers and connects the other chamber to the pump reservoir. This flow creates a pressure difference P_L across the piston. This pressure difference multiplied by the piston area A_p provides the active force F_A for the suspension system.

Dynamics for the hydraulic actuator [11] valve are given below. Parameter definitions and experimental values are given in Appendix 1.C. The change in force is proportional to the position of the spool with respect to center, the relative velocity of the piston, and the leakage through the piston seals. A second input u_2 may be used to bypass the piston component by connecting the piston chambers.

$$\dot{F}_A = A_p \alpha [C_{d1} w u_1 \sqrt{\frac{P_s - \text{sgn}(u_1) P_L}{\rho}} - C_{d2} u_2 \, \text{sgn}(P_L) \sqrt{\frac{2|P_L|}{\rho}} \\ - C_{tm} P_L - A_p(\dot{x}_s - \dot{x}_u)] \quad (1.3)$$

The bypass valve u_2 could be used to reduce the energy consumed by the system. If the spool position u_1 is set to zero, the bypass valve and actuator will behave in a similar fashion to a variable orifice damper. For the purposes of this work, the bypass valve input u_2 is set to zero during experiments.

Spool valve positions u_1 and u_2 are controlled by a current-position feedback loop. The essential dynamics of the spool, Equation (1.4), have been shown to resemble a first order system forced by a voltage for frequencies less than 15 Hz [4].

$$\tau \dot{u} + u = kv \tag{1.4}$$

Complete system. The system to be controlled by the FTC is the combined quarter-car plant and hydraulic actuator; spool voltage, V, is the control input. Defining the state $x_5 = P_L = F_A/A_p$ and choosing the state vector of Equation (1.5), the state space representation of the system can be written as in Equation (1.6). Suspension friction and road disturbance are considered model errors and are not shown.

$$X = [r - x_u, \dot{x}_u, x_u - x_s, \dot{x}_s, \frac{F_A}{A_p}, u_1] \tag{1.5}$$

$$\dot{X} = \begin{bmatrix} 0 & -1 & 0 & 0 & 0 & 0 \\ \frac{k_t}{m_u} & -\frac{c_t+c_s}{m_u} & \frac{-k_s}{m_u} & \frac{c_s}{m_u} & \frac{-A_p}{m_u} & 0 \\ 0 & 1 & 0 & -1 & 0 & 0 \\ 0 & \frac{c_s}{m_s} & \frac{k_s}{m_s} & \frac{-c_s}{m_s} & \frac{A_p}{m_s} & 0 \\ 0 & A_p\alpha & 0 & -A_p\alpha & -\alpha C_{tm} & 0 \\ 0 & 0 & 0 & 0 & 0 & \frac{-1}{\tau} \end{bmatrix} X + \begin{bmatrix} 0 \\ 0 \\ 0 \\ 0 \\ \Phi \\ 0 \end{bmatrix} + \begin{bmatrix} 0 \\ 0 \\ 0 \\ 0 \\ 0 \\ \frac{k}{\tau} \end{bmatrix} V \tag{1.6}$$

where $\Phi = \alpha C_d w x_6 \sqrt{\frac{P_s - \operatorname{sgn}(x_6) x_5}{\rho}}$

Control Algorithms

As seen in Equation (1.6), there is a severe non-linearity Φ in the dynamic behavior of the system. The most direct approach to solving this problem is dynamic surface control [2].

Dynamic surface control. For the system in Equation (1.6), the control enters through the spool voltage. Applying dynamic surface control as described by Slotine and Li [14], the output F_A was differentiated with respect to time until the control input appeared. The resulting system has relative degree 2 and 4 internal dynamic states. The controller surfaces are

$$P_L = \frac{F_A}{A_p} \quad \text{and} \quad u_{spool} \tag{1.7}$$

For the P_L surface, an integral term was added to the standard definition of s. The integral term, weighted by $0 < \lambda_1 < 1$, slightly attenuates control noise.

$$s_1 = \tilde{x}_5 + \lambda_1 \int \tilde{x}_5 dt \qquad \text{where } \tilde{x}_5 = x_5 - x_{5d} \tag{1.8}$$

Applying the sliding surface approach, the control law must satisfy the condition in Equation (1.9) to ensure asymptotic tracking of F_{des}.

$$s_1 \dot{s}_1 = s_1(\dot{\tilde{x}}_5 + \lambda \tilde{x}_5) \leq -\eta_1 s_1^2 \tag{1.9}$$

Plugging in the equation of dynamics for $\dot{x}_{5d}$ and solving for u_{des}:

$$u_{des} = \frac{1}{\Phi}\{A_p \alpha(x_4 - x_2) + \alpha C_{tm} x_5 + \dot{x}_{5d} - \lambda_1 \tilde{x}_5 - \eta_1 s_1\} \tag{1.10}$$

In Equation (1.10), the desired force profile enters through the terms $\dot{x}_{5d}$ and s_1. Because the time derivative of the desired force is used in control computation, it is important for the force profile to be smooth.

Following the method used for the P_L surface, the equation for control input V can be obtained as follows:

$$\begin{aligned} s_2 &= u - u_{des} \\ s_2 \dot{s}_2 &= s_2(\dot{u} - \dot{u}_{des}) \leq -\eta_2 s_2^2 \end{aligned} \tag{1.11}$$

Substituting the equation of dynamics for $\dot{u}$ into Equation (1.11) and solving, the control input is thus:

$$V = \frac{1}{k}\{u + \tau \dot{u}_{des} - \eta_2 \tau s_2\} \tag{1.12}$$

The time derivative of u_{des} is needed to compute the control input V. Using the filter of Equation (1.13) allows theoretical proof that the resulting controller is asymptotically stable.

$$\dot{\Psi} = \frac{u_{des} - \Psi}{\tau} \tag{1.13}$$

Note that $\dot{\Psi}$ is used in place of $\dot{u}_{des}$ in Equation (1.12). The state Ψ is maintained via forward Euler integration of $\dot{\Psi}$.

In theory, choosing the sliding surface gains to overcome the worst-case model and disturbance errors ensures asymptotic tracking of the desired profile.

Output redefinition reduces model errors by directly considering the lack of a pure damping element in the system [10]. Other methods that afford increased gains are presented in the following section.

Implementation

The desired spool position command output by the first surface, P_L, is very noisy. The second surface amplifies the noise and causally decreases sliding mode gains. It was empirically determined that the filter in Equation (1.14) reduced control noise and improved controller performance. With this filter,

u'_{des} replaces the u_{des} command sent to the second surface in Equation (1.11).

$$u'_{des} = \frac{u_{des}(k-1) + u_{des}(k-2)}{2} \quad (1.14)$$

Another empirical study showed that numerical differentiation, Equation (1.15), of $\dot{u}_{des}$ worked better than the sliding mode filter described in Equation (1.13).

$$\dot{w}_{des}(k) = \frac{w_{des}(k) - w_{des}(k-1)}{\Delta t} \quad (1.15)$$

Model error filters. The FTC formulation above treats the full car dynamics as a disturbance. Results indicate that FTC performance around the resonant chassis modes is poor. Resonant frequency for the pitch and heave modes is around 2 Hz and around 4 Hz for the roll mode. Attenuating the u_{des} command inputs near these frequencies improves force tracking. To implement these filters with high-level controller force generation a heave, pitch, and roll quantification scheme was used. Ultimately, FTC tuning was sufficient as to eliminate the need for these filters. Moreover, a model predictive control (MPC) formulation considers these resonant frequencies when computing F_{des}.

Higher-level controller filters. Hierarchical control inputs are generated at a slower sampling rate (30 ms due to processor constraints) than the 1 ms FTC task. A 1 ms sampling rate is necessary to ensure good tracking up to 8 Hz as dictated by the system time constants. For smooth convergence to F_{des}, considering the derivative terms in Equation (1.10), the desired force was filtered by Equation (1.16). A plot of the filter step response is shown in Figure 1.4, the rise time is approximately 27 ms.

$$\hat{F}_{des}(s) = \frac{6.4e9}{s^4 + 950s^3 + 385625s^2 + 7.6e7s + 6.4e9} \quad (1.16)$$

1.4 Higher-level Controllers

Higher-level controllers compute the desired force for the four independent FTCs. Controllers have access to the vehicle information listed below. The bold items are states provided by a 14 state Kalman filter (KF) operating on a 7 degree-of-freedom (DOF) full car model.

- **Suspension Expansion**, Suspension velocity;
- **Hub Velocity**;
- **Tire Deflection**;
- **Chassis Pitch and Roll rate**, Chassis HPR.

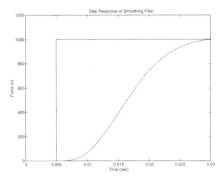

Figure 1.4 Plot of F_{des} smoothing filter step response

Zero Force

The desired force is set to zero. This controller is a very simple attempt to reduce suspension acceleration.

Skyhook Damping Controller

Four independent skyhook dampers are implemented on the HMMWV, one for each wheel. The plant dynamics are derived from those of a modified quarter-car model. For skyhook damping, a theoretical damper is used to reduce the velocity of the sprung mass. The control law is given by Equation (1.17), [1].

$$F_{des} = -B_{sky}\dot{x}_s + K_{vel}(\dot{x}_u - \dot{x}_s) \tag{1.17}$$

Controller gains are chosen to adjust the pole locations of the original system. The gain set $\{B_{sky}, K_{vel}\} = \{2000, 1000\}$ is used on the HMMWV.

Linear Quadratic Regulator

A standard linear quadratic regulator (LQR) formulation for suspension systems is implemented. The plant dynamics are of the form used by the Kalman filter. Thus, the cost function includes

{Chassis accel, Susp travel, Tire deflection, Pitch & roll rates, Hub vel, Control usage}

Some transformations are required to put the associated cost function into standard form and obtain the Riccati equation. Consult Thompson *et al* [16] for more details. MATLAB^{TM} is used to generate the LQR optimal matrix gain K. Chassis acceleration, pitch and roll rates have the highest costs.

Model Predictive Controller

The MPC was designed and coded for the HMMWV environment by Scientific Systems Inc., source code and libraries are implemented in SIMULINK

Table 1.1 LQR weighting gains

Parameter	Weight
Chassis acceleration	10
Pitch & roll rate	10
Suspension travel	1
Hub velocity	1
Tire deflection	0.1
Control usage	$1e^{-4}$

via S-function. MPC is the primary computation for the 300 MHz Alpha processor, at a Δt of 30 ms.

At each sampling instant, the MPC computes a finite number of future control moves such that a cost function, over a finite horizon, is minimized. The first control output is fed to the FTC. The exact workings of the MPC involve output prediction (based on a system model) and a receding-horizon approach. For more information on MPC formulation, consult Gopalasamy et al [6]. Therein, they describe how to recast the MPC problem to a constrained quadratic programming (QP) problem and how to select an appropriate real-time algorithm.

The superscript p denotes the usage of preview information. In other words, MPC$_p$ enhances MPC by considering $\dot{Z}_{road}$ and relative road heights for the desired preview horizon (pH) at each wheel.

Of interest to this project are the cost function weighting parameters and the physical constraint set, Table 1.2 and Table 1.3, respectively. Field testing of the MPC and skyhook controllers motivated the addition of an "optimal" skyhook (suspension velocity) damping term in the MPC cost function. The absorbed power term will be explained in Section 1.6.

Table 1.2 MPC weighting gains

Parameter	Weight
Absorbed power	23
Suspension travel	0.02
Suspension velocity	192
Tire deflection	0.08
Control usage	$1.1e^{-6}$

Table 1.3 MPC constraint values

Constraint	Value
Force	± 8000 N
Force rate	± 5000 N/s
Suspension travel	± 0.06 m

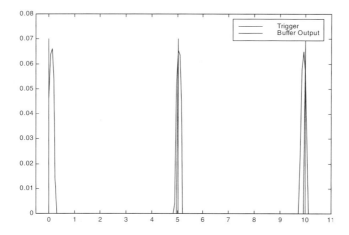

Figure 1.5 Buffered, generated preview data matched peak loads

1.5 Preview Information

The MPC_p requires the road profile, Z_{road}, and the rate that Z_{road} is changing with respect to time $\dot{Z}_{road}$ for n preview steps, pH, at each wheel. Road profiles for each side of the car are stored in a buffer. When extracting preview data, the buffer is parsed and information of the current vehicle velocity and Z_{road} are combined to create $\dot{Z}_{road}$. The HMMWV system has two methods to obtain Z_{road}.

Preview generation. Preview generation is used on courses with a known road profile, such as the test track, Appendix 1.B. The preview buffer is fed a pre-stored profile in place of the sensor preview data. The digital profile is synchronized to the actual profile using HMMWV sensors. Figure 1.5 shows a sample buffer output matched with peaks from the suspension load cells. The load triggered spikes indicate the most probable location of the actual bump. Preview generation relies on absolute position and is susceptible to error accumulation; in Figure 1.9 at 10 m the predicted bump location is no longer accurate. To contrast, preview sensor data requires at most 4 m, slightly more than the length of the vehicle.

Preview sensors. On unknown terrain, we obtain preview information via sensors that measure the range to ground. Sensor measurements are converted to road height by Equation (1.18), see variable definitions in Figure 1.6. The preview sensors are rigidly attached to the chassis and have the same heave, pitch and roll (HPR). HPR are measured much faster than the rate of change of HPR and the chassis is assumed to have negligible

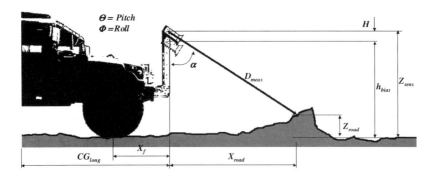

Figure 1.6 Diagram and nomenclature definition for preview correction computation

warp. Thus, it is reasonable to directly apply trigonometry.

$$\begin{aligned} Z_{road} &= Z_{sens} - D_{meas}\cos(\alpha - \theta) \\ Z_{sens} &= H - CG_{long}\sin\theta - CG_{lat}\sin\phi + D_{bias}\cos\alpha \\ X_{road} &= D_{meas}\sin(\alpha - \theta) + X_f \end{aligned} \qquad (1.18)$$

The set of values X_{road} and Z_{road} are fed into the buffer and used to attain preview information for the MPC_p. Current values of HPR are obtained from the Kalman filter (KF). For accurate KF estimates, road disturbance information from the preview buffer is input into the KF. This interdependence, coupled with processing delays, created unstable preview dynamics.

For robustness, tire dynamics compensation was done externally from the KF. The free response of a quarter-car model, Equation (1.1), to the buffered road disturbance was used to modify the HPR input to Equation (1.18). Figure 1.7 shows the results—raw preview sensor data is shown in the top plot; buffered and corrected road information is shown in the bottom two plots. The plots denoted "original" (dashed lines) are the output of Equation (1.18) using KF estimates without disturbance information; "Tire Comp." includes tire compensation done external to the KF. Observe the negative bump just after the actual bump (at 4.3 s) in the Z_r plot. The negative impression is removed by accounting for tire dynamics. MPC_p places the most weight on $\dot{Z}_{road}$; in the final plot we see a tremendous improvement over the original $\dot{Z}_{road}$. A more robust solution is to resolve issues encountered when incorporating road disturbance information into the Kalman Filter computation.

Preview Buffer

Incoming, consecutive road data is not guaranteed to have an equal spacing or even a consistent order. New road information is sorted and stored to the buffer with respect to X_{road}. Interpolated data is retrieved for the requested pH for each wheel.

Active Suspension, Preview Controller for Improved Ride Comfort 13

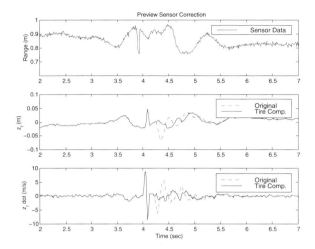

Figure 1.7 Sample sensor data with HPR correction

The buffer is fixed length, circulating memory. An integer increment in the array pointer corresponds to a fixed increment in the physical distance. Relative distance travelled is maintained by integrating vehicle speed. To improve the stochastic properties of the buffer, new information is interpolated and updated, if necessary, with a forgetting factor.

A standard velocity sensor is used to measure V_k for the experimental HMMWV. In final implementation, an accurate estimation of the ground speed is required to avoid errors introduced by wheel slip, by wheel liftoff, and by loss of traction.

1.6 Experimental Results

Experimental set-up. The specially equipped HMMWV (Appendix 1.A) is repeatedly driven over the test track, see Appendix 1.B, at a speed of $20 mph$[1]. In subsequent trials, the different higher-level controllers are used. Realistic performance data is collected on an off-road, natural terrain. Prior to system testing, a generated F_{des} profile is fed to the FTCs to verify proper FTC performance; the vehicle is stationary.

Performance criterion. The US Army TARDEC has empirically developed a criterion known as 'absorbed power' to quantify ride comfort. This formulation filters the sprung mass acceleration through a human response filter (HRF) that represents the frequency range most undesirable by a human driver. A second order approximation of the HRF is given in Equation

---

[1] 20 mph is the limiting speed for a passive suspension. All experiments were conducted at this speed for consistency

(1.19), the units of input acceleration are m/s². Output from the filter is squared and time averaged over a moving window to produce the absorbed power measure, also known as the cumulative absorbed power (CAP). Over a given terrain the CAP should remain less than 6 W for driver comfort. Drivers inherently slow down when the CAP persistently exceeds the 6 W limit.

$$\widehat{HRF}(s) = \frac{12s}{s^2 + 30.02s + 901.3} \tag{1.19}$$

Force Tracking Controller

Figure 1.8 depicts nominal FTC performance across the 1 Hz to 10 Hz frequency range as well as the response to a filtered square wave. There is no compensation for model error, Section 1.3, or ORD [10]. If present, ORD would lessen the dip at each square wave peak. Figure 1.8 also shows FTC performance while tracking a discrete F_{des}, Section 1.3.

High-level Controllers

Figure 1.9 depicts typical results. All of the higher-level controllers perform similarly; the MPC is slightly better than the rest. There is a better than twofold improvement in the absorbed power criterion when compared to the passive suspension[2]. LQR and skyhook performance, not shown, are comparable to the MPC performance. This is true provided the system remains within constraints, Table 1.3, and preview information is not used. For the MPC trial, the FTC performance is shown. Observe the two spikes corresponding to the test track bumps (the first bump is larger than the second). At these instances, there is saturation of the control input u_1. In theory, the MPC_p reduces the amount of saturation.

Using the controllers off-road, the results of Figure 1.10 are generated. Now, MPC handedly beats the other higher-level controllers. Moreover, the off-road results show that only the MPC maintained a CAP < 6 W. By military standards, this terrain is only drivable at this speed, 20 mph, if the MPC is used. To better understand the improvements of Figure 1.10, the suspension expansion (LVDT) measurements are shown as well. The MPC reduced suspension travel and the likelihood of suspension saturation, which occurs at approximately ± 0.06 m.

MPC Preview Controller

Finally, the preview information is added. Only access to the generated preview is available. Figure 1.11 shows MPC verses MPC_p performance for one big bump. There is 15% improvement over normal MPC control. For persistently exciting road profiles this added improvement will greatly improve

---

[2]Passive suspension is emulated with the hydraulic pump turned off and unpowered spool valves.

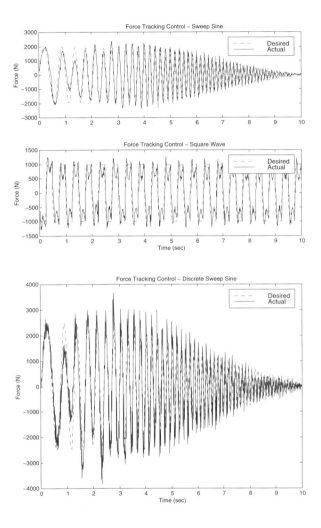

Figure 1.8 Nominal FTC tracking of generated control signals F_{des}. frequency sweep and filtered square wave (*top*) and discrete frequency sweep (*bottom*).

ride comfort. Comparing the preview results with those of Figure 1.9, we see a near threefold improvement over the passive suspension when preview control is utilized.

1.7 Conclusions

Practical, implementation-oriented, modifications to dynamic surface control theory were successfully employed. Modifications involved adding filters at various levels of the control computation. To the end of realizing full functionality of model predictive control (MPC) using preview infor-

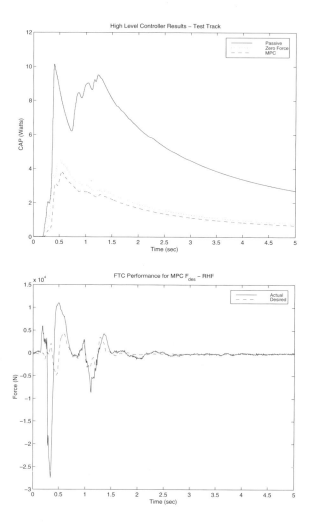

Figure 1.9 Test track, higher-level controller performance (*top*) and *FTC* tracking for the MPC F_{des} (*bottom*)

mation, numerous subsystems were designed. All subsystems work well. A skyhook damping controller and a linear quadratic regulator were developed to benchmark the performance of the MPC without preview. For the non-preview controllers, more than a twofold increase in ride comfort over passive suspension was obtained. With preview control, the ride comfort was improved threefold. This resulted in an increased drivable speed for rough terrain. In particular, the MPC allowed for the fastest speed over off-road terrain. All of the infrastructure is in place to use the experimental HMMWV as a test bed for future control algorithms.

Future work is needed on the preview correction algorithm. We feel that

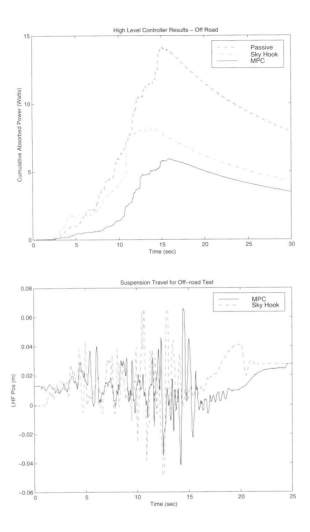

Figure 1.10 Off-road, higher-level controller performance (*top*) and suspension travels (*bottom*), respectively.

our performance was limited by hydraulic actuator capabilities. Moreover, hydraulics related problems were a nuisance throughout the project. Alternative actuators need be explored for the ultimate realization of the active suspension system.

Acknowledgments

This work was made possible by the prior research of Professor Andrew Alleyne, University of Illinois at Urbana Champagne, and Carlos F. Osorio of the University of California at Berkeley. The technical support and software expertise of Jayesh Amin from Scientific Systems Company Inc. enabled

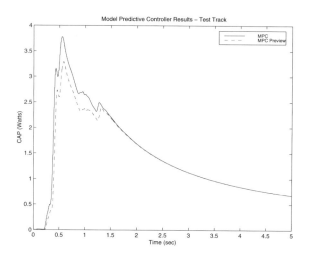

Figure 1.11 MPC performance using generated preview data

much of the higher-level control theory and implementation. Scientific Systems Company Inc., the SBIR Project Office, and the US Army TARDEC sponsored the project. This work was completed under Phase II of the SBIR contract number DAAE07-96-C-X007.

1.8 References

[1] A. Alleyne. *Nonlinear and Adaptive Control with Applications to Active Suspensions*. Ph.D. thesis, Department of Mechanical Engineering, University of California at Berkeley, Berkeley, CA, 1994.

[2] A. Alleyne and J. K. Hedrick. Nonlinear control of a quarter car active suspension. In *Proceedings of the 1992 American Control Conference*, Chicago, IL, 1992.

[3] A. Alleyne and J. K. Hedrick. Nonlinear adaptive control of active suspensions. *IEEE Transactions on Control Systems Technology*, 3(1):94–102, 1995.

[4] A. Alleyne and R. Liu. On the limitations of force tracking control for hydraulic servosystems. *ASME Journal of Dynamic Systems, Measurement and Control*, 1999.

[5] G. H. Engleman and G. Rizzoni. Including the force generation process in active suspension control formulation. In *Proceedings of the 1993 American Controls Conference*, 701–705, San Francisco, CA, 1993.

[6] S. Gopalasamy, J. K. Hedrick, C. Osorio, and R. Rajamani. Model predictive control for active suspensions - controller design and experimental study. *ASME J. Dynamic Systems and Control*, 61:725–733, 1997.

[7] A. Hac. Suspension optimization of a 2-dof vehicle model using stochastic optimal control technique. *Journal of Sound and Vibration*, 100(3):343–357, 1985.

Table 1.4 Important Vickers PV3-115 hydraulic pump specifications

Specification	Value
Supply pressure	3000 psi
Flow rate	45 l/min
Power consumption	< 32 hp at < 3500 rpm

[8] A. Hac. Optimal linear preview control of active vehicle suspension. *Vehicle System Dynamics*, 21:167–195, 1992.

[9] D. Hrovat. Survey of advanced suspension developments and related optimal control applications. *Automatica*, 33(10):1781–1817, 1997.

[10] C. Osorio, S. Gopalasamy, and J. K. Hedrick. Force tracking control for electro hydraulic active suspensions using output redefinition. In *Proceedings of the ASME Winter Annual Meeting*, Nashville, TN, 1999.

[11] R. Rajamani. *Observers for Nonlinear Systems, with application to Automotive Active Suspensions*. Ph.D. thesis, Department of Mechanical Engineering, University of California at Berkeley, Berkeley, CA, 1993.

[12] R. Rajamani and J. K. Hedrick. Observer-based control of an active suspension. In *IEEE Conference on Control Applications*, Dayton, OH, 1992.

[13] R. S. Sharp and S. A. Hassan. The relative performance capabilities of passive, active, and semi-active car suspension systems. In *Proceedings of the Institution of Mechanical Engineers*, volume 203-3 of *D*, 219–228, 1986.

[14] J. J. Slotine and W. P. Li. *Applied Nonlinear Control*. Prentice Hall, 1991.

[15] A. G. Thompson and P. M. Chaplin. Force control in electrohydraulic active suspensions. *Vehicle System Dynamics*, 25:185–202, 1996.

[16] A. G. Thompson, B. R. Davis, and C. E. M. Pearce. An optimal linear active suspension with finite road preview. Paper 0148-7191/80/0225-0520(800520), Society of Automotive Engineers, 1980.

[17] M. Tomizuka. Optimum linear preview control with application to vehicle suspension revisited. *Trans. ASME, J. Dynamic Systems, Measurement and Control*, 98(3):309–315, September 1976.

1.A HMMWV Equipment

Lotus Engineering completed original instrumentation of the HMMWV. The University of California at Berkeley added additional sensors and a new computer. Provided below are tables detailing important information regarding the sensor and actuator suites.

Table 1.5 Essential HMMWV sensors

Qty	Sensor Type	Location	Measurement
4	Load cell	Top mount of each actuator	Actuator forces
4	LVDT	Inside each actuator	Actuator displacement
4	Hub accelerometer	On each wheel hub	Axle vertical acceleration
2	Chassis accelerometer	Opposite corners of chassis	Chassis vertical acceleration
2	Rate gyro	Center console	Pitch and roll rates
2	Range	Front of vehicle	Preview, distance to ground
1	Speedometer	Engine compartment	Vehicle speed

Digital signal processing boards. The processor suite of choice is the dSpace Autobox components identified below
   DS1003: TI TMS320C40 Parallel 60 MHz DSP board
   DS1004: DEC Alpha AXP21164 300 MHz DSP board

Preview sensors. For relatively straight path motion or for uniform, wide bumps in the road profile, it is sufficient to use only range finding sensors to obtain preview information. This assumption simplifies the sensor requirements and, as shown in Section 1.5, the preview processing algorithm. Two types of sensors were explored, see comparison in Table 1.6:

1. Frequency modulated continuous wave (FMCW) radar by O'Conner Engineering. The FMCW radar has a central frequency of 24.5 GHz and scanning range of 0.5 GHz.

2. WTA24-P5401 LED optical sensor by Sick Optic. The WTA optical sensor consists of a modulated infrared LED, with an average life of 100,000 hours at 25 °C, and precision reflectors all mounted inside a

Table 1.6 Comparison of HMMWV preview sensors

Specification	FMCW Radar Nominal Value	WTA24-P5401 Nominal Value
Range	1.0 m – 5.0 m	0.6 m – 1.2 m
Resolution	0.01 m	0.02 m
Light spot	0.3 m – 0.6 m	0.02 m – 0.03 m
Response time	1.1 ms	5.0 ms

rugged diecast metal housing. The unit meets or exceeds shock and vibration standards: IEC 68-2-27/IEC 68-2-6.

1.B Test Track

The HMMWV is stored and tested at the University of California's Richmond Field Station (RFS) in Richmond, CA. The available testing paradigms are: 1) a custom test track, described below, 2) gravel roads, 3) dirt roads, and 4. off-road, grassy terrain. The test track is asphalt to limit the effects of weather, erosion, and wear. We use six standard, hard rubber speed bumps, manufactured by Scientific Developments Inc.. There are 10 possible temporary locations for the bumps along the 32' test region. Bump raisers are used to increase the height of the bumps in 1.5" increments.

1.C Nomenclature

A_p	Piston area	$0.0044 \, \text{m}^2$
C_{d1}	Discharge coefficient	0.7
C_{tm}	Leakage coefficient $\dfrac{C_{im} + C_{em}}{2}$	$15e - 12$
c_s	Suspension damping	$12000 \, \text{Ns/m}$
c_t	Tire damping	$200 \, \text{Ns/m}$
F_A	Actuator force	
F_f	Friction force	$120 \, \text{N}$
k	Voltage to position conversion factor	$1481 \, \text{V/m}$
k_s	Suspension spring stiffness	$240 \, \text{kN/m}$
k_t	Tire spring stiffness	$1000 \, \text{kN/m}$
k_v	Relative velocity, chassis $\to$ axle	2.1
m_s	Sprung mass	$2800 \, \text{kg}$

m_u	Unsprung mass	270 kg
m_{eq}	Equivalent mass $(\dfrac{1}{m_u}+\dfrac{1}{m_s})$	
P_s	Supply pressure	20684 kN/m²
P_L	Pressure induced by load	
u_1	Spool valve position	
u_2	Bypass valve area	
V	Input voltage command	
V_t	Total volume of actuator cylinder chamber	
w	Spool valve width	0.008 m
x_s	Sprung mass position	
x_u	Unsprung mass position	
α	Hydraulic coefficient, $4\beta/V_t$	2.273e9 N/m⁵
β	Bulk modulus of hydraulic fluid	
ρ	Specific gravity of hydraulic fluid	3500
τ	Spool valve time constant	0.003 s

2

Active and Passive Suspension Control for Vehicle Dive and Squat

Fu-Cheng Wang M.C. Smith

Abstract

Performance capabilities of passive and active vehicle suspension systems will be examined from a mechanical networks point of view. It is known that the reduction of effects of road disturbances is a conflicting requirement with the reduction of effects of inertial loads in a quarter-car model when passive control is used, but not with active control of suitable structure. The extension of these ideas to a half-car trailing-arm model will be considered. It will be shown that the choice of suspension geometry does not remove the basic trade-offs for passive suspensions. An active control structure to allow the road and load transmission paths to be optimised independently will be presented. The design approach will be applied to a non-linear trailing-arm vehicle model to demonstrate good anti-dive and anti-squat behaviour together with a soft ride in response to road disturbances. The performance of the controller will be demonstrated using the multi-body simulation code *AutoSim*.

2.1 Introduction

This paper studies the problem of vehicle active suspension control where a key feature is the need to insulate the vehicle body from both road irregularities and load disturbances (*e.g.*, inertial loads induced by braking and cornering). In the context of a quarter-car model, it was shown that these are conflicting requirements when passive suspensions are used [10], but the conflict may be removed when active control is employed with appropriate hardware structure, *e.g.*, choice of sensor location, number and type [9]. In this paper, we will generalise the result of [10] to show that the conflict remains for passive suspensions applied to a simple half-car model. We will also discuss why the optimisation of suspension geometry does not remove the problem in the context of a half-car trailing-arm model.

For the active suspension problem, once a suitable hardware structure is selected, there remains the problem of designing the controller to achieve satisfactory responses for all of the disturbance transmission paths. We will summarise some results from [11], which considers the problem of parameterising the set of all stabilising controllers for a given plant. This leaves the transfer function for a given disturbance transmission path the same as when some nominal stabilising controller is employed. In this way, the design for each disturbance path can be carried out successively, providing there is sufficient freedom to adjust the responses independently. The design approach will be applied to a half-car trailing-arm model to demonstrate good anti-dive and anti-squat behaviour in response to braking and accelerating while retaining a soft ride in response to road disturbances. The multi-body simulation code *AutoSim* is then employed to demonstrate the behaviour of the non-linear vehicle model.

2.2 Limitations Imposed by Passivity in Vehicle Suspension Design

From a systems design point of view, there are two main categories of disturbances on a vehicle, namely road and load disturbances. Road disturbances have the characteristics of large magnitude in low frequency (such as hills) and small magnitude in high frequency (such as road roughness). Load disturbances include the variation of loads due to accelerating, braking and cornering. Therefore, a good suspension design relates to disturbance rejection from these disturbances to the outputs (*e.g.*, vehicle height, rotation, *etc.*) in which we are interested. In ordinary passive suspensions, good rejection of road disturbances is achieved by a "soft" suspension, while good rejection of load disturbance is achieved by a "hard" suspension. In this section, we will show that passivity imposes a fundamental restriction on the independent design of these disturbance transmission paths. To this end, we will make use of the classical characterisation of passivity from electrical network theory together with an electrical-mechanical analogue.

Network Theorems

There is a standard analogy between electrical and mechanical systems in which force and velocity are analogous to current and voltage, respectively. A *port* in a mechanical network is a pair of points to which an equal and opposite force F can be applied and which experience a relative velocity v. More generally, an n-port network has n pairs of port variables $\{F_i, v_i\}$, $i = 1, 2, \ldots, n$. Such a network is termed *passive* if for all admissible vectors of admissible port variables v and F that are square integrable on $(-\infty, t]$

$$\int_{-\infty}^{t} F(\tau)^* v(\tau) d\tau \geq 0 \tag{2.1}$$

The quantity on the left-hand side of (2.1) is the total energy delivered to the network up to time t. Thus, a passive network cannot deliver energy to the environment.

For a linear network, a matrix $Z(s)$ (or $Y(s)$) for which $\hat{v} = Z(s)\hat{F}$ (or $\hat{F} = Y(s)\hat{v}$) is termed the impedance (or admittance) matrix. It is also possible to define a mixed immittance matrix relating a vector containing both forces and velocities to a vector containing the complementary variables.

THEOREM 2.1—[6, CHAPTERS 4, 5], [1, CHAPTER 5]
Consider a multi-port network for which the impedance matrix $Z(s)$ exists and is real-rational. The network is passive if and only if one of the following two equivalent conditions is satisfied:

1. $Z(s)$ is analytic and $Z(s) + Z(s)^* \geq 0$ in $\text{Re}(s) > 0$.[1]

2. $Z(s)$ is analytic in $\text{Re}(s) > 0$, $Z(j\omega) + Z(j\omega)^* \geq 0$ for all ω at which $Z(j\omega)$ is finite, and any poles of $Z(s)$ on the imaginary axis or at infinity are simple and have a non-negative definite, Hermitian residue.

□

Any real-rational matrix $Z(s)$ satisfying Conditions 1. or 2. above is called *positive real*. Theorem 2.1 also holds with $Z(s)$ replaced by $Y(s)$ or a mixed immittance $G(s)$.

The Half-car Model

In this section, we shall generalise the procedure of [10, Example 3] to the half-car model. Figure 2.1 shows a standard half-car model together with an abstract representation as a four-port mechanical network. The model consists of the vehicle body of mass m_s and moment of inertia I_ψ, the unsprung (point) masses m_1 and m_2 and tyre springs with vertical stiffness k_{t_1} and k_{t_2}. The suspension elements are assumed to provide equal and opposite forces u_1 and u_2 on the vehicle body and unsprung masses,

---
[1] * denotes complex conjugate transpose and ≥ 0 means non-negative definite Hermitian.

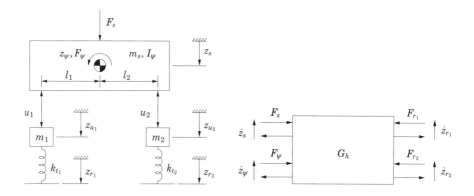

Figure 2.1 The half-car model, and as a four-port network

and to have negligible masses. The applied force and torque on the vehicle body are F_s and F_ψ and the road disturbance displacements are z_{r_1} and z_{r_2}. The vertical and angular displacements of the vehicle body are z_s and z_ψ, and the vertical displacements of the unsprung masses are z_{u_1} and z_{u_2}. The linearised dynamic equations can be expressed as follows:

$$m_s \ddot{z}_s = F_s - u_1 - u_2$$
$$I_\psi \ddot{z}_\psi = F_\psi - u_1 l_1 + u_2 l_2$$
$$m_1 \ddot{z}_{u_1} = u_1 + F_{r_1}$$
$$m_2 \ddot{z}_{u_2} = u_2 + F_{r_2}$$

where the tyre forces F_{r_1}, F_{r_2} are given by:

$$F_{r_1} = k_{t_1}(z_{r_1} - z_{u_1})$$
$$F_{r_2} = k_{t_2}(z_{r_2} - z_{u_2})$$

and $u_1 = c_1(\dot{z}_s + l_1 \dot{z}_\psi - \dot{z}_{u_1}) + k_1(z_{a_1} - z_{u_1})$, $u_2 = c_2(\dot{z}_s - l_2 \dot{z}_\psi - \dot{z}_{u_2}) + k_2(z_{a_2} - z_{u_2})$ for passive suspensions.

Let us suppose the performance requirements are that the system should be "soft" to road disturbances z_{r_1}, z_{r_2} corresponding to passive elements c_1^s, c_2^s, k_1^s, k_2^s, and "hard" to load disturbances F_s, F_ψ corresponding to passive elements c_1^h, c_2^h, k_1^h, k_2^h. Then we require a mixed immittance matrix of the form:

$$\begin{bmatrix} s\hat{z}_s \\ s\hat{z}_\psi \\ \hat{F}_{r_1} \\ \hat{F}_{r_2} \end{bmatrix} = \begin{bmatrix} G_{11}^h & G_{12}^h & G_{13}^s & G_{14}^s \\ G_{21}^h & G_{22}^h & G_{23}^s & G_{24}^s \\ G_{31}^h & G_{32}^h & G_{33}^s & G_{34}^s \\ G_{41}^h & G_{42}^h & G_{43}^s & G_{44}^s \end{bmatrix} \begin{bmatrix} \hat{F}_s \\ \hat{F}_\psi \\ s\hat{z}_{r_1} \\ s\hat{z}_{r_2} \end{bmatrix} \quad (2.2)$$

The following parameters are used for the half-car model: $m_s = 1600\,\text{kg}$, $I_\psi = 1000\,\text{kg m}^2$, $m_1 = m_2 = 100\,\text{kg}$, $k_{t_1} = k_{t_2} = 500\,\text{kN/m}$, $l_1 = 1.15\,\text{m}$,

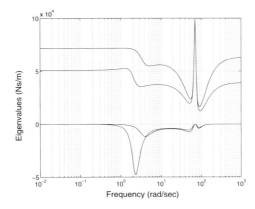

Figure 2.2 Eigenvalues of $(Y+Y^*)(j\omega)$ for the half-car model

$l_2 = 1.35$ m. As the suspension settings, we choose $c_1^h = c_2^h = 10$ kNs/m, $k_1^h = k_2^h = 40$ kN/m, $c_1^s = c_2^s = 1$ kNs/m, $k_1^s = k_2^s = 4$ kN/m. In order to get a suitable scale, we check the positive realness of another transformation matrix Y, where Y is determined directly from (2.2) as $[\hat{F}_s, \hat{F}_\psi, \hat{F}_{r_1}, \hat{F}_{r_2}]' = Y[s\hat{z}_s, s\hat{z}_\psi, s\hat{z}_{r_1}, s\hat{z}_{r_2}]'$. As shown in Figure 2.2, there are two negative eigenvalues of $(Y+Y^*)(j\omega)$ in the region around 3 rad/sec. These numerical results were obtained from the multi-body dynamics package *AutoSim* by finding state-space realisations of the linearised model of Figure 2.1. Therefore, we know that the above suspension behaviour cannot be implemented passively.

2.3 Suspension Geometry in the Half-car Trailing-arm Model

In the classical approach to suspension design, the optimisation of suspension geometry is used to get better compromises in the responses to the various disturbance forces acting on a vehicle. We consider this approach in this section and show that, although improved performance is possible, the basic trade-off between road and load disturbances remains for passive suspensions.

Anti-dive and Anti-squat Suspension Geometry

We now explain and slightly generalise the classical reasoning for the choice of suspension geometry in a trailing-arm vehicle model, as shown in Figure 2.3. To simulate acceleration and braking, torques T_1 and T_2 are applied to the two wheels, and corresponding forces F_1 and F_2 are induced at the tyre contact patch to maintain a no-slip rolling contact. For the acceleration scenario (anti-squat case), the reaction torques are applied to the sprung mass in pitch motion. The basic idea (following [4] and [5]) is to derive

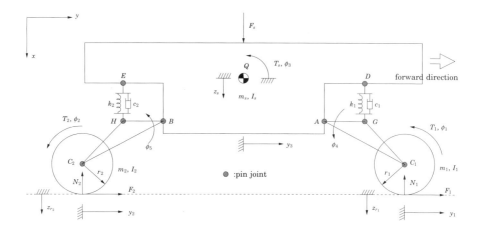

Figure 2.3 A half-car trailing-arm model with passive suspensions

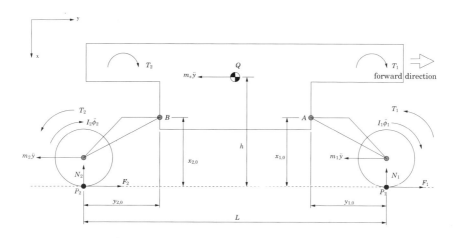

Figure 2.4 Anti-squat design

conditions for the geometrical arrangement to remain invariant (*i.e.*, no pitching motion of the sprung mass or trailing arms) under an acceleration that occurs with a preassigned constant ratio of forces at the tyre contact patches. Our approach is a slight generalisation in that we allow non-zero values for the mass and inertia of the wheels. We now make use of the simplified model of Figure 2.4 in which the suspension struts are removed. We introduce D'Alembert forces and torques $m_1\ddot{y}$, $m_2\ddot{y}$, $m_s\ddot{y}$, $I_1\ddot{\phi}_1$ and $I_2\ddot{\phi}_2$ and write down the corresponding equilibrium conditions. This assumes that the geometrical configuration remains invariant, *i.e.*, there is no pitching or vertical motion of the sprung mass.

1. Suppose the accelerating force ratio (F_1/F_2) is $p/(1-p)$, then the acceleration $\ddot{y}$ is given by:

$$\ddot{y} = \frac{F_1 + F_2}{m_1 + m_2 + m_s} \tag{2.3}$$

so that F_1, F_2 can be expressed as:

$$F_1 = pM_x\ddot{y}, \qquad F_2 = (1-p)M_x\ddot{y}$$

where M_x is defined as:

$$M_x = m_1 + m_2 + m_s. \tag{2.4}$$

2. Taking moments about P_2 for the whole system results in:

$$N_1 L + m_s \ddot{y} h + m_1 \ddot{y} r_1 + m_2 \ddot{y} r_2 + \frac{I_1}{r_1}\ddot{y} + \frac{I_2}{r_2}\ddot{y} = 0,$$

which gives N_1 as:

$$N_1 = -M_y \ddot{y} \tag{2.5}$$

where M_y is defined as:

$$M_y = \frac{1}{L}(m_s h + m_1 r_1 + m_2 r_2 + \frac{I_1}{r_1} + \frac{I_2}{r_2}) \tag{2.6}$$

3. Applying Newton's second law in the vertical direction for the whole system gives:

$$N_2 = -N_1 = M_y \ddot{y} \tag{2.7}$$

4. For the front wheel, taking moments about A results in:

$$F_1 x_{1,0} + N_1 y_{1,0} - m_1 \ddot{y}(x_{1,0} - r_1) - I_1 \ddot{\phi}_1 + T_1 = 0,$$

which is equivalent to

$$pM_x(x_{1,0} - r_1) - M_y y_{1,0} - m_1(x_{1,0} - r_1) = 0 \tag{2.8}$$

where M_x, M_y are given by (2.4), (2.6), respectively. Here, we have used the following equation for the rotational motion of the front wheel:

$$T_1 + F_1 r_1 - I_1 \ddot{\phi}_1 = 0 \tag{2.9}$$

5. For the rear wheel, taking moments about B gives:

$$F_2 x_{2,0} - N_2 y_{2,0} - m_2 \ddot{y}(x_{2,0} - r_2) - I_2 \ddot{\phi}_2 + T_2 = 0,$$

which gives

$$(1-p)M_x(x_{2,0} - r_2) - M_y y_{2,0} - m_2(x_{2,0} - r_2) = 0, \qquad (2.10)$$

where M_x, M_y are given by (2.4), (2.6), respectively. Here we have used the following equation for the rotational motion of the rear wheel:

$$T_2 + F_2 r_2 - I_2 \ddot{\phi}_2 = 0. \qquad (2.11)$$

The above derivation shows that the trailing-arm vehicle model satisfies an anti-squat property for tractive forces in the ratio $p : (1-p)$ only if (2.8) and (2.10) hold. Conversely, we can also show that conditions (2.8) and (2.10) are sufficient for the anti-squat property. The proof of this direction is straightforward through the analysis of each element of the model, allowing the possibility of general motion of the sprung mass, but showing that no pitching or vertical motion actually occurs.

We can carry out a similar analysis for the case of anti-dive under braking. If we have braking forces in the ratio $F_1/F_2 = q/(1-q)$ and consider the case of inboard brakes (i.e., reaction torques applied to the car body) then Equations (2.8) and (2.10) with p replaced by q are the necessary and sufficient conditions for anti-dive. For the case of outboard brakes, a similar analysis gives the following necessary and sufficient conditions:

$$qM_x x_{1,0} - M_y y_{1,0} - m_1(x_{1,0} - r_1) + \frac{I_1}{r_1} = 0 \qquad (2.12)$$

$$(1-q)M_x x_{2,0} - M_y y_{2,0} - m_2(x_{2,0} - r_2) + \frac{I_2}{r_2} = 0 \qquad (2.13)$$

where M_x, M_y are given by (2.4) (2.6), respectively.

Classical Conditions for Anti-dive and Anti-squat

The necessary and sufficient conditions (2.8), (2.10) for perfect anti-squat (or anti-dive with inboard braking), and (2.12), (2.13) for perfect anti-dive with outboard braking were derived for non-zero values of m_1, m_2, I_1 and I_2. We now show that these conditions reduce to the standard geometrical conditions if the masses and inertias of the wheels, m_1, m_2, I_1, I_2, are negligible. In such a case, (2.8) and (2.10) reduce to:

$$\frac{x_{1,0} - r_1}{y_{1,0}} = \frac{h}{pL}, \qquad \frac{x_{2,0} - r_2}{y_{2,0}} = \frac{h}{(1-p)L}$$

and (2.12) and (2.13) reduce to:

$$\frac{x_{1,0}}{y_{1,0}} = \frac{h}{qL}, \qquad \frac{x_{2,0}}{y_{2,0}} = \frac{h}{(1-q)L}$$

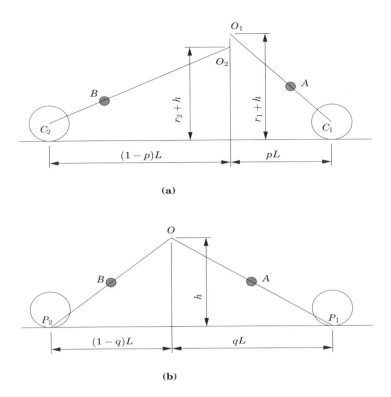

Figure 2.5 Simplified trailing-arm geometry: (a) anti-squat and anti-dive with inboard braking; (b) anti-dive with outboard braking

The geometrical interpretation of these conditions is illustrated in Figure 2.5. For the anti-squat case or anti-dive with inboard braking, the trailing-arm joint A (and B) should be on the line C_1O_1 (and C_2O_2), where O_1 is at a height h above C_1 and a horizontal distance pL from C_1, and similarly for O_2. For the anti-dive case with outboard braking, the trailing-arm joint A (and B) should be on the line P_1O (and P_2O). These results agree with the standard conditions in the literature (see [4] and [5]) when we set $r_1 = r_2$. An interesting alternative approach to suspension geometry, in which the geometry of suspension linkages is defined in terms of mathematical relationships rather than specific linkages is given in [7].

Passivity Analysis

We observe that in each case the general conditions for anti-dive (2.8), (2.10) and anti-squat (2.12), (2.13) amount to two equations in four unknowns ($x_{1,0}$, $y_{1,0}$, $x_{2,0}$, $y_{2,0}$). However, these conditions can rarely be satisfied all together in practice for the usual case when p and q are unequal. For example, in the case of inboard brakes, the only solution is $x_{1,0} = r_1$,

$x_{2,0} = r_2$, and $y_{1,0} = y_{2,0} = 0$, which is nonsensical.

Typically then, if the suspension geometry is chosen to achieve full anti-dive control of vehicle squat will need to be achieved through the suspension struts, and similarly if full anti-squat is chosen. If a high resistance to squat is provided by the suspension forces, then an unsatisfactory response to road disturbances results if passive suspension is used. If a "soft" response is required from road disturbances, then active suspension is needed, as is shown by the following example.

EXAMPLE 2.1
Consider a half-car trailing-arm model, as shown in Figure 2.3, with the following coefficients: $m_s = 1600\,\text{kg}$, $I_s = 1000\,\text{kg m}^2$, $m_1 = m_2 = 100\,\text{kg}$, $I_1 = I_2 = 1\,\text{kg m}^2$, $r_1 = r_2 = 0.15\,\text{m}$, and model the front (rear) tyre vertical force as a spring with constant $k_{t_1} = 500\,\text{kN/m}$ ($k_{t_2} = 500\,\text{kN/m}$). Suppose the vehicle is four-wheel drive, with $T_1/T_2 = 2/1$ (or $p = 0.6749$), and the design braking force ratio is $F_1/F_2 = 4/6$ (or $q = 0.4$). The trailing arms are regarded as massless and without inertia. A no-slip constraints is set for both wheels. Suppose the coordinates in the nominal configuration are given by: $Q = (0,0)$, $A = (0.15, 0.6126)$, $B = (0.15, -1.0008)$, $C_1 = (0.35, 1.3)$, $C_2 = (0.35, -1.3)$, $D = (0.1, 1.3)$, $E = (0.1, -1.3)$, $G = (0.35, 1.3)$, $H = (0.35, -1.3)$ in the units of meters. The trailing-arm geometry arrangement satisfies the anti-squat conditions (2.8) and (2.10). The performance requirements are that it should be soft to the road disturbances z_{r_1} and z_{r_2} as with a soft passive suspension with parameters $c_1^s = c_2^s = 1\,\text{kNs/m}$, $k_1^s = k_2^s = 4\,\text{kN/m}$ is used. At the same time, it should also be stiff to the accelerating/braking torques T_1, T_2 as when a hard passive suspension with parameters $c_1^h = c_2^h = 100\,\text{kNs/m}$, $k_1^h = k_2^h = 400$ is used. A corresponding network for the outboard braking model is shown in Figure 2.6(a). The transformation matrix G can be expressed as:

$$\left[s(\hat{\phi}_1 - \hat{\phi}_4), s(\hat{\phi}_2 - \hat{\phi}_5), -\hat{N}_1, -\hat{N}_2\right]' = [G^h, G^s]\left[\hat{T}_1, \hat{T}_2, s\hat{z}_{r_1}, s\hat{z}_{r_2}\right]'.$$

In order to get a suitable scale, we check the positive realness of Y instead of G, where Y is defined as:

$$\left[\hat{T}_1, \hat{T}_2, -\hat{N}_1, -\hat{N}_2\right]' = Y\left[s(\hat{\phi}_1 - \hat{\phi}_4), s(\hat{\phi}_2 - \hat{\phi}_5), s\hat{z}_{r_1}, s\hat{z}_{r_2}\right]'$$

The results in Figure 2.6(b) show that the linearised vehicle dynamics fail to be passive by virtue of the fact that some eigenvalues of $(Y + Y^*)$ become negative.

For the inboard braking and accelerating cases, similar results can also be obtained by changing the cross variables ϕ_4, ϕ_5 to be ϕ_3 in Figure 2.6(a).

□

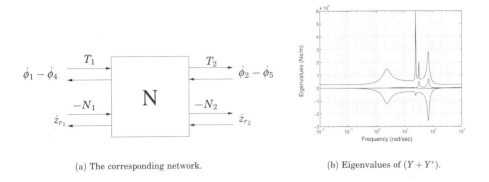

(a) The corresponding network. (b) Eigenvalues of $(Y+Y^*)$.

Figure 2.6 The passivity analysis of the half-car trailing-arm model with out-board braking

2.4 Active Suspension Design for Independence of Disturbance Responses

Motivated by the problem of achieving the separate requirements in active suspension systems for road and load disturbances, the authors derived in [11] a parameterisation of the set of all stabilising controllers for a given plant, which leaves some pre-specified closed-loop transfer function fixed. This allows the disturbance responses to be designed successively. We summarise these results here.

We consider the LFT (linear fractional transformation) model in Figure 2.7, where the Laplace transfer function of the generalised plant P is partitioned as:

$$P = \begin{bmatrix} P_{11} & P_{12} \\ P_{21} & P_{22} \end{bmatrix}$$

and further partitioned conformably with the disturbance signals as:

$$\begin{bmatrix} \begin{bmatrix} \hat{z}_1 \\ \hat{z}_2 \end{bmatrix} \\ \hat{y} \end{bmatrix} = \begin{bmatrix} \begin{bmatrix} P_{11,11} & P_{11,12} \\ P_{11,21} & P_{11,22} \end{bmatrix} & \begin{bmatrix} P_{12,1} \\ P_{12,2} \end{bmatrix} \\ \begin{bmatrix} P_{21,1} & P_{21,2} \end{bmatrix} & P_{22} \end{bmatrix} \begin{bmatrix} \begin{bmatrix} \hat{w}_1 \\ \hat{w}_2 \end{bmatrix} \\ \hat{u} \end{bmatrix} \quad (2.14)$$

where $w_1 \in \mathbb{R}^{m_1}$, $w_2 \in \mathbb{R}^{m_2}$, $u \in \mathbb{R}^{m_3}$, $z_1 \in \mathbb{R}^{p_1}$, $z_2 \in \mathbb{R}^{p_2}$, $y \in \mathbb{R}^{p_3}$ at any time instant and $\hat{u}$ denotes the Laplace transform of $u(t)$ etc. We consider the problem of parametrising all stabilising controllers which leave $T_{\hat{w}_1 \to \hat{z}_1}$ (the transfer function from $\hat{w}_1$ to $\hat{z}_1$) the same as for some given stabilising controller K_0.

We summarise the result for the case where P is stable and $K_0 = 0$. Suppose r_2 and r_3 are the normalrank of $P_{12,1}$ and $P_{21,1}$, respectively. It can be shown [11] that there always exists a factorisation $P_{21,1} = U_1 F$ where

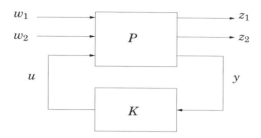

Figure 2.7 Generalised model in LFT form

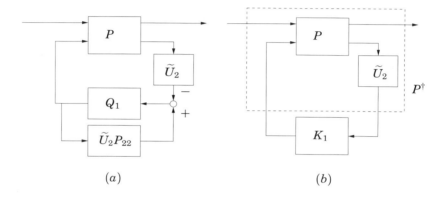

Figure 2.8 General controller structure and equivalent controller

$F \in \mathbb{RH}_\infty^{r_3 \times m_1}$, $U_1 \in \mathbb{RH}_\infty^{p_3 \times r_3}$, $U = (U_1, U_2)$ over $\mathbb{RH}_\infty$ is unimodular and F is of full row normalrank r_3. We make a further assumption that $m_3 = r_2$. Consider the following partition of U^{-1}:

$$U^{-1} = (U_1, U_2)^{-1} = \begin{pmatrix} \widetilde{U}_1 \\ \widetilde{U}_2 \end{pmatrix}$$

THEOREM 2.2
The set of all stabilising controllers for which $T_{\hat{w}_1 \to \hat{z}_1} = P_{11,11}$ can be parameterised as:

$$K = -(I - Q_1 \widetilde{U}_2 P_{22})^{-1} Q_1 \widetilde{U}_2 \qquad (2.15)$$

for $Q_1 \in \mathbb{RH}_\infty^{r_2 \times (p_3 - r_3)}$ (see Figure 2.8 (a)). Furthermore, the controller structure given in Figure 2.8 (b), where K_1 ranges over the set of stabilising controllers for $\widetilde{U}_2 P_{22}$, also parameterises all stabilising controllers for which $T_{\hat{w}_1 \to \hat{z}_1} = P_{11,11}$. □

In [11], the authors derived explicit parameterisations of the set of all stabilising controllers that leave the road disturbance responses fixed in

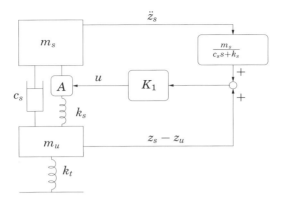

Figure 2.9 Controller structure for the quarter-car with two feedbacks

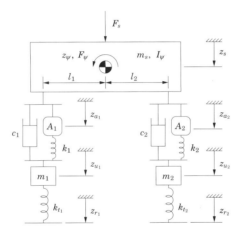

Figure 2.10 The half-car model

a variety of vehicle models incorporating a "Sharp" actuator system [8], [13]. The controller structure for the quarter-car model is shown in Figure 2.9, where K_1 is any stabilising controller for the system. This turns out to be a generalisation of a structure used in [12].

The simple half-car model shown in Figure 2.10 with four measurements: $[\ddot{z}_s, \ddot{z}_\psi, D_1, D_2]$, where $D_1 = z_s + l_1 z_\psi - z_{u1}$, $D_2 = z_s - l_2 z_\psi - z_{u2}$ are strut deflections, was also discussed in [11]. The controller structure, which leaves the road responses the same as the passive system is then found as in Figure 2.11.

A key step in the method described in this section is the computation of the matrix $\widetilde{U}_2$, which determines the required controller structure. When a symbolic calculation of $\widetilde{U}_2$ is not feasible, the following numerical procedure

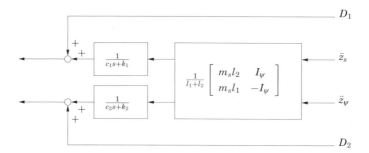

Figure 2.11 Controller structure ($\widetilde{U}_2$) for the half-car model

can be followed: 1) partition $T_{3,1} = [B', A']'$, where B is square; 2) find a minimal realisation of $L := AB^{-1}$; (3) find a left coprime factorisation $L = \widetilde{B}^{-1}\widetilde{A}$ and set $\widetilde{U}_2 = [-\widetilde{A}, \widetilde{B}]$ (e.g., see [14, Theorem 12.19]). Such an approach will be used in Section 2.5.

2.5 Active Suspension Design for the Trailing-arm Model

In this section, we numerically calculate an active controller that improves the load responses, while keeping the road responses the same as in the passive case, for the nonlinear half-car trailing-arm model of Figure 2.3 with the suspension struts replaced by "Sharp" actuators as in Figure 2.10. The approach makes use of Theorem 2.2. We will use the coefficients in Example 2.1, and compare the responses of the active and passive system.

Controller Design

We first set up a trailing-arm model in *AutoSim*, and obtained a linearised model for the controller design. The suspension geometry was chosen to give perfect anti-squat for $T_1/T_2 = 2$: the heights of joints A and B were fixed as $x_{1,0} = x_{2,0} = 0.35$ m, then (2.8) gives $y_{1,0} = 0.6874$ m and (2.10) gives $y_{2,0} = 0.2992$ m, which give the same coordinates as in Example 2.1. The linearised half-car model had four measurements: $[\ddot{z}_s, \ddot{z}_\psi, D_1, D_2]$, as the linear half-car model in Section 2.4. For disturbance response decoupling design, a soft passive suspension with $k_1 = k_2 = 40$ kN/m, $c_1 = c_2 = 10$ kNs/m was chosen to give satisfactory responses to road disturbances, and the active controller was designed to stiffen the responses to load disturbances and braking torques without changing the road responses. After the numerical design, the controller structure $\widetilde{U}_2$ after model reduction became:

$$\widetilde{U}_2 = \begin{bmatrix} \dfrac{0.0788(s+3.8402)}{(s+3.9793)(s+3.7744)} & \dfrac{-0.0385(s+3.7307)}{(s+3.9793)(s+3.7744)} & -1 & 0 \\ \dfrac{0.0766(s+4.0033)}{(s+3.9793)(s+3.7744)} & \dfrac{0.0356(s+3.9341)}{(s+3.9793)(s+3.7744)} & 0 & -1 \end{bmatrix}$$
(2.16)

Choosing a weighting function as follows:

$$W_1 = \dfrac{8(s+80)}{(s+2)} \begin{bmatrix} 1 & 0 \\ 0 & 1 \end{bmatrix}$$

an H_∞ loop-shaping controller after model reduction was obtained as:

$$K_1 = \begin{bmatrix} \dfrac{-1.16(s+4.88 \cdot 10^3)(s+33.7 \pm 17.1j)}{(s+2.13)(s+152.8 \pm 109.8j)} & \dfrac{-5.32(s-0.57)(s+40.1 \pm 35.8j)}{(s+2.13)(s+152.8 \pm 109.8j)} \\ \dfrac{-4.92(s-3.02)(s-11.4 \pm 76.1j)}{(s+2.1)(s+104.8 \pm 114.4j)} & \dfrac{-22.6(s+29.4)(s+86.9 \pm 20.2j)}{(s+2.1)(s+104.8 \pm 114.4j)} \end{bmatrix}$$

with the controller layout as shown in Figure 2.8 (b).

Simulation Results

The closed-loop responses of the active system reduced the d.c. gain of $T_{\hat{F}_s \to \hat{z}_s}$ from 1.35×10^{-5} to 1.12×10^{-6}, $T_{\hat{T}_s \to \hat{\phi}_3}$ from 7.99×10^{-6} to 6.60×10^{-7}, $T_{\hat{T}_1 \to \hat{z}_s}$ from -2.66×10^{-5} to -2.47×10^{-7}, $T_{\hat{T}_2 \to \hat{z}_s}$ from 5.33×10^{-5} to 4.94×10^{-7}, $T_{\hat{T}_1 \to \hat{\phi}_3}$ from -9.83×10^{-6} to -1.84×10^{-6}, $T_{\hat{T}_2 \to \hat{\phi}_3}$ from 1.44×10^{-5} to -1.61×10^{-6}. As expected, the active controller significantly reduces the load and torque responses without changing the road responses. The active controller was implemented in *AutoSim* for accelerating and braking scenarios. Figure 2.12 (a)–(c) show the applied torques, forward velocity and pitch behaviour under acceleration and braking. Figure 2.12(c) shows that perfect anti-squat behaviour is achieved in acceleration as expected. (A more practical approach would be to select the suspension geometry to give a suitable compromise between partial anti-squat and anti-dive, but the present choice was made to better illustrate the theory.) A further simulation was carried out incorporating a vertical tyre spring and a relaxation of the no-slip condition. To model a rolling wheel with tyre the following method was employed: the unsprung masses were taken to be discs of mass 100 kg and inertia 1 kg m² with a no-slip rolling contact on a virtual road; the instantaneous vertical position of the virtual road was determined by the point being connected to the true road's vertical position at each wheel station by a spring of the same stiffness as the tyre spring. (This is not dissimilar to the method used to produce a tyre spring effect on the ThrustSSC supersonic car [2], which used aluminium forged wheels

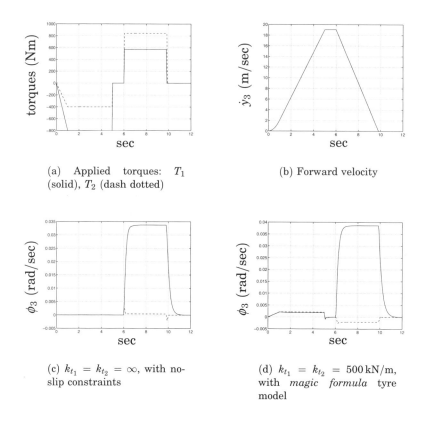

Figure 2.12 (a) Applied torques; (b) velocity; (c) the pitch motion of the active (dashed) and passive (solid) systems with solid wheels and no-slip condition; (d) same as (c) but with tyre spring and slip

without tyres running on a water-laid desert surface with a natural 'yield'!) The relaxation of the no-slip condition made use of the *magic formula* of [3]. The simulation results are shown in Figure 2.12(d).

2.6 References

[1] B. D. O. Anderson and S. Vongpanitlerd. *Network Analysis and Synthesis: A Modern Systems Theory Approach*. Prentice Hall, Englewood Cliffs, NJ, 1973.

[2] R. Ayres. Thrust ssc environmental testing. *Environmental Engineering*, 9(2):28–30, June, 1996.

[3] E. Bakker, L. Nyborg, and H. B. Pacejka. Tyre modelling for use in vehicle dynamics studies. *Society of Automotive Engineerings Transactions*, 96(2):190–204, 1988.

[4] C. Campbell. *Automobile Suspensions*. Chapman and Hall, 1981.

[5] J.C. Dixon. *Tyres, Suspension and Handling*. Cambridge University Press, first edition, 1991.

[6] R.W. Newcomb. *Linear Multiport synthesis*. McGraw-Hill, 1966.

[7] R.S. Sharp. Influences of suspension kinematics on pitching dynamics of cars in longitudinal manoeuvring. *Vehicle System Dynamics Supplement*, 33:23–36, 1999.

[8] R.S. Sharp and S.A. Hassan. On the performance capabilities of active automobile suspension systems of limited bandwidth. *Vehicle System Dynamics*, 16:213–225, 1987.

[9] M.C. Smith. Achievable dynamic response for automotive active suspension. *Vehicle System Dynamics*, 24:1–33, 1995.

[10] M.C. Smith and G.W. Walker. Performance limitations and constraints for active and passive suspension: a mechanical multi-port approach. *Vehicle System Dynamics*, 33:137–168, 2000.

[11] M.C. Smith and F-C. Wang. Controller parametrisation for disturbance response decoupling: Application to vehicle active suspension control. *IEEE Transactions on Control Systems Technology*, to be published, 2001.

[12] R.A. Williams, A. Best, and I.L. Crawford. Refined low frequency active suspension. *Int. Conf. on Vehicle Ride and Handling, Proc. ImechE*, C466/028:285–300, 1993.

[13] P.G. Wright and D.A. Williams. The case for an irreversible active suspension system. *SAE Transactions, J. of Passenger Cars, Sect. 6,*, 83–90, 1989.

[14] K. Zhou, J.C. Doyle, and K. Glover. *Robust and Optimal Control*. Prentice Hall, Englewood Cliffs, NJ, 1996.

3

Modeling of Drivers' Longitudinal Behavior

J. Bengtsson   R. Johansson   A. Sjögren

Abstract

In the last few years, many vehicle manufacturers have introduced advance driver support in some of their automobiles. One of those new features is adaptive cruise control (ACC), which extends the conventional cruise control system to control of relative speed and distance to other vehicles. In order to design an ACC controller, it is suitable to have a model of drivers' behavior. Our approach to find dynamical models of the drivers' behavior was to use system identification. Basic data analysis is made by means of system identification methodology, and several models of drivers' longitudinal behavior are proposed, including both linear regression models and subspace-based models. In various situations, detection for when a driver's behavior changes or deviates from the normal is useful. To that purpose, a GARCH (generalized autoregressive conditional heteroskedasticity) model was used to model the driver in situations such as arousal.

Figure 3.1 Radar mounted on the vehicle (left) with working range, laser mounted on vehicle with working range (middle), the laser device used from IBEO (right)

3.1 Introduction

Systems that support a driver in traffic situations and reduce the total driver workload, have been studied since the 1950s. Several of these support systems aim towards fully or partially automatic driver assistance system such as those for longitudinal control, which are often called ACC systems [11], [12], [13], [19]. Much attention has also been paid to ACC devices in the PATH project [10]. The motivation for these systems is that they are aiming to increase the driving comfort, reduce the traffic accidents and increase the flow throughput. These ACC systems autonomously adjust the vehicle's speed according to current driving conditions. In order to accomplish driver comfort the system must resemble driver behavior in traffic and the system must avoid irritation of the driver and the surrounding traffic. Therefore, to design a system that resembles the natural longitudinal behavior a good model of a driver is needed. There exist several models of the drivers' longitudinal behavior, which all aim to describe various parts of the drivers' behavior. The model structures are different, some are based on cognitive models [5], [7], [16], [20] or are general longitudinal models [2], [17], [24] or only car-following models [4], [1], [3]. Most of them have one thing in common in that they are using static models.

3.2 Material and Methods

Experimental platform—Vehicles. The cars used were two Volvo 850s. One car was used as a leading vehicle and the other as a following vehicle. The car used as the leading vehicle had the property that it was possible to program the car to drive along a trajectory. By using that feature, it was possible to reproduce the driving situation and to let all the different drivers drive the same situation.

Table 3.1 Autoliv-CelsiusTech Electronics

Modulation characteristics	Modulation type FMCW
Radar scanning principle	Mechanical scanning
Frequency	76-77 GHz
Transmitted power	10 mW
Minimum tracking distance	2 m
Maximum tracking distance	200 m
Update rate of radar	10 Hz
Field of view	24°
Angle resolution	0.1°
Distance resolution	1 m

Table 3.2 IBEO Laser scanner LD Automotive

Minimum tracking distance	0.4 m
Maximum tracking distance	100 m
Update rate of laser	10 Hz
Field of view	up to 270°
Angle resolution	0.25°
Distance resolution	0.004 m

Sensor equipment. A radar from Autoliv-CelsiusTech Electronics was used to measure the distance to the front vehicle ΔY and it's relative speed Δv. Some technical radar information is given in Table 3.1. A practical difficulty was that the radar must have good resolution, also at small distances, and that the relative speed measured with high resolution. A laser from IBEO was used to measure ΔY and Δv—see Figure 3.1 and Table 3.2. The reason to use both radar and a laser is their different working ranges, the radar having a narrow but long working range and the laser having a wide but short working range (Figure 3.1).

ACC device of Volvo team. Some work on ACC at Volvo Technical Development was reported in [18]. In this study, a stop-and-go controller for ACC was designed and implemented. As both cars used in this study were equipped with ACC, it was possible to let the cars drive along a specified trajectory. This property was used in the experiment protocol in order to reproduce the same driving situations for all participating drivers.

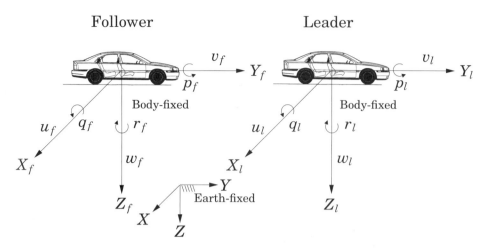

Figure 3.2 Body-fixed and earth-fixed reference frames

Data acquisition. The data was collected with a sampling rate of 10 Hz. The collected data variables are space headway (ΔY), differential velocity (Δv), velocity (v_f), throttle angle (α_t), and brake pressure (p_b). A natural choice of inputs to the driver model are ΔY, Δv, and v_f. The outputs are then α_t, p_b. In the test situations, the front car was programmed to drive along a predefined trajectory. The driver in the front car only had to steer the car, whereas the throttle and the brakes were feedback-controlled to keep the trajectory.

Experimental Design

Figure 3.2 shows a car-following situation with the speed of the preceding and following vehicle denoted v_l and v_f, respectively, the distance between the vehicles being denoted as ΔY, headway $\Delta Y = y_l - y_f$ and the relative speed being defined as:

$$\Delta v = v_l - v_f = \frac{d}{dt}\Delta Y \tag{3.1}$$

There are four types of situations where data have been collected: following, cut-in, braking, and mode changing. Cut-in situations describe a scenario wherein a vehicle cuts in front of the driver's vehicle from another lane. In the braking situations, the headway distance decreased under the individual minimal headway distance, and the driver braked to re-establish the headway distance. In mode-changing situations, the driver shifted from uninfluenced driving to car following. Data collection of various situations have been done on public roads as well as on test tracks. Seven different drivers of various sex and age (23–35) participated in the data collection. The data acquisition was performed in the summer of 2000 during good weather conditions.

Modeling of Drivers' Longitudinal Behavior

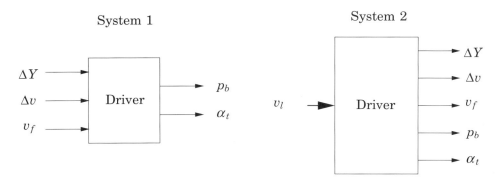

Figure 3.3 Two different input and output separations—one with ΔY, Δv, and v_f as inputs to the driver *(left)* and another one with the velocity v_l of the leader as independent variable

Data Analysis

There are at least two possible separation of variables into inputs and outputs, the first one being to choose ΔY, Δv, and v_f as inputs and the outputs as α_t, p_b (Figure 3.3). This is the standard separation. The second approach is to let the velocity of the leader, v_l be the input and ΔY, Δv, v_f, α_t, and p_b to be the output (Figure 3.3). This decomposition is useful in understanding the interaction between the driver and the vehicle. Figures 3.4 and 3.5 show data from one of the following situations in which seven different drivers participated. There are individual differences between the drivers, but also large similarities among their behavior. The major differences among the drivers consist of the choice of space headway and safety distance. In these situations, the front car drove along a predefined trajectory. This allowed all the drivers to drive the same test situation. The fact that the drivers all drove the exact same situation makes it very easy to compare the drivers against each other. Some of the drivers drove with caution and kept a long headway distance. These drivers also only have to used a small brake pressure whereas those drivers that drove more aggressively and kept a short headway distance needed to use higher brake pressure.

Data analysis was made by means of system identification methodology [14]. Autospectra, cross-spectra and coherence spectra of the inputs (ΔY, Δv, and v_f) and outputs (α_t and p_b), were made for assessment of the various signals levels and relationships.

In Figure 3.6, the coherence spectra among inputs and outputs are shown. The coherence among inputs and outputs is high, which can be interpreted as an indication that there exists a linear relationship between the inputs and the outputs. Note that the coherence for p_b is higher than for α_t.

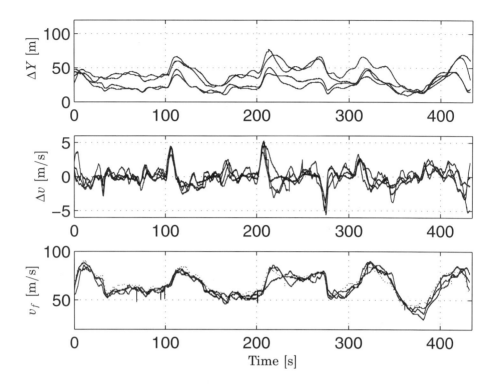

Figure 3.4 Data acquisition of the input data from one following situation. Data are from different drivers

Modeling

The human driver can be viewed as a closed-loop system, with feedback from the front vehicle's velocity v_l (Figure 3.7). All the experiments were performed in closed-loop feedback operation and there may be systematic problems in how to obtain relevant information from this type of experiment [14, Chapter 8]. As the experiment condition implies feedback operation, the data may not be informative enough to establish a valid model of the driver. Note that the system possesses a multi-input multi-output structure.

Linear regression models. To find out if there is some relationship between the input data and the output, a linear regression model was estimated [14]. The linear regression model takes on the format

$$y_k = [\Delta Y_k ... \Delta Y_{k-n} \Delta v_k ... \Delta v_{k-n} ... v_{f_k} ... v_{f_{k-n}}] \theta + e_k \qquad (3.2)$$

where n is the estimated order and the noise sequence $\{e_k\}$ represents additive errors. A linear regression model of high order was estimated. Since the model order is high, it may be assumed that the computed residual ε_k is a good approximation of the noise e_k. The residual sequence was used in pseudolinear regression to estimate a model of lower order.

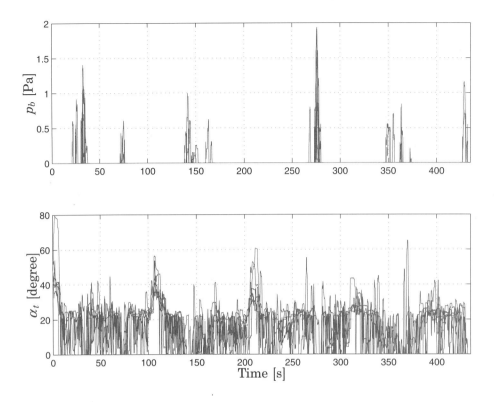

Figure 3.5 Data acquisition of the output data—i.e., brake pressure p_b and throttle angle α_t—from one following situation. Data are from different drivers

State-space models using subspace-based identification. A discrete-time time-invariant system in state-space realization:

$$x_{k+1} = Ax_k + Bu_k + Kw_k$$
$$y_k = Cx_k + Du_k + w_k$$

where $\{w_k\}$ is an noise sequence on innovations form.

The problem is to estimate the order n of the system and the system matrices A, B, C, D. In Figure 3.8 is a schematic representation of the identification problem. The subspace method is well suited to modeling of multivariable systems [14]. To determine the model order, a Hankel matrix of input-output data is constructed and the choice of model order is based on the singular values of the Hankel matrix [22], [21]. However, if there is strong noise influence then this criterion degrades and becomes non-conclusive.

Behavioral model. Behavioral model identification may be suggested in cases without clear-cut distinction of signals as inputs or outputs [23],

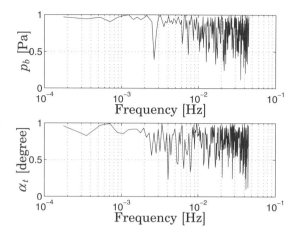

Figure 3.6 Coherence spectra between the inputs and the outputs. The upper figure: coherence between inputs $[\Delta Y \quad \Delta v \quad v_f]$ and the output p_b. The lower figure: coherence between inputs $[\Delta Y \quad \Delta v \quad v_f]$ and the output α_t

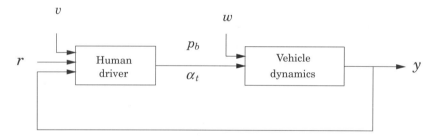

Figure 3.7 Structure of a human driver in car-following. r is the inputs to the driver from the lead vehicle. v is the observation noise, w is the motor noise, and y is the car position and velocity

[15]. This may be preferable since there is feedback interaction between the driver and the car. There are also interactions between the driver and the other vehicle, for example, in cut-in situations. The behavioral method has great similarities with the subspace method, but differs in its absence of explicit separation among inputs and outputs. Thus, the estimated state-space model represents all the dynamics, both for the inputs and for the outputs. Then, by matrix fraction description an input-output model can be obtained.

3.3 Results and Validation

In all cases, identification accuracy was measured using the variance-ac-

Modeling of Drivers' Longitudinal Behavior

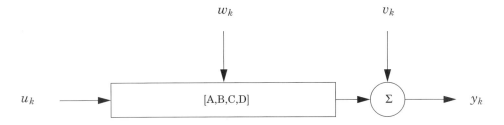

Figure 3.8 Schematic representation of the innovations model identification problem

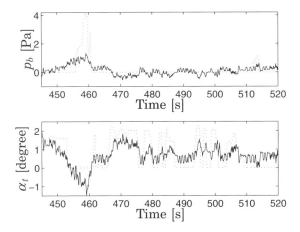

Figure 3.9 Data (*dotted*) and simulated output data from a linear regression model of order $n = 30$ (*solid*)

counted-for (VAF) as test quantity

$$\text{VAF} = 1 - \frac{\text{Var}(y - \hat{y})}{\text{Var}(y)} \times 100\,\% \tag{3.3}$$

In the model estimation, the normalized ΔY, Δv, v_f, α_t, and p_t were used.

Linear regression

A linear regression model of order $n = 30$, with input and output chosen as System 1 (Figure 3.3) was estimated and is shown in Figure 3.9. The model captures some of the driver's behavior. One reason why not even this high-order model succeeds in modeling the driver may be that the experiment set-up is a closed-loop system. The model is better in predicting the driver's throttle angle α_t behavior than the brake pressure p_b behavior. A possible background would be that the acceleration and deceleration have different explanations, for example that deceleration could be explained by air resistance or topography. The residual analysis of the model is shown in

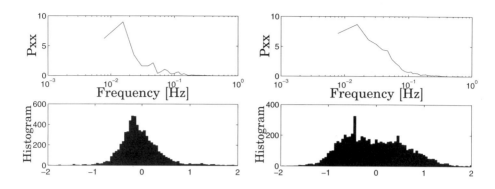

Figure 3.10 Histogram and auto-correlation of the residuals from a thirtieth order linear regression model. To the left is the residuals of output p_b. To the right is the residuals of output α_t

Figure 3.10 and it is found that the residual from output α_t and the output p_b have different distributions. The residuals of this high-order model were further used to estimate a pseudolinear regression model. The result is shown in Figure 3.11, and the model captures most of the driver's behavior, even the braking behavior. The VAF scores for the linear regression model are 41.1 % (p_b) and 46.2 % (α_t) whereas the VAF scores for the pseudolinear regression model: 89.9 % (p_b) and 73.2 % (α_t), respectively.

Subspace-based Identification

The state-space models using subspace methods have been designed by using the SMI Toolbox [9] in Matlab. In Figure 3.12 the data and simulated output data have been compared and it was found that lower order models of state-space models have problems to capture the behavior of the driver. The state-space models captured some of the driver's behavior (Figure 3.12). The best result was obtained for the model order $n = 15$, with input and output chosen as System (Figure 3.3) but there are some possible time delays. The estimated model $n = 15$ is better at capturing the driver's throttle behavior than the brake behavior with VAF scores: 44.3 % (p_b) and 48.7 % (α_t). Residual analysis of the model is shown in Figure 3.13. Similar to the result for linear regression models, the residual sequences for α_t and p_b have different empirical distributions.

Behavioral Model

A behavioral model of order $n = 30$, with input-output representation chosen as System 2 of Figure 3.3 was estimated and model response is shown in Figure 3.14. The model captures the driver behavior very well, it captures both the braking behavior and the throttle behavior. The residual analysis of the car-following model based on the behavior method is shown in Figure 3.15 and they both have empirical distribution comparable with

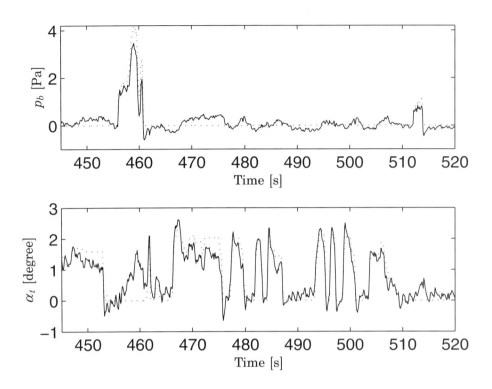

Figure 3.11 Data (*grey*) and simulated output data from a pseudolinear regression model of first order (*black*)

normal distributions. VAF scores for the behavioral model are 81.9% (p_b) and 92.2% (α_t).

The Arousal Behavior

We notice that the residual ε from the seventh order behavior model for p_b becomes large when the braking starts. We may call this phenomenon "arousal behavior" and estimate a generalized autoregressive conditional heteroskedasticity (GARCH) model [6], [8]. A GARCH(r,m) model is:

$$u_t = v_t \sqrt{h_t} \tag{3.4}$$

where v_t is an independently distributed Gaussian sequence with zero mean and unit variance and h_t is:

$$h_t = \kappa + \delta_1 h_t + \delta_2 h_{t-2} + \cdots + \delta_t h_{t-r} + \alpha_1 u_{t-1}^2 + \alpha_2 u_{t-2}^2 + \cdots + \alpha_m u_{t-m}^2$$

where $\kappa \equiv [1 - \delta_1 - \delta_2 - \cdots - \delta_r]\zeta$. In Figure 3.16, the squared residual sequence is shown, and the residual of p_b seems to increase linearly during the brake part.

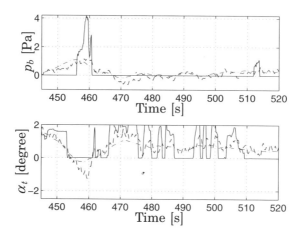

Figure 3.12 Data (*grey*) and simulated output data from state-space models using subspace methods, n=5 (*dashdot*), and n=15(*solid*)

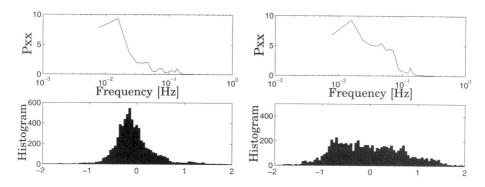

Figure 3.13 Histogram and auto-correlation of the residuals from a fifteen order sub-spaced model. To the left is the residuals of output p_b. To the right is the residuals of output α_t

The estimated third order linear regression models for the different drivers capture the behavior of the residual well, Figure 3.16, 3.17, and 3.18. In Figure 3.19, the impulse response from the linear regression of the second driver is shown. Notice that there is an association between the arousal behavior in the brake situation and the throttle behavior, respectively.

3.4 Conclusion

We have studied the dynamical longitudinal behavior of drivers' and models designed with various model structure. The design approach to use system identification, such as linear regression, state-space models using subspace

Modeling of Drivers' Longitudinal Behavior

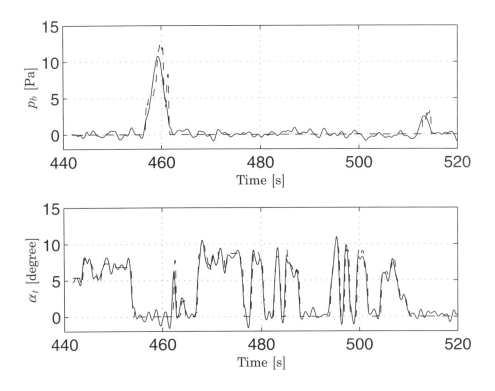

Figure 3.14 Data (*grey*) and simulated output data from a behavioral model (*black*)

methods, and behavioral models was found to work fairly well, especially with the behavioral models. The accuracy differs between the various model structures, and the best VAF scores are achieved by the behavioral model. Progress of the VAF scores for increasing model orders and for various model structures is shown in Figure 3.20. The modeling of the arousal behavior was found to work well. The proposed model captures the deviating behavior in arousal situations.

Acknowledgments

This work was a joint project with partners of Volvo Technological Development Corporation and the Department of Automatic Control, Lund University. The authors thank NUTEK for financial support.

3.5 References

[1] P. S. Addison and D. J. Low. A novel nonlinear car-following model. *Chaos*, 8(4):791–799, 1998.

[2] K. I. Ahmed. *Modeling Drivers' Acceleration and Lane Changing Behavior*. Phd thesis, Massachusetts Institute of Technology, Cambridge, MA, 1999.

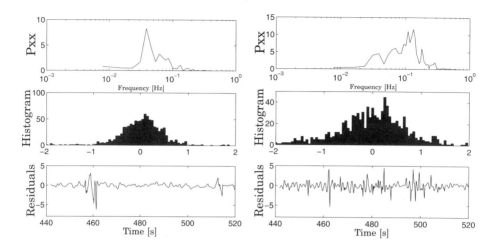

Figure 3.15 Residuals of output p_b from a behavioral model $n = 30$ (*left*). Residuals of output α_t from a behavior model $n = 30$ (*right*). Notice that the residual of the output p_b becomes large when the time is around 460 s

[3] R. F. Benekohal and J. Treiterer. Carsim: Car-following model for simulation of traffic in normal and stop-and-go conditions. *Transportation Research Record 1194*, 99–111, 1998.

[4] T. Bleile. A new microscopic model for car-following behavior in urban traffic. In *Proc. 4th World Congress on Intelligent Transport Systems*, Berlin, 1997.

[5] E. R. Boer and M. Hoedemaeker. Modeling driver behavior with different degrees of automation: A hierarchical decision framework of interacting mental models. In *Proc. XVIIth European Conference on Human Decision Making and Manual Control*, Valenciennes, France, 1998.

[6] T. Bollerslev. Generalized autoregressive conditional heteroskedasticity. *J. Econometrics*, 31:307–327, 1986.

[7] M. A. Goodrich and E. R. Boer. Semiotics and mental models: Modeling automobile driver behavior. In *Proc. 1998 IEEE ISIC/CIRA/ISAS Joint Conference*, 771–776, Gaithersburg, MD, 1998.

[8] J. D. Hamilton. *Time Series Analysis*. Princeton Univ. Press, Princeton, NJ, 1994.

[9] B. Haverkamp. *Subspace Model Identification, Theory and Practice*. Ph. D. thesis, Delft University of Technology, Delft, NL, 2001.

[10] J. K. Hedrick, V. Garg, J. C. Gerdes, D. B. Maciua, and D. Swaroop. Longitudinal control development for IVHS fully automated and semi-automated system phase 3. Technical Report UCB-ITS-PRR-97-20, Dept. Mechanical Engineering, Univ. California Berkeley, Berkeley, CA, 1997.

[11] J. Hitz, J. Koziol, and A. Lam. Safety evaluation results from the field operational test of an intelligent cruise control (ICC) system. *Number 2000-011352 in SAE Technical Paper Series*, 2000.

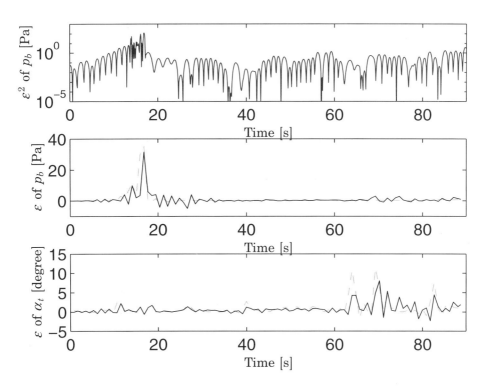

Figure 3.16 Upper figure shows squared residuals ε from the behavioral model $n = 30$, illustrating the heteroskedasticity properties according to a GARCH model. Center and lower figures show residuals ε from driver 1 (*dashed*) and computed residuals ε from a linear regression model (*solid*)

[12] T. Iijima, A. Higashimata, S. Tange, K. Mizoguchi, H. Kamiyama, K. Iwasaki, and K. Egawa. Development of an adaptive cruise control system with brake actuation. *Number 2000-01-1353 in SAE Technical Paper Series*, 2000.

[13] P. A. Ioannou and C. C. Chen. Autonomous intelligent cruise control. *IEEE Transactions on Vehicular Technology*, 42(4), 1993.

[14] R. Johansson. *System Modeling and Identification*. Prentice Hall, Englewood Cliffs, NJ, 1993.

[15] R. Johansson, M. Verhaegen, C. T. Chou, and A. Robertsson. Behavioral model identification. In *IEEE Conf. Decision and Control (CDC'98)*, 126–131, Tampa, FL, December 1998.

[16] N. Kuge, T. Yamamura, O. Shimoyama, and A. Liu. A driver behavior recognition method based on a driver model framework. *Number 2000-01-0349 in SAE Technical Paper Series*, 2000.

[17] W. Leutzbach. *Introduction to the Theory of Traffic Flow*. Springer-Verlag, Berlin, Germany, 1988.

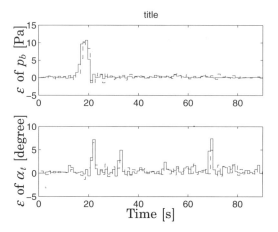

Figure 3.17 Residuals ε from driver 2 (*dashed*) and computed residuals ε from a linear regression model (*solid*). Heteroskedasticity properties according to a GARCH model

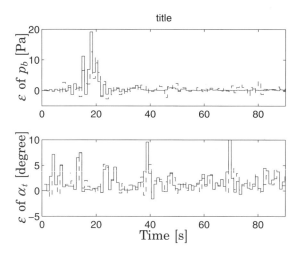

Figure 3.18 Residuals ε from driver 3 (*dashed*) and computed residuals ε from a linear regression model (*solid*). Heteroskedasticity properties according to a GARCH model

[18] M. Persson, F. Botling, E. Hesslow, and R. Johansson. Stop-and-go controller for adaptive cruise control. In *Proc. 1999 IEEE Int. Conf. Control Applications*, volume 2, 1692–1697, Kohala Coast, HI, 1999.

[19] W. Prestl, T. Sauer, J. Steinle, and O. Tschernoster. The BMW active cruise control ACC. *Number 2000-01-0344 in SAE Technical Paper Series*, 2000.

[20] T. A. Ranney. Models of driving behavior: A review of their evolution. *Accid. Anal. and Prev*, 26(6):733–750, 1994.

Modeling of Drivers' Longitudinal Behavior

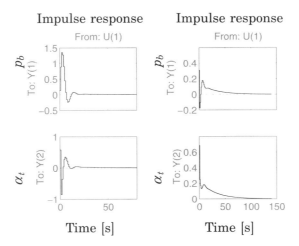

Figure 3.19 Impulse response where input is residual from p_b (*left*) and impulse response where input is residual from α_t (*right*). Heteroskedasticity properties according to a GARCH model

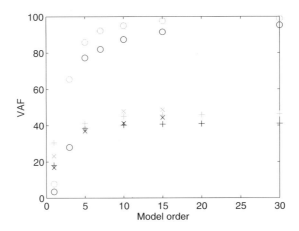

Figure 3.20 VAF scores (p_b (*grey*) and α_t (*black*)) for various model structures and model order (*linear regression model +, state-space model using subspace methods x, and behavioral model o*)

[21] M. Verhaegen and P. Dewilde. Subspace model identification—analysis of the elementary output-error state-space model identification algorithm. *Int. J. Control*, 56(5):1211–1241, 1992.

[22] M. Verhaegen and P. Dewilde. Subspace model identification—the output-error state-space model identification class of algorithms. *Int. J. Control*, 56:1187–1210, 1992.

[23] J.C. Willems. From time series to linear systems. Part I: Finit dimensional

linear time invariant systems, and Part II: Exact modelling. *Automatica*, 22:561–80 and 675–694, 1986.

[24] Q. Yang and H. N. Koutsopoulos. A microscopic traffic simulator for evaluation of dynamic traffic management systems. *Transportation Research C*, 4:113–129, 1996.

4

Nonlinear Adaptive Backstepping with Estimator Resetting using Multiple Observers

J. Kalkkuhl T. A. Johansen J. Lüdemann

Abstract

A multiple model based observer/estimator for the estimation of parameters is used to reset the parameter estimation in a conventional Lyapunov-based nonlinear adaptive controller. Transient performance can be improved without increasing the gain of the controller or estimator. This allows performance to be tuned without compromising robustness and sensitivity to noise and disturbances. The advantages of the scheme are demonstrated in an automotive wheel slip controller.

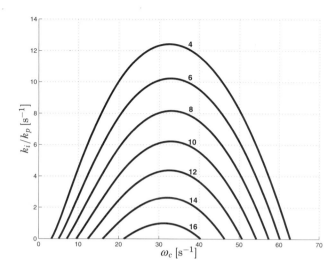

Figure 4.1 Regions of robust stability of open loop system (4.1) for different plant poles a, a time delay of $\tau = 0.014$ s and a desired phase margin of 36 degrees

4.1 Introduction

This paper addresses some performance issues in nonlinear adaptive control of a class of systems with unknown but fixed parameters. It is well known, that transient performance in such control systems is conditioned by the set of controller and adaptation gains. The higher the controller and/or adaptation gain, the faster the error transients will be. However, when it comes to implementation, performance limitations due to sampling, computational delay, unmodelled dynamics, and noise almost always arise. This is, in particular, crucial for unstable nonlinear systems where nonlinear control design would be most beneficial, but non-robustness may arise from excessively high gain. For linear systems, these limitations have been well investigated and can be quantified easily. Consider, for instance, the PI-controller design for a first order unstable system with a small computational delay τ in the controller, where the open-loop transfer function is given by

$$G(s) = \frac{k_i + k_p s}{s} \mathrm{e}^{-s\tau} \cdot \frac{a}{s-a}, \qquad a > 0, \quad k_i > 0, \quad k_p > 0 \qquad (4.1)$$

In Figure 4.1, the regions of parametric robust stability in terms of crossover frequency ω_c and controller zero $z = -\frac{k_i}{k_p}$ are shown. Obviously, for given ω_c, robust stability can only be achieved if the integral gain of the controller is below a certain limit. For nonlinear systems, robustness results with respect to unmodelled dynamics are scarce. However, the linear example clearly shows the practical benefits of having low gains, in particular, low adaptation gains for robustness with respect to unmodelled dynamics.

Therefore, the focus of this paper will be a multiple model-based method for achieving fast transients in adaptive control without employing high adaptation gains.

The use of multiple models to switch or reset parameter estimators has originally been proposed for certainty equivalence adaptive control of linear systems [13], [17], [18], [19], [15], [16], [20], [27]. External disturbances, changes in the system's dynamics parameter variations, *etc.*, are treated as unknown different environments in which the system has to operate. Multiple linear models, which may either be fixed or adaptive, are used to describe those different environments [17], [18], [19], [20]. All models operate in parallel and, corresponding to each model there exists a parameterized controller. The optimal model is chosen among the set of multiple models on the basis of some measure of the identification errors. Various strategies have been proposed to prevent the system from arbitrarily fast switching [13], [15].

For nonlinear systems a similar overall structure as for linear systems has been proposed [19] whereby the choice of models, the controllers and the adaptive algorithms were adopted to the nonlinear case.

Similarly, the supervisory control approach in [15], [16], [5] compares in real-time the norm-squared output prediction errors of a number of available estimators instead of models, and uses the estimator with the least prediction error.

Both approaches base decision making on the principle of certainty equivalence [14], [3], [11], that is, an adaptive feedback controller is designed on the basis of current values of plant parameter estimates with the understanding that these estimates are to be viewed as correct, even though they may not be. To prevent the system from chattering, a dwell time switching logic [15] is applied. The system is prevented from switching until the dwell time has been reached. After the switching, the time is set to zero again.

The approach in [27] differs from the approaches above. The methodology is to design stable linear controllers for linear time-invariant (LTI) and linear time-varying (LTV) systems. The state space is decomposed into subsets and a common quadratic Lyapunov function is derived for each subset giving the admissible controller.

In this paper, we present a hybrid approach to speed up transients in continuous Lyapunov-based nonlinear adaptive control systems. Hereby, a multiple model observer (MMO) is used to reset the parameter estimation in a nonlinear adaptive controller that is not based on certainty equivalence. The advantage of combining both approaches is that transients due to adaptation can be damped out while the robustness and performance of the controller with respect to disturbances can be improved. As a consequence, the gain of the continuous adaptive controller can be considerably lowered, thus increasing the robustness with respect to noise and unmodelled dynamics. The parameter resetting is based on a control Lyapunov function (CLF) and can guarantee asymptotic stability. The approach will be applied to hybrid friction estimation and wheel slip control design. The improvement of the transient performance of nonlinear adaptive backstepping using estimator resetting based on multiple models has been presented in previous papers [9], [10]. The main contributions of the present paper consists of:

- the use of multiple model-based observers instead of estimators;
- the formulation of a set of sufficient closed loop stability conditions for resetting tuning function nonlinear adaptive controllers;
- introduction of reference trajectory resetting strategies to obtain tighter performance bounds;
- the introduction of a fast multiple model observer, from which even under transient conditions a parameter estimate can be obtained;
- the application of the technique to friction estimation in wheel slip control.

The paper is organised as follows: In Section 4.2, some results of constructive nonlinear adaptive control are briefly reviewed and a motivation for discontinuous parameter resetting is given. This is followed by an analysis of the closed-loop stability implications of resetting parameter estimates (Section 4.3) where a first-order and a second-order example are used to illustrate the results. Section 4.4 describes the concept of multiple model observers and gives for a special plant structure sufficient conditions for stability of parameter resetting. In Section 4.5, the combination of tuning function approach and multiple observer-based resetting is applied to a second order benchmark system. At the end, Section 4.6 discusses wheel slip control as an application of the method and gives some preliminary experimental results.

4.2 Nonlinear Adaptive Backstepping

Consider the adaptive tracking problem for a parametric strict-feedback system [11]

$$\dot{x}_1 = x_2 + \varphi_1(x_1)^T \theta$$
$$\dot{x}_2 = x_3 + \varphi_2(x_1, x_2)^T \theta$$
$$\vdots$$
$$\dot{x}_{n-1} = x_n + \varphi_{n-1}(x_1, x_2, \ldots, x_{n-1})^T \theta$$
$$\dot{x}_n = \beta(x)u + \varphi_n(x)^T \theta$$

where $\theta \in \mathbb{R}$ is a vector of unknown constant parameters, β, and

$$F = [\varphi_1, \ldots, \varphi_n]$$

are smooth nonlinear functions taking arguments in $\mathbb{R}^n$. It has been shown that in a tuning function adaptive controller for such a system, adaptive control law and parameter update law take the following form:

$$u = \frac{1}{\beta(x)}\left[\alpha_n(x, \hat{\theta}, \bar{y}_r^{(n-1)}) + y_r^{(n)}\right] \quad (4.2)$$

$$\dot{\hat{\theta}} = \Gamma \tau_n(x, \hat{\theta}, \bar{y}_r^{(n-1)}) \quad (4.3)$$

where y_r is the reference signal to be tracked by the output y

$$\bar{y}_r^{(i)} = \left(y_r, \dot{y}_r, \ldots, y_r^{(i)}\right)$$

and the control law and the tuning functions are given recursively by

$$z_i = x_i - y_r^{(i-1)} - \alpha_{i-1} \quad (4.4)$$

$$\alpha_i(\bar{x}_i, \hat{\theta}, \bar{y}_r^{(i-1)}) = -z_{i-1} - c_i z_i - w_i^T \hat{\theta}$$
$$+ \sum_{k=1}^{i-1}\left(\frac{\partial \alpha_{i-1}}{\partial x_k}x_{k+1} + \frac{\partial \alpha_{i-1}}{\partial y_r^{(k-1)}}y_r^{(k)}\right) \quad (4.5)$$
$$- \kappa_i |w_i|^2 z_i + \frac{\partial \alpha_{i-1}}{\partial \hat{\theta}}\Gamma \tau_i + \sum_{k=2}^{i-1}\frac{\partial \alpha_{k-1}}{\partial \hat{\theta}}\Gamma w_i z_k$$

$$\tau_i(\bar{x}_i, \hat{\theta}, \bar{y}_r^{(i-1)}) = \tau_{i-1} + w_i z_i \quad (4.6)$$

$$w_i(\bar{x}_i, \hat{\theta}, \bar{y}_r^{(i-1)}) = \varphi_i - \sum_{k=1}^{i-1}\frac{\partial \alpha_{i-1}}{\partial x_k}\varphi_k \quad (4.7)$$

$$i = 1 \ldots n$$

$$\bar{x}_i = (x_1, \ldots, x_i), \quad \bar{y}_r^{(i)} = (y_r, \dot{y}_r, \ldots, y_r^{(i)})$$

$$\alpha_0 = 0, \quad \tau_0 = 0, \quad c_i > 0$$

The control law together with the parameter update law render the time derivative of the Lyapunov function

$$V_n = \frac{1}{2}z^T z + \frac{1}{2}\tilde{\theta}^T \Gamma^{-1}\tilde{\theta} \quad \text{with} \quad \tilde{\theta} = \theta - \hat{\theta} \quad (4.8)$$

negative semi-definite:

$$\dot{V}_n = -\sum_{k=1}^{n}c_k z_k^2 - \sum_{k=i}^{n}\kappa_i|w_i|^2 z_i^2 \leq -c_0 |z|^2 \quad \text{where} \quad c_0 = \min_{1\leq i\leq n} c_i \quad (4.9)$$

Furthermore, the resulting closed-loop system is of the form

$$\dot{z} = Az + W\tilde{\theta} \tag{4.10}$$

$$A = \begin{bmatrix} -c_1 - \kappa_1|w_1|^2 & 1 & 0 & \cdots & 0 \\ -1 & -c_2 - \kappa_2|w_2|^2 & 1+\sigma_{23} & & \sigma_{2n} \\ 0 & -1-\sigma_{23} & \ddots & \ddots & \vdots \\ \vdots & \vdots & \ddots & \ddots & 1+\sigma_{n-1,n} \\ 0 & -\sigma_{2n} & \cdots & -1-\sigma_{n-1,n} & -c_n - \kappa_n|w_n|^2 \end{bmatrix}$$

$$W = \begin{bmatrix} w_1^T \\ \vdots \\ w_n^T \end{bmatrix}, \quad \sigma_{ik} = -\frac{\partial \alpha_{i-1}}{\partial \hat{\theta}} \Gamma w_k$$

where

$$\tilde{\theta} = \theta - \hat{\theta} \tag{4.11}$$

denotes the parameter estimation error.

Our main objective is to improve the transient performance of the closed-loop system described above, in particular, with respect to adaptation of the unknown parameter vector θ, which is assumed to be constant with respect to time. Assuming for simplicity the adaptation weights to be a diagonal matrix $\Gamma = \gamma I$, it can be shown [11] that the following inequalities hold for the $\mathcal{L}_2$ and $\mathcal{L}_\infty$ transient error as a function of the initial values of the control error and the parameter error:

$$\|z\|_2 \leq \frac{|\tilde{\theta}(0)|}{\sqrt{2c_0\gamma}} + \frac{1}{\sqrt{2c_0}}|z(0)| \tag{4.12}$$

$$|z(t)| \leq \frac{1}{\sqrt{\gamma}}|\tilde{\theta}(0)| + |z(0)| \tag{4.13}$$

$$|z(t)| \leq \frac{1}{2\sqrt{c_0\kappa_0}}\sqrt{|\tilde{\theta}(0)|^2 + \gamma|z(0)|^2} + |z(0)|e^{-c_0 t} \tag{4.14}$$

where

$$\kappa_0 = \left(\sum_{i=1}^n \frac{1}{\kappa_i}\right)^{-1}$$

and c_0 is given in (4.9). It will be shown later that a zero initial control error $z(0) = 0$ can always be achieved by reference initialisation, in which case the corresponding bounds are given by

$$\|z\|_2 \leq \frac{|\tilde{\theta}(0)|}{\sqrt{2c_0\gamma}} \tag{4.15}$$

$$|z(t)| \leq \min\left\{\frac{1}{2\sqrt{c_0\kappa_0}}, \frac{1}{\sqrt{\gamma}}\right\} |\tilde{\theta}(0)| \tag{4.16}$$

Hence, the transient performance can be improved by increasing any of the design parameters c_i, κ_i and γ. The higher the gain, the faster the transient response of the control systems. In practical applications, however, high gain should be avoided as there are always unmodelled dynamics or even time delays (related to computer implementation) in the system which may lead to instability if the loop-gain is too high. Thus, other strategies of counteracting uncertainties are highly desirable.

Such a strategy is provided by the multiple model switching and tuning approach, where the estimates are taken from a finite set

$$\hat{\theta}_i, \quad i = 1, \ldots, N$$

thus, trading in accuracy for speed. The multiple model observer provides additional information on parameter uncertainies, which can then be used to instantaneously reset the parameter estimate $\hat{\theta}$ of the controller. Suppose the best estimate of the multiple model observer with respect to *modelling performance* is

$$\hat{\theta}^+ = \hat{\theta}_j$$

Then, a decision has to be made whether or not to use this additional information. In the case when the multiple model estimate is used the current continuous estimate $\hat{\theta}^-$ will be discarded and the corresponding value *reset* to the new value $\hat{\theta}^+$. This resetting decision should not be based on the modelling performance alone. It should also be guaranteed that global stability of the control system is preserved and the control performance, in particular, the transient behaviour, is improved via resetting.

In between the resetting events, the parameter estimate will still be governed by the adaptation law and it will, thus, be piecewise continuous. This will result in discontinuous control and adaptation laws. Since the state transformation in Equation (4.4) is parameterised by $\hat{\theta}$, the states $z_2, \ldots, z_n$ will be discontinuous in time.

In the remainder of the paper, the switching conditions and the implications of such a resetting strategy will be studied.

4.3 Stability Analysis of Parameter Resetting

Sufficient Conditions for Stability

Stability results for discontinuous Lyapunov functions are reported in [26]. For stability, it is sufficient that:

1. $V(x)$ be continuous with respect to its arguments;
2. $V(x)$ is non-decreasing along trajectories in between switching events;
3. $V(x^+) \leq V(x^-)$ whenever there is a jump from x^- to x^+ at some time instant t.

Here, we denote

$$x^- = x(t^-) := \lim_{\tau \downarrow t} x(\tau)$$
$$x^+ = x(t^+) := \lim_{\tau \uparrow t} x(\tau) \qquad (4.17)$$

the one-sided limits. Consider the Lyapunov function (4.8) of the tuning function approach

$$V_n(z, \theta, \hat{\theta}) = \frac{1}{2} z^T z + \frac{1}{2} \tilde{\theta}^T \Gamma^{-1} \tilde{\theta} \qquad \text{with } \tilde{\theta} = \theta - \hat{\theta} \qquad (4.18)$$

For the tuning function approach, it can be easily shown that properties 1 and 2 hold due to the stability of the closed-loop system when no resetting is applied. When the parameter estimate $\hat{\theta}$ is reset, the state variable z depending on $\hat{\theta}$ changes discontinuously with time. Then, to obtain a sufficient condition for stability, it remains to be analysed whether

$$\Delta V_n = V_n(z(\hat{\theta}^+), \theta, \hat{\theta}^+) - V_n(z(\hat{\theta}^-), \theta, \hat{\theta}^-) \leq 0 \qquad (4.19)$$

holds. If this is the case, then a resetting of $\hat{\theta}$ from $\hat{\theta}^-$ to $\hat{\theta}^+$ is admissible. In general, the state vector z will depend on $\hat{\theta}$ in a nonlinear way and can be computed from Equations (4.4)–(4.6).

Denote the step change in parameter

$$\Delta\hat{\theta} = \hat{\theta}^+ - \hat{\theta}^- \qquad (4.20)$$

The set of admissible parameter changes depends on the state z and on the parameter error $\tilde{\theta}$. Thus, in general, additional information on the estimation error is necessary to check the admissibility of $\Delta\hat{\theta}$. It can be easily verified that even in the case when $\hat{\theta}$ steps from $\hat{\theta}^-$ to the *correct parameter value* $\hat{\theta}^+ = \theta$, the Lyapunov function could actually *increase* due to its dependency on z.

Reference trajectory resetting

The condition (4.19) on $\Delta\hat{\theta}$ can be considerably simplified when resetting of the reference trajectory y_r is used in combination with parameter resetting. Reference trajectory resetting can be applied most easily in the case where y_r and its derivatives are generated by a linear reference model, which is driven by some external reference input signal $r(t)$. For the following considerations, we assume the existence of a reference model since the states of such a system can be reset directly.

Reference trajectory initialisation was originally developed for improving the transients in adaptive tuning function control systems [11]. In fact, by resetting the n values $y_r(t^+), \dot{y}_r(t^+), \ldots, y_r^{(n-1)}(t^+)$, an additional degree of

Nonlinear Adaptive Backstepping using Multiple Observers

freedom is obtained which enables us to set $z^+ = 0$. From Equation (4.4), it can be seen that $z^+ = 0$ requires the solution of the set of equations

$$y_r^{(i-1)}(t^+) = x_i - \alpha_{i-1}(\bar{x}_i, \hat{\theta}^+, y_r(t^+), \ldots, y_r^{(n-1)}(t^+) y_r^{(i-2)}(t^+))$$
$$i = 1, \ldots, n \quad (4.21)$$

It can be shown [11] that the solution to these equations does not depend on the controller parameters.

The corresponding step change in the Lyapunov function with reference trajectory resetting is

$$\begin{aligned}\Delta V_n &= (\theta - \hat{\theta}^+)^T \Gamma^{-1} (\theta - \hat{\theta}^+) \\ &\quad - (z^-)^T (z^-) - (\theta - \hat{\theta}^-)^T \Gamma^{-1} (\theta - \hat{\theta}^-) \\ &= \Delta\hat{\theta}^T \Gamma^{-1} \Delta\hat{\theta} - 2(\tilde{\theta}^-)^T \Gamma^{-1} \Delta\hat{\theta} - (z^-)^T(z^-)\end{aligned} \quad (4.22)$$

for which we can obtain a controller-independent upper bound

$$\Delta V_n \le \Delta\hat{\theta}^T \Gamma^{-1} \Delta\hat{\theta} - 2(\tilde{\theta}^-)^T \Gamma^{-1} \Delta\hat{\theta} \quad (4.23)$$

Moreover, after each resetting step, the transient error will be bounded by:

$$\|z(t-t^+)\|_2 \le \frac{|\tilde{\theta}^+|}{\sqrt{2c_0\gamma}} \quad (4.24)$$

$$|z(t-t^+)| \le \min\left\{\frac{1}{2\sqrt{c_0\kappa_0}}, \frac{1}{\sqrt{\gamma}}\right\} |\tilde{\theta}^+| \quad (4.25)$$

In comparison, the transient errors without resetting are given by

$$\|z^o(t-t^+)\|_2 \le \frac{|\tilde{\theta}^-|}{\sqrt{2c_0\gamma}} + \frac{1}{\sqrt{2c_0}} |z^-| \quad (4.26)$$

$$|z^o(t-t^+)| \le \frac{1}{\sqrt{\gamma}} |\tilde{\theta}^-| + |z^-| \quad (4.27)$$

$$|z^o(t-t^+)| \le \frac{1}{2\sqrt{c_0\kappa_0}} \sqrt{|\tilde{\theta}^-|^2 + \gamma |z^-|^2} + |z^-| e^{-c_0 t} \quad (4.28)$$

This shows clearly that whenever $|\tilde{\theta}^+| < |\tilde{\theta}^-|$ holds, a reduction of the transient error bound will be obtained. Moreover, the use of reference trajectory resetting guarantees that the value of the Lyapunov function V_n^+ after resetting is actually decreasing with decreasing parameter estimation error $\tilde{\theta}^+$, and for $\hat{\theta}^+ = \theta$ we would achieve $V_n^+ = 0$ after one step.

Application to Low-order Systems

Sufficient stability conditions for first order systems. Consider the tracking control of the first order system

$$\dot{x}_1 = \varphi_1(x_1)\theta + u \quad (4.29)$$

An adaptive tuning function controller is simply

$$u = -\varphi_1(x_1)\hat{\theta} - c_1 z_1 - \dot{y}_r \tag{4.30}$$
$$\dot{\hat{\theta}} = \gamma z_1 \varphi_1(x_1) = \gamma \tau_1 \tag{4.31}$$
$$z_1 = x_1 - y_r$$

This controller based on the control Lyapunov function

$$V = \frac{1}{2}z_1^2 + \frac{1}{2\gamma}\left(\theta - \hat{\theta}\right)^2 \tag{4.32}$$

renders the derivative of the Lyapunov function negative semi-definite

$$\dot{V} = -c_1 z_1^2 \leq 0$$

The closed-loop system is given by

$$\dot{z}_1 = -c_1 z_1 + \varphi_1(x_1)\tilde{\theta} \tag{4.33}$$

The time derivative of the squared error along the solution of (4.33) is

$$\frac{d}{dt}\left(\frac{1}{2}z_1^2\right) = z_1 \dot{z}_1$$
$$= -c_1 z_1^2 + z_1 \varphi_1(x_1)\tilde{\theta} \tag{4.34}$$

For the rest of the discussion of the first-order case, we assume that $\varphi_1(x_1) > 0$. This assumption is not necessary for the approach in general but it simplifies the resetting conditions considerably.

For the first order system (4.29) and the Lyapunov function (4.32) we obtain by use of Equation (4.19) the following sufficient stability condition:

$$\Delta V = V^+ - V^-$$
$$= \frac{1}{2\gamma}\left(\hat{\theta}^+ - \hat{\theta}^-\right)^2 - \frac{1}{\gamma}\left(\left(\theta - \hat{\theta}^-\right)\left(\hat{\theta}^+ - \hat{\theta}^-\right)\right) \leq 0 \tag{4.35}$$

This gives the following conditions on the step change in the parameter estimate:

$$\text{sgn}\left(\Delta\hat{\theta}\right) = \text{sgn}\left(\tilde{\theta}^-\right) \tag{4.36}$$
$$|\Delta\hat{\theta}| \leq 2|\tilde{\theta}^-| \tag{4.37}$$

In general, Condition (4.37) cannot be verified without additional information on the parameter estimate. However, a parameter resetting law can be designed such that condition (4.36) holds. Due to the assumption that φ is always positive using the closed-loop error equation (4.33), we obtain:

$$\dot{x}_1 x_1 > 0 \quad \text{implies} \quad \text{sgn}(\dot{x}_1) = \text{sgn}(\tilde{\theta}^-) \tag{4.38}$$

Thus, from the signs of the closed-loop error and its derivative, we can determine the sign of the parameter estimation error. Since these two

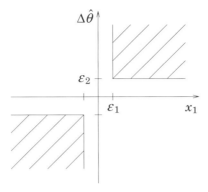

Figure 4.2 Parameter resetting law

signals might be noisy and generate chattering, the following parameter resetting condition with threshold and a hysteresis is derived:

$$x_1 \text{sgn}(\dot{x}_1) = \varepsilon_1 \bigwedge x_1 \Delta\hat{\theta} > \varepsilon_1 \varepsilon_2, \quad \varepsilon_1 > 0, \, \varepsilon_2 > 0 \tag{4.39}$$

The resetting condition is depicted in Figure 4.2. Condition (4.39) states that resetting occurs whenever the magnitude of the control error x_1 exceeds some threshold, and at the same time there is a significant discrepancy between the continuous parameter estimate and the multiple observer parameter estimate having the same sign as the control error. Thus, provided that $|x_1|$ is increasing while it crosses the threshold ε_1, the sign of $\dot{x}_1$ is a direct indicator of the sign of the parameter error $\tilde{\theta}^-$. In the general case, the sign of φ will be known and the resetting law can be modified accordingly. Since Condition (4.39) only satisfies (4.36) but not (4.37), it is neither necessary nor sufficient. A sufficient condition is given by the following theorem:

THEOREM 4.1
1. Consider the first order system (4.29) together with the continuous control law (4.30) and the update law (4.31). If the parameter $\hat{\theta}$ is reset under Condition (4.39), then the sign condition (4.36) is satisfied.

2. Provided the sign condition is satisfied, then a decrease of V in Equation (4.32) at the switching instant is obtained, provided that

$$|\Delta\hat{\theta}| < 2|\tilde{\theta}^-| \tag{4.40}$$

   holds. Thus, a sufficient condition for stability is satisfied.

3. If to the contrary

$$|\Delta\hat{\theta}| \geq 2|\tilde{\theta}^-| \tag{4.41}$$

holds, then the control error x_1 is driven towards zero as long as $|x_1| > \varepsilon_1$ despite of the increase in value of V.

Proof. The first and second part of the lemma has been proven above.

If the assumptions of the third part of the theorem hold, then, outside the ball $|x_1| > \varepsilon_1$ we have along the solutions of the closed-loop equation:

$$\begin{aligned}
\frac{d}{dt}\left(\frac{1}{2}x_1^2\right) &= x_1\dot{x}_1 \\
&= x_1\left(-c_1 x_1 + \varphi_1 \tilde{\theta}\right) \\
&= -c_1 x_1^2 + x_1 \varphi_1(x_1)\tilde{\theta} \\
&= -c_1 x_1^2 + x_1 \varphi_1(x_1)\left[\tilde{\theta}^- - \Delta\hat{\theta} S(y, \Delta\hat{\theta})\right] \\
&\leq -c_1 x_1^2 + |x_1 \varphi_1(x_1)|\left[|\tilde{\theta}^-| - |\Delta\hat{\theta}|\right] < 0
\end{aligned} \quad (4.42)$$

due to (4.41) which implies that x_1 is driven towards the origin □

Information concerning condition (4.37) will be derived later in Section 4.4 using the properties of the multiple model observer. Thus, by combining the parameter resetting law (4.39) with the multiple observer approach, we will arrive at a sufficient condition for stability.

Second-order systems. Consider the second-order system with one parameter

$$\begin{aligned}
\dot{x}_1 &= x_2 + \varphi(x_1)\theta \\
\dot{x}_2 &= u
\end{aligned} \quad (4.43)$$

Designing the tuning function controller for such a system requires one backstep. Setting

$$\begin{aligned}
z_1 &= x_1 - y_r \\
z_2 &= x_2 - \alpha_1(x_1, \hat{\theta}, x_r, \dot{x}_r) - \dot{y}_r
\end{aligned} \quad (4.44)$$

and assuming that the parameter estimate $\hat{\theta}$ can vary discontinuously with time, we will thus have also discontinuous changes with time in α_1 and z_2 and in the corresponding Lyapunov function

$$V = \frac{1}{2}z_1^2 + \frac{1}{2}z_2^2 + \frac{1}{2\gamma}\tilde{\theta}^2 \quad (4.45)$$

If reference trajectory resetting is used, the step change in the Lyapunov function can be expressed according to (4.22) and (4.23) as

$$\begin{aligned}
\Delta V = V^+ - V^- &= \frac{1}{\gamma}\Delta\hat{\theta}^2 - \frac{2}{\gamma}\tilde{\theta}^- \Delta\hat{\theta} - (z_1^-)^2 - (z_2^-)^2 \\
&\leq \frac{1}{\gamma}\Delta\hat{\theta}^2 - \frac{2}{\gamma}\tilde{\theta}^- \Delta\hat{\theta}
\end{aligned} \quad (4.46)$$

from which we obtain

$$\Delta \hat{\theta}^2 - 2\tilde{\theta}^- \Delta \hat{\theta} \leq 0 \tag{4.47}$$

This is identical to the first-order case (4.35), and thus, Conditions (4.36)-(4.37) are required to hold. Similar to the first-order case, a switching law can be derived from the closed-loop error equations

$$\begin{aligned}
\dot{z}_1 &= -c_1 z_1 + z_2 + w_1 \tilde{\theta} \\
\dot{z}_2 &= -z_1 - c_2 z_2 + w_2 \tilde{\theta}
\end{aligned}$$

Under the condition $w_i \neq 0$, the following is obtained:

$$\begin{array}{c}
[\dot{z}_1 > 0 \wedge (-c_1 z_1 + z_2^-)\text{sign}(w_1) < 0] \\
\vee \\
[\dot{z}_2 > 0 \wedge (-c_2 z_2^- + z_1^-)\text{sign}(w_1) < 0]
\end{array} \quad \text{implies} \quad \tilde{\theta} > 0$$

$$\begin{array}{c}
[\dot{z}_1 < 0 \wedge (-c_1 z_1 + z_2^-)\text{sign}(w_2) > 0] \\
\vee \\
[\dot{z}_2 < 0 \wedge (-c_2 z_2^- + z_1^-)\text{sign}(w_2) > 0]
\end{array} \quad \text{implies} \quad \tilde{\theta} < 0 \tag{4.48}$$

4.4 Multiple Model Observer (MMO)

In this section, a multiple model observer structure for fast parameter estimation will be derived. It will be shown that for special cases, an additional bound on the estimation error can be given, even under transients. The MMO will then be combined with the tuning function controller and sufficient conditions for stability of the overall control system will be given.

The MMO Structure

Quite similar to the multiple model estimation described in [17], [18], [19], [20], the basic idea of the MMO is to construct a finite set of parallel observers, each of which is designed for a fixed parameter value or a fixed plant structure, respectively. In its simplest form, the MMO consists of a set O of N individual observers o_i. Each individual observer stands for a fixed parameter θ_i, $i = 1, \ldots, N$ covering the range of admissible parameter values. Figure (4.3) shows the structure of a multiple observer parameter estimation. Each of the N observer estimates the states of the system and is driven by the output error $e_{1i} = x_1 - \hat{x}_{1i}$, where x_1 denotes the vector of measurable states. Since any mismatch between a single observer and the physical system will in general lead to a steady-state estimation error, this error can be used to determine the best observer for the actual system.

Using discontinuous output injection functions is common in sliding mode observers [25]. A hybrid observer using convergence information to

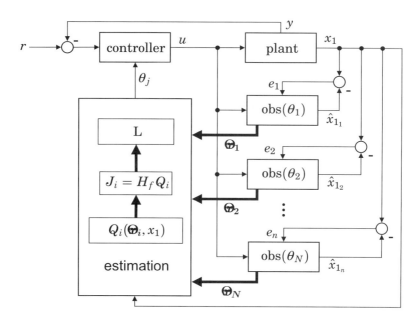

Figure 4.3 Multiple observer-based parameter estimation

switch between several discontinuous output injection functions for nonlinear systems has been reported in [12]. Here, we propose instead to use a set of observers with fixed output injection functions that are run in parallel, and thus can have considerably faster transients. A performance index $Q_i(\boldsymbol{\theta}_i, x_1)$ is defined for each individual observer of the set O. The performance index weighs the output error of the observer, and thus quantifies the mismatch between the plant and the observer. The performance index is low-pass filtered to make it less sensitive to noise:

$$J_i = H_f Q_i$$

A switching logic L is used to select an optimal estimate θ_j among the parameters $\{\theta_1, \ldots, \theta_N\}$ of the multiple observer O. L satisfies two purposes:

1. Selecting the parameter θ_j corresponding to the observer o_j with the best performance:

$$j = \underset{i}{\operatorname{argmin}}\, J_i$$

2. Providing a mechanism that ensures a convergence of the estimator after a finite number of switches.

In order to prevent the optimal parameter estimate from chattering, two different approaches have been suggested in the literature:

- dwell time switching [15] where after each switch for a certain period of time, switching to a new estimate is prohibited;
- hysteresis switching [13], [8]: Let o_p be the valid observer at time t^-, then a switch to a new observer o_j occurs only if $Q_j(t^+)(1+h) < Q_p(t^+)$, where $Q_p(t^+)$ is the current performance of the observer o_p and $h > 0$ is the hysteresis. Otherwise, no switching will occur and o_p will remain valid.

Construction of the Individual Observers in the First-order Case

Consider the system (4.29) where the parameter θ is treated as an augmented state

$$\begin{aligned}
\dot{x}_1 &= \varphi_1(y)\theta + u \\
\dot{\theta} &= 0 \\
y &= x_1 \\
\varphi_1(y) &> 0
\end{aligned} \tag{4.49}$$

It is assumed that the parameter θ is contained in a closed interval $[\theta_{min}, \theta_{max}]$. The interval is discretised using a set N parameter values

$$\theta_{min} < \theta_1 < \theta_2 < \ldots < \theta_N < \theta_{max}$$

Each of the N individual observers of the multiple model observer will centered around one of the discrete parameter values θ_i. For this purpose, Equation (4.49) is rewritten into

$$\begin{aligned}
\dot{x}_1 &= \varphi_1(y)\theta_i + \varphi_1(y)x_{2i} + u \\
\dot{x}_{2i} &= 0 \\
y &= x_1
\end{aligned} \tag{4.50}$$

where the system has been extended by the state $x_{2i} = \theta - \theta_i$. Following the Lyapunov-based observer design in [7], we propose to use the following individual nonlinear observer:

$$\begin{aligned}
\dot{\hat{x}}_{1i} &= \varphi_1(y)\theta_i + 2\omega\varphi_1(y)(y - \hat{y}_i) + u + \varphi_1(y)x_{2i} \\
\dot{\hat{x}}_{2i} &= \omega^2 \varphi_1(y)(y - \hat{y}_i), \quad \omega > 0
\end{aligned} \tag{4.51}$$

Defining the error $e_i = [e_{1i}, e_{2i}]^T = [x_1 - \hat{x}_{1i}, x_{2i} - \hat{x}_{2i}]^T$, the observer will result in the bilinear error dynamics

$$\dot{e}_i = \varphi(y) A e_i \tag{4.52}$$

where the matrix

$$A = \begin{pmatrix} -2\omega & 1 \\ -\omega^2 & 0 \end{pmatrix}$$

is Hurwitz and $\varphi(y)$ represents the nonlinearity in the system output. The observer design renders the derivative of the Lyapunov function

$$V_i(e_i) = \frac{1}{2} e_i^T \begin{pmatrix} 1 & 0 \\ 0 & \omega^{-2} \end{pmatrix} e_i \qquad (4.53)$$

negative semi-definite

$$\dot{V}_i = -2\omega\varphi(y)e_{1i}^2 \leq 0 \qquad (4.54)$$

ASSUMPTION 4.1
In order to establish a lower bound on the convergence rate of the observer, an additional assumption on φ is necessary:

$$\varphi(y) \geq c_\varphi > 0 \qquad (4.55)$$

where c_φ is a positive constant. □

An important property of the error differential equation (4.52) is that its solution can be explicitly given. Defining

$$y^*(t,0) = \int_0^t \varphi_1(y(\tau))d\tau \geq c_\varphi \cdot t > 0, \quad \text{for} \quad t > 0 \qquad (4.56)$$

we obtain

$$e_i(t) = \exp(-\omega y^*(t,0)) \cdot \begin{pmatrix} 1 - \omega y^*(t,0) & y^*(t,0) \\ -\omega^2 y^*(t,0) & 1 + y^*(t,0) \end{pmatrix} e_i(0) \qquad (4.57)$$

Knowing the measurable output error $e_{1i}(0)$ and $e_{1i}(t)$ at some time instant t, Equation (4.57) can be used to determine the parameter estimation error $e_{2i}(t)$ according to

$$\begin{aligned} e_{2i}(t) &= h(y^*, e_{1i}(t), e_{1i}(0)) \\ &= \frac{1}{y^*(t,0)} \left[(1 + \omega y^*(t,0)) e_{1i}(t) - e^{-\omega y^*(t,0)} e_{1i}(0) \right] \end{aligned} \qquad (4.58)$$

Thus, even under observer transients a parameter estimate

$$\hat{\theta}_i = \theta_i + \hat{x}_{2i}(t) + e_{2i}(t) \qquad (4.59)$$

can be computed as shown in Figure 4.4. Anti-windup is introduced for the observer state $\hat{x}_{2i}$ by defining the local bounds

$$\bar{\theta}_i = \begin{cases} \theta_{min} & \text{for} \quad i = 0 \\ \frac{1}{2}(\theta_{i+1} + \theta_i) & \text{for} \quad 1 \leq i \leq N-1 \\ \theta_{max} & \text{for} \quad i = N \end{cases} \qquad (4.60)$$

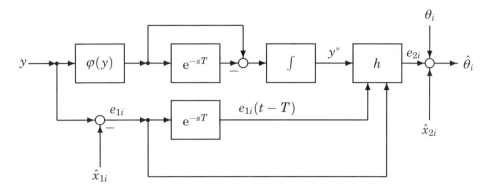

Figure 4.4 Computation of $\hat{\theta}_i$

and setting the state equation

$$\dot{\hat{x}}_{2i} = \begin{cases} 0 & \text{for } x_{2i} \notin [\bar{\theta}_{i-1}, \bar{\theta}_i] \text{ and } (y - \hat{y}_i)x_{2i} > 0 \\ \omega^2 \varphi_1(y)(y - \hat{y}_i) & \text{elsewhere} \end{cases} \quad (4.61)$$

When x_{2i} is in saturation, then e_{2i} will be constant and the error dynamics equation, Equation (4.52), simplifies to the stable first-order system

$$\dot{e}_{1i} = \varphi_1(y)\left[-2\omega e_{1i} + e_{2i}\right] \quad (4.62)$$

driven by a constant input e_{2i}. Due to Assumption 4.1, the solution to this equation

$$e_{1i} = \frac{1}{2\omega}\left[(2\omega e_{1i}(0) - e_{2i})\,e^{-2\omega y^*} - e_{2i}\right] \quad (4.63)$$

will eventually converge to the finite value

$$e_{1i}(\infty) = -\frac{e_{2i}}{2\omega} \quad (4.64)$$

Hence, only one individual observer will have an output error converging to zero and consequently, a cost index Q_i converging to zero independently of the particular cost index that is used.

The properties of the MMO can be used to derive the following resetting law:

THEOREM 4.2
Consider the control system (4.29) together with the control law (4.30), the parameter update law (4.31) and the MMO (4.51). Suppose that o_i is the observer that has been selected as the optimal one according to the cost index and the switching logic. Then, setting $\hat{\theta}^+ = \bar{\theta}_i$ will result in a negative step of the Lyapunov function (4.32) if:

1. $x_{2i}(\tau)$ does not saturate within the time interval $\tau \in [t-T, t]$.
2. $\bar{\theta}_{i-1} < \hat{\theta}_i < \bar{\theta}_i$
3. either

$$\hat{\theta} - \bar{\theta}_i > \bar{\theta}_i - \theta_i \tag{4.65}$$

or

$$\bar{\theta}_{i-1} - \hat{\theta} > \theta_i - \bar{\theta}_{i-1} \tag{4.66}$$

Proof. Due to Condition 1 of the theorem, according to Equations (4.58) and (4.59) we have

$$\hat{\theta}_i = \theta_i + \hat{x}_{2i}(t) + h(y^*(t, t-T), e_{1i}(t), e_{1i}(t-T)). \tag{4.67}$$

In addition to this, due to Condition 2 it can be implied that the real parameter is contained in

$$\bar{\theta}_{i-1} < \theta < \bar{\theta}_i \tag{4.68}$$

From Condition 3, it follows that either (4.65) is satisfied, in which case we obtain by adding $\hat{\theta}$ to both sides and rearranging and employing (4.68)

$$-\Delta\hat{\theta} = \hat{\theta} - \theta_i < 2(\hat{\theta} - \bar{\theta}_i) \leq 2(\hat{\theta} - \theta) = -2\tilde{\theta} \tag{4.69}$$

If on the other hand (4.66) is satisfied then by subtracting $\hat{\theta}$ from both sides and employing (4.68)

$$\Delta\hat{\theta} = \theta_i - \hat{\theta} < 2(\bar{\theta}_{i-1} - \hat{\theta}) \leq 2(\theta - \hat{\theta}) = 2\tilde{\theta} \tag{4.70}$$

Consequently, Conditions (4.38) and (4.40) are satisfied, which is sufficient for stability. □

4.5 A Second Order Benchmark System

The approach will now be demonstrated using the second order benchmark system

$$\dot{x}_1 = x_2 + x_1^2 \theta \tag{4.71}$$
$$\dot{x}_2 = u \tag{4.72}$$
$$y = x_1 \tag{4.73}$$

For the system, a tuning function controller with the parameters $c_1 = 1$, $c_2 = 20$, $\gamma = 10$, $\kappa_i = 0$ was designed. A multiple model observer with 10 models at the parameter values

$$\{\theta_i\} = \{-10, -8, -6, -4, -2, 2, 4, 6, 8, 10\}$$

was used. Since full state feedback is applied, the observer design problem can be reduced to first order, with x_2 being treated as an input into the observer. The observer error injection parameter is $\omega = 1000$ and the filtering time for computation of $\hat{\theta}_i$ is $T = 0.3$. The cost index $Q_i = e_{1i}^2 + e_{2i}^2$ was filtered with the low-pass filter

$$H_f = \frac{10}{s+10}$$

In Figure 4.5, simulations of the tracking of a sinusoidal reference input r are shown. The reference output y_r is generated using the second order reference model

$$y_r(s) = \frac{36}{(s+6)^2} r(s)$$

During the tracking experiment, the system parameter was changed stepwise from a value of $\theta = 9$ to $\theta = -8$ and afterwards to $\theta = 4$. It can be seen in Figure 4.5(c) that the MMO estimate (thin solid line) follows those step changes almost immediately. The performance of the control system with and without resetting of the tuning function controller is compared. In Figure 4.5(c), the controller parameter estimates with resetting (dashed line) and without resetting (dotted line) are shown. It can be seen that due to the low adaptation gain, the controller without resetting is unable to track the parameter changes, while the estimate of the controller with resetting converges very fast. Consequently, the tracking performance of the controller with resetting (dashed line in Figure 4.5(a)) is much better then the one without resetting (dotted line). To emphasise this, the corresponding tracking errors are also displayed in Figure 4.5(b).

4.6 Wheel Slip Control

Wheel slip control in automotive systems is a challenging problem for a number of reasons:

- the friction between tyre and road is highly nonlinear, uncertain and time varying due to changing road conditions;
- slip dynamics can exhibit an unstable behaviour;
- additional (unmodelled) dynamics such as actuator dynamics, suspension dynamics and computational delays impose significant constraints on the loop bandwidth and require a robust controller design.

The simplest model used in wheel slip control is shown in Figure 4.6. The model consists of a single wheel attached to a mass m. The wheel moves, driven by inertia of the mass m in the direction of the velocity vector v. A tyre reaction force F_x is generated by the friction between the tyre surface and the road surface. The tyre reaction force will generate a torque that

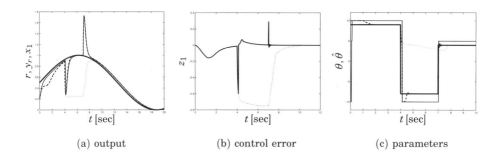

(a) output   (b) control error   (c) parameters

Figure 4.5 Performance with and without MMO for the second order benchmark system. Plot (a): reference input r (thick solid line), reference output y_r (thin solid line), output y for system with MMO (dashed) and without MMO (dotted); Plot (b): control error z_1 with MMO (solid) and without MMO (dotted); Plot (c): real parameter value θ (thick solid line), MMO estimation $\hat{\theta}_i$ (thin solid), controller parameter estimate $\hat{\theta}$ with MMO (dashed line) and without MMO (dotted line)

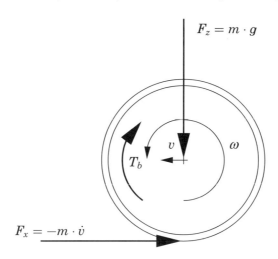

Figure 4.6 Quarter-car slip model

initiates a rolling motion of the wheel, causing an angular velocity ω. A brake torque applied to the wheel will act against the spinning of the wheel, causing a negative angular acceleration. The equations of motion of the quarter-car are

$$m\dot{v} = -F_x \qquad (4.74)$$
$$J\dot{\omega} = r F_x - T_b \, \text{sign}(\omega) \qquad (4.75)$$

where

Nonlinear Adaptive Backstepping using Multiple Observers

v horizontal speed at which the car travels
ω angular speed of the wheel
F_z vertical force
F_x tyre friction force
T_b brake torque
r wheel radius
J wheel inertia

The tyre reaction force F_x is given by

$$F_x = F_z \cdot \mu(\lambda, \mu_H, \alpha, F_z) \tag{4.76}$$

where the friction coefficient μ is a nonlinear function of

λ tyre slip
μ_H friction coefficient between tyre and road
α slip angle of the wheel

The slip λ is the output to be controlled. It describes the normalised difference between horizontal speed v and speed of the wheel perimeter ωr

$$\lambda = \frac{v - \omega r}{v} \tag{4.77}$$

The slip value of $\lambda = 0$ characterises the free motion of the wheel, where zero friction force F_x is exerted. If the slip attains the value $\lambda = 1$, then the wheel is locked which means that it has come to a standstill.

The tyre road dynamics depend nonlinearly on the friction between tyre and the uncertain road characteristic. In the literature, numerous models describing tyre road friction can be found [22], [4], [6].

The friction coefficient μ can vary in a very wide range. Its qualitative dependence on slip λ is shown in Figure 4.7. Calculating the time derivative of Equation (4.77) results in the two equations

$$\dot{\lambda} = -\frac{1}{v}\left\{\frac{1}{m}(1-\lambda) + \frac{r^2}{J}\right\} F_z \mu(\lambda, \mu_H) + \frac{1}{v} \cdot \frac{r}{J} T_b \tag{4.78}$$

$$\dot{v} = -\frac{1}{m} F_z \mu(\lambda, \mu_H) \tag{4.79}$$

When considering the slip as the output $y = \lambda$ and the brake torque as an input $u = T_b$ then Equation (4.79) describes the zero dynamics of the system. Reformulation of Equations (4.78) and (4.79) results in the following structure:

$$\begin{aligned}\dot{x}_1 &= -\frac{1}{x_2}\varphi(x_1, \theta) + \frac{1}{x_2} u \\ \dot{x}_2 &= -\frac{\varphi(x_1, \theta)}{1 - x_1 + \frac{mr^2}{J}}\end{aligned} \tag{4.80}$$

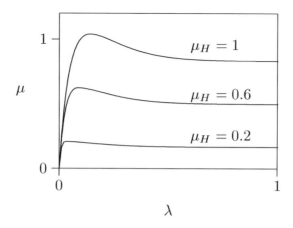

Figure 4.7 Tyre friction

where

$$\varphi(x_1, \theta) = \left\{\frac{J}{mr}(1-x_1) + r\right\} F_z \mu(x_1, \theta) > 0$$

$$0 < u \leq u_{max}$$

$$\left.\frac{\partial \varphi}{\partial \theta}\right|_{x_1} > 0$$

where $x_1 = \lambda$ is the slip, $x_2 = \frac{Jv}{r}$, $\varphi(x_1, \theta)$ is a nonlinear function of slip, and the uncertain tyre road friction coefficient $\theta = \mu_H$, u is the brake torque. The function φ depends nonlinearly on the parameter θ. However, everywhere except for $x_1 = 0$, we have

$$\frac{\partial \varphi(x_1, \theta)}{\partial \theta} > 0 \tag{4.81}$$

as shown in Figure 4.8. Hence, $\varphi(x_1, \theta)$ is a non-decreasing function of θ and consequently, gradient-based parameter estimation methods can be applied in controller and observer design [2], [21], [1].

The control problem is stabilising the slip around a constant set point $y_r = x_1^0$ where $\dot{y}_r = 0$. Depending on the set point and the road conditions, the slip dynamics can be highly unstable. With respect to the output x_1, the system is of relative degree one with the vehicle velocity x_2 as the zero dynamics state. From physical considerations, it is clear that the zero dynamics of the system are stable since the velocity will be driven to zero while braking. Also, x_2 can be estimated from measurement data. Thus, the system can be treated like a first order system.

A nonlinear adaptive slip controller was designed for system (4.80). The

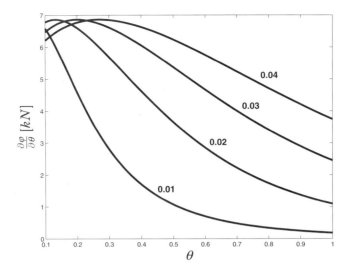

Figure 4.8 Derivative of the tyre friction force F_x with respect to the friction coefficient θ with slip x_1 as a parameter

controller design was based on the Lyapunov function

$$V = \frac{1}{2}x_2^2 \cdot (x_1 - x_1^0)^2 + \frac{1}{2\gamma}\tilde{\theta}^2 \tag{4.82}$$

and uses an inverse optimal control law [24], [23] in order to avoid cancelling of nonlinear terms. The control law is required to be robust with respect to the unmodelled actuator dynamics and an additional computational delay of 0.014 s. The controller parameters are selected in such a way that around the set point x_1^0, a bandwidth of the linearised system of about 7 Hz is achieved. At the set point, the system is open-loop unstable and thus, as explained in Section 4.1, high adaptation gain would compromise robust stability.

An MMO with six models for the parameter values $\theta_i = \{\frac{1}{6}, \frac{2}{6}, \ldots, 1\}$ was also designed and combined with the adaptive controller using the resetting conditions derived earlier.

Some experimental results of slip control for an experimental car with electromechanical brakes are shown in Figure 4.9. While braking on snow at an initial speed of 16 m/s, the slip is controlled to a set point of $\lambda_0 = 0.07$. It can be seen that the friction coefficient is about $\mu_H = 0.5$. The plots of Figure 4.9(a) show the experimental result without resetting. Starting at a value of $\hat{\theta}(0) = 0.8$, the controller estimate (lower plot solid line) converges slowly. In Fig. 4.9(b), the corresponding results with resetting are shown. Although the initial value of the estimate $\hat{\theta}(0) = 1$ is much further away from the true friction coefficient (resulting in a slightly higher peaking of the slip), it can be seen that the transients are much faster.

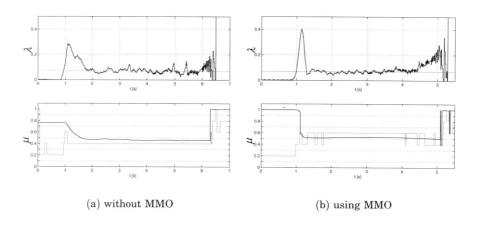

(a) without MMO (b) using MMO

Figure 4.9 Experiment with the tyre slip controller. The upper plots show the controlled variable $y = \lambda$. The lower plots show the friction coefficient estimate $\hat{\theta}$ (solid line) the MMO etimate θ_i (dashed line).

4.7 Conclusions

The presented paper provides an extension of multiple model-based adaptive control to the class of strict feedback nonlinear systems. As main contribution, a set of sufficient closed loop stability conditions for resetting tuning function-based nonlinear adaptive controllers was given. Also, a fast multiple model observer was introduced, from which even under transient conditions a parameter estimate can be obtained. A wheel slip control example showed that recovering of wheel slip can be improved after instantaneous changes of road friction.

Acknowledgments

This work has been supported by the European Union under the ESPRIT LTR funding scheme, project number 28104.

4.8 References

[1] A. M. Annaswamy, F. P. Skantze, and Ai-Poh Loh. Adaptive control of continuous time systems with convex/concave parameterisation. *Automatica*, 33–49, 1998.

[2] M. Arcak and P. Kokotović. Robust output-feedback design using a new class of nonlinear observers. In *Proceedings of the 39th IEEE Conference on Decision and Control, Sydney*, 752–755, Australia, 2000.

[3] K. J. Åström and B. Wittenmark. *Adaptive Control*. Addison Wesley, second edition, 1995.

[4] M. Burckhardt. *Radschlupf-Regelsysteme*. Vogel Buchverlag, Würzburg, Germany, 1993.

[5] G. W. Chang, J. P. Hespanha, A. S. Morse, M. S. Netto, and R. Ortega. Supervisory field-oriented control of induction motors with uncertain rotor resitance. Submitted for publication, 2000.

[6] A. Daiß and U. Kiencke. Estimation of tyre slip during combined cornering and braking observer supported fuzzy estimation. In *13th Triennal World Congress*, 41–46, San Francisco, 1996.

[7] C. Canudas de Wit, R. Horowitz, and P. Tsiotras. Model-based observers for tire/road contact friction prediction. In Nijmeier and T. I. Fossen., editors, *New Directions in Nonlinear Observer Design, Springer Lecture Notes on Control and Information Science No. 244*, 23–42. Springer-Verlag, 1999.

[8] J. P. Hespanha. *Logic-Based Switching Algorithms in Control*. Dissertation, Graduate School, Yale University, December 1998.

[9] J. Kalkkuhl, T. A. Johansen, and J. Lüdemann. Improved transient performance of nonlinear adaptive backstepping using estimator resetting based on multiple models. *IEEE Transactions on Automatic Control*, 2001.

[10] J. Kalkkuhl, T. A. Johansen, and J. Lüdemann. Nonlinear adaptive backstepping with estimator resetting using multiple observers. In M. D. DiBenedetto and A. Sangiovanni-Vincentelli, editors, *Hybrid Systems: Computation and Control, Proc. 4th Int. Workshop, HSCC2001*, March 2001, 319–332, Rome, Italy. Springer-Verlag, 2001.

[11] M. Krstić, I. Kanellakopoulos, and P. Kokotović. *Nonlinear Adaptive Control Design*. John Wiley & Sons Inc., 1995.

[12] Yong Liu. Switching observer design for uncertain nonlinear system. In *Proceedings of the 34th IEEE CDC*, 1756–1761, New Orleans, 1995.

[13] R. H. Middleton, G. C. Goodwin, D. J. Hill, and D. Q. Mayne. Design issues in adaptive control. *IEEE Transactions on Automatic Control*, 33(1):50–58, 1 1988.

[14] A. S. Morse. Towards a unified theory of parameter adaptive control–partii: Certainty equivalence and implicit tuning. *IEEE Transactions on Automatic Control*, 37(1):15–29, January 1992.

[15] A. S. Morse. Supervisory control of families of linear set-point controllers–Part 1: Exact matching. *IEEE Transactions on Automatic Control*, 41(10):1413–1431, October 1996.

[16] A. S. Morse. Supervisory control of families of linear set-point controllers–Part 2: Robustness. *IEEE Transactions on Automatic Control*, 42(11):1500–1515, 10, 1997.

[17] K. S. Narendra and J. Balakrishnan. Improving transient response of adaptive control systems using multiple models and switching. *IEEE Transactions on Automatic Control*, 39(9):1861–1866, 1994.

[18] K. S. Narendra and J. Balakrishnan. Intelligent control using fixed and adaptive models. In *Proc. 33rd CDC*, 1680–1685, December 1994, Lake Buena vista, FL, 1994.

[19] K. S. Narendra, J. Balakrishnan, and M. K. Ciliz. Adaption and learning using multiple models, switching and tuning. *IEEE Control Systems Magazine*, 15(3):37–51, 1995.

[20] K. S. Narendra and C. Xiang. Adaptive control of discrete-time systems using multiple models. In *Proc. 37th CDC*, 3978–3983, December 1998, Tampa, FL, 1998.

[21] K. S. Narendra and A. M. Annaswamy. *Stable Adaptive Systems*. Prentice Hall, Englewood Cliffs, NJ, 1989.

[22] H. B. Pacejka and R. S. Sharp. Shear force developments by pneumatic tires in steady-state conditions: A review of modeling aspects. *Vehicle Systems Dynamics*, 29:409–422, 1991.

[23] R. Sepulchre, M. Janković, and P. Kokotović. *Constructive Nonlinear Control*. Springer-Verlag, London, 1997.

[24] E. D. Sontag. A "universal" construction of Artstein's theorem on nonlinear stabilization. 1989.

[25] V. I. Utkin. *Sliding Modes in Control and Optimization*. Communication and Control Engineering Series. Springer-Verlag, Berlin, third edition, 1992.

[26] A. van der Schaft and H. Schumacher. *An Introduction to Hybrid Dynamical Systems*. Springer-Verlag, London, 2000.

[27] P. V. Zhivoglyadov, R. H. Middleton, and Minyue Fun. Localisation based switching adaptive control for time-varying discrete time systems. *IEEE Transactions on Automatic Control*, 45(4):752–755, 2000.

5

ABS Control—A Design Model and Control Structure

S. Solyom A. Rantzer

Abstract

The anti-lock braking system (ABS) is an important component of a complex steering system for the modern car. Most of the ABS controllers available on the market are table-based on-off controllers. In the latest generation of "brake-by-wire" systems, the performance requirements of the ABS are much higher. The control objective shifts to maintain a specified tire slip for each wheel during braking. The authors propose a design model and based on that a gain-scheduled controller that regulates the tire-slip. Simulation and test results are presented.

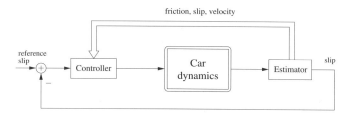

Figure 5.1 ABS control

5.1 Introduction

The first ABS systems in cars were implemented in the late 70s, the main objective of the control system being prevention of wheel-lock while braking.

Most of the ABS controllers available on the market are table and relay-feedback-based, making use of hydraulic actuators to deliver the braking force [5], [10], [9], [13]. One approach used with hydraulic brakes is to measure the wheel rotational velocity and compute the wheel deceleration. Then, given some thresholds for the wheel deceleration, the pressure is increased, held or decreased while trying to maintain the slip around a value that determines the point of maximum friction [7]. This kind of algorithm will have as a side effect vibrations that are noticeable while braking.

In the literature, another approach is presented for hydraulic brakes, where the maximum friction point is reached by measuring the angular velocity of the wheel and the brake pressure [4]. This is a model-based approach and uses a sliding mode to reach and track the maximum friction during emergency braking.

In [8], feedback linearization is used to design a slip controller and gain scheduling to handle variation with speed of the tire friction curve.

In the latest generation of "brake-by-wire" systems, electromechanic actuators are used, which are capable of delivering continuously varying and different brake forces on each of the four wheels. This way, the control objective shifts to maintain a specified tire slip for each of the four wheels. The set-point slip is supposed to be provided by a higher level in the hierarchy (e.g., an ESP system), and can be used for stabilizing the steering dynamics of the car while braking. This might imply different reference values for the slip of each wheel.

The control problem is highly uncertain and nonlinear, mainly due to the tire friction characteristics. An important limiting factor is time delay due to sampling and communication.

This article proposes a synthesis method that handles uncertainties induced, mainly by the friction curve, while the system has to operate in a noisy environment. A simple static model of friction is used. Based on this, we develop a gain-scheduled controller that switches between local controllers (Figure 5.1).

The layout of the paper is the following: in the next section, the proposed design model is presented and the fundamental limitations in control performance are shown. Section 5.3 describes the control structure, the local designs and scheduling. Section 5.4 presents simulation and experimental results done in cooperation with DaimlerChrysler. Section 5.5 presents some conclusions.

5.2 The Design Problem

The control objective is, as mentioned above, to follow a reference trajectory for the tire slip on each of the four wheels. Only longitudinal motion is considered. Nevertheless, there are differences between the front and the back wheels as well as between the right and the left wheels. This is mainly due to pitching of the car body while braking.During the design, only a simplified model of the quarter car will be considered, that is, one wheel with a mass where no variations of the normal force while braking is taken into account, and furthermore, no suspension dynamics are explicitly considered. Finally, the resulting controller is applied to each wheel, irrespective of the wheel location (left or right).

Some of the robustness requirements can be identified due to:

- the feedback signal (λ) is not measurable but results from estimation and the signal quality is rather poor;
- time delay due to sampling and communication;
- high uncertainty in the tire-friction curve, especially in the nonlinear region.

On the other hand, fast response time is imperative, which is obviously contradictory to the above-mentioned robustness requirements.

As pointed out in the previous section, the proposed controller is model based, therefore a natural point to start with is the quarter-car model.

The Quarter-car Model

The equations of motion of the quarter car are given by:

$$J\dot{\omega} = rF_x - T_b$$
$$m\dot{v} = -F_x$$

where:

m mass of the quarter-car

v velocity over ground of the car

ω angular velocity of the wheel

F_z vertical force

F_x tire friction force

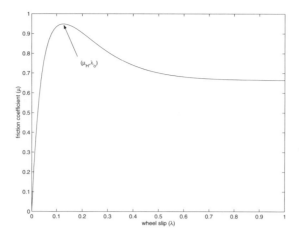

Figure 5.2 Tire friction curve

T_b brake torque

r wheel radius

J wheel inertia

The tire slip is defined as:

$$\lambda = \frac{v - \omega r}{v}$$

hence, a locked wheel is described by $\lambda = 1$, while the free motion of the wheel is described by $\lambda = 0$.

The tire friction force, F_x, is determined by:

$$F_x = F_z \mu(\lambda, \mu_H, \alpha, F_z, v)$$

where $\mu(\lambda, \mu_H, \alpha, F_z)$ is the road-tire friction coefficient, a nonlinear function with a typical dependence on the slip shown in Figure 5.2 [11]. This function also depends on the normal force (F_z), steering angle (α), road surface, tire characteristics, and car velocity. Depending on the road condition and the tire characteristics, the peak of the friction curve will be more or less pronounced and the value of the maximum friction coefficient (μ_H) will be different.

In the following, we consider only the case when no steering is present (i.e., $\alpha = 0$) and no side slip occurs.

Proposed Design Model

From the equations of motion, taking into account that the velocity of the car varies much slower than the other variables involved, one obtains the

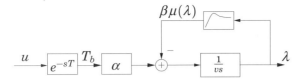

Figure 5.3 Design model for an ABS

dynamics of the tire slip:

$$\dot{\lambda}v = \frac{r}{J}T_b - \frac{r^2 F_z}{J}\mu \qquad (5.1)$$

Relation (5.1) is a first order nonlinear differential equation due to the tire friction coefficient function. Denoting

$$\beta \triangleq -\frac{r^2 F_z}{J}, \quad \alpha \triangleq \frac{r}{J}$$

and adding a time delay T, the proposed design model (see Figure 5.3) can be synthesized in the following:

$$\dot{\lambda}(t)v = -\beta\mu(\lambda(t)) + \alpha u(t-T) \qquad (5.2)$$

where v is considered constant but uncertain.

This model captures the main control difficulties of an anti-lock braking system. Notice that in addition to those pointed out at the beginning of this section, velocity dependence of the system is also included.

If the above-presented model is linearized around an operating point, the resulting model is of the form:

$$\dot{\lambda}(t)v = \beta(m_i\lambda(t) + \Psi) + \alpha u(t-T) \qquad (5.3)$$

where m_i is the slope of the tire-friction curve at the considered operating point. Then locally the slip dynamics is given by a first order system, which are stable or unstable depending on the slope m_i.

Fundamental limitations

If the slope m_i resulting from the linearization (see Section 5.2) is negative, one obtains locally an unstable system, which in conjuncture with a time delay, will give rise to fundamental limitations in control performance [1]. In the following a local analysis of the system in the mentioned situation will be carried out.

Consider that there are no other unstable nor non-minimum-phase dynamics in the system. Then the unstable pole in question is:

$$p = \beta\frac{|m_i|}{v}$$

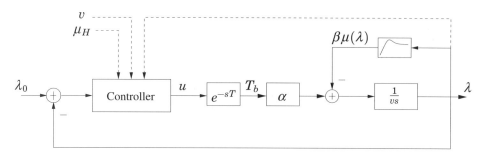

Figure 5.4 ABS control scheme

with $m_i < 0$. According to rules of thumb in [1], satisfactory control performance with a phase margin $\varphi_m = \pi/4$ requires

$$pT \leq 0.3$$

where T is the time delay, and the resulting crossover frequency is:

$$\omega_c = p\sqrt{\frac{2}{pT} - 1}$$

Consider process parameters such that $\beta \approx 440$, a time delay of $T = 14\,\text{ms}$ and a friction curve with local negative slope of -0.5 (that is, a deflection from the horizontal of $-26°$). Then satisfactory control performance can be obtained until a velocity over ground not less than $v = 10\,\text{m/s}$. While for a local negative slope of -0.05 (that is, a deflection of approximately $-3°$) the same performance can be obtained up to a velocity over ground not less than $v = 1\,\text{ms}$. The crossover frequency where this performance can be achieved is $\omega_c \approx 50\,\text{rad/s}$.

Thus, the time delay plays an important role in the investigated system.

5.3 The Control Structure

The proposed control structure (Figure 5.4) is a gain-scheduling scheme, based on tire-slip value, velocity over ground (v), and the maximum friction coefficient μ_H (i.e., friction coefficient at the top of the friction curve). The tire-slip and the maximum-friction coefficient are not directly measurable, but they are estimated. A Kalman filter is used to estimate the velocity of the car based on the speed of the wheels and acceleration measurements. A fast estimate of the friction coefficient at each wheel is obtained by a nonlinear multiple observer as described in a different paper of this book [6]. This paper is focused on the slip control loop based on given estimates of velocity and friction. No dynamics of the estimators are taken into account.

The main idea behind the slip control design is to use a few local controllers that locally robustly stabilize the system for different slopes

ABS Control—A Design Model and Control Structure

of the friction curve and which tolerate the time variations due to the decreasing velocity over ground of the car (v). Switching between the linear controllers is done according to the estimated friction and slip, which define the operating point on the friction curve.

Design of the Local Linear Controllers

Due to high uncertainty in the real process, it is natural to look for a simple robust controller that can easily be tuned in the test vehicle. Therefore, PI-controllers are used and the gains are scheduled based on the three variables mentioned above.

Consider a linearized model as in (5.3). The bandwidth of this model depends on v, i.e., the bandwidth is smaller for high car velocities than for low car velocities. Therefore, it is natural to design the controller to counteract this variation. The controller is scaled by velocity (v) to ensure a higher gain for high velocities.

Note that scaling the controller by velocity over ground (v) will, in turn, scale the relative uncertainty caused by the friction curve by v^{-1}. This scaled uncertainty remains in feedback with a nominal plant as shown in Figure 5.3. Considering the uncertainty as being cone bounded, this scaling will "tighten" the cone with increasing velocity. In particular, when the system is operating at maximum friction, that is, at the top of the friction curve, this scaling will theoretically remove the dependence on velocity over ground.

The chosen local controllers are of the form:

$$u(t) = K\left(\lambda_0(t) - \lambda(t)\right)v(t) + \int K_i\left(\lambda_0(t) - \lambda(t)\right)v(t)dt \qquad (5.4)$$

and can be viewed as PI-controllers scaled by the velocity over ground.

As seen in (5.3), in stationarity the slip dynamics do not depend on v. Hence, in stationarity the control output should not be affected by the velocity scaling. This can be achieved by moving the velocity inside the integral, thus the integral term is kept constant as long as the slip error is zero. Also, the gain K_i is inside the integral in order to obtain a smooth transition while switching between parameters [2].

Another important issue in ABS control is to prevent wheel-lock in case of changes in surface condition (e.g., a transition from dry to wet surface). A change in surface condition (e.g., transition from dry to wet surface) will act as a load disturbance of magnitude $\beta(\Psi_1 - \Psi_2)$ according to (5.3). Thus, it is important that the controller minimizes the effect of load disturbances on the system. On the other hand, as seen in relation (5.3), the slope of the approximating line (resulting from the linearization of the friction curve) affects the pole of the linear system. Furthermore, this is scaled by the velocity as a consequence of (5.4). Then, the local control problem is to robustly stabilize the system while minimizing the effects of a load disturbance. The main uncertainty comes from one pole and the gain of the plant.

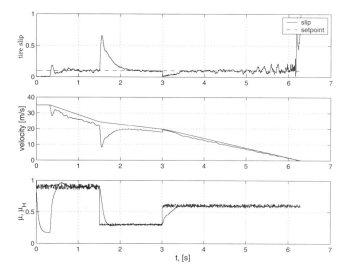

Figure 5.5 Simulation results for left front-wheel

To synthesize a PI-controller that minimizes the effects of load disturbances, one can solve a constrained optimization problem as suggested in [3]. In order to guarantee additional robustness against the uncertainty in the plant, it is possible to add a further inequality constraint based on the circle criterion as described in [12]. For improved accuracy, a model of the actuators can also be introduced in the optimization.

PID controllers can be designed in the same way, with the arising design difficulties described in [12]. The main potential advantage of using PID- instead of PI-controllers for the above-described system is the ability to significantly increase the integral gain while inactive. Simulations have been encouraging.

Scheduling

As described above, local robust controllers have been designed to handle different slopes on the friction curve at different velocities. By scheduling the gains K, K_i, the controller can be adapted to the current operating mode/position on the estimated friction curve. In the results presented below, only two local PI-controllers are used, hence four parameters.

The restriction to two local controllers is based on the observation that usually there is a maximum on the friction curve and to the left of this there is a positive slope region, while to the right of the top (tire slips up to 0.5 are considered) there is a region with negative slope that tends to flatten out for higher slips. For this reason, one of the scheduling variables is the slip value where the assumed maximum is located (λ_H). To the left of this, a controller is used which is tuned for relatively high positive slopes, while to the right a controller that can handle negative slopes is used.

ABS Control—A Design Model and Control Structure

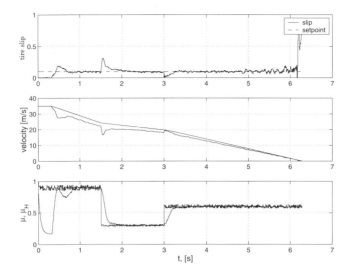

Figure 5.6 Simulation results for right back-wheel

The coordinate λ_H changes with the friction curve, thus a new scheduling variable, the maximum friction coefficient (μ_H), is introduced and estimated. Accordingly, a new λ_H is considered.

Due to the robustness of local designs, it is enough to use the same λ_H for a family of friction curves. This is a point where trade-off between robustness and performance will lay a mark on the controller's complexity.

In order to have a fast response at the beginning of the braking action, an initial braking force is applied by initializing the controller state at soon as the ABS is switched on. In this way, fast response times are possible while the controller's robustness is maintained.

5.4 Simulation and Experimental Results

The simulation environment consists of a four wheel model with pitch dynamics as well as estimators for μ_H and v. This simulator has been provided by DaimlerChrysler. Figure 5.5 and Figure 5.6 show simulation results for a front and a back wheel, respectively. Note the influence of the pitching dynamics, which makes it harder to control the slip for the front wheel. This is due to a higher normal force F_z, which in turn, translates into a faster plant. There is also some minor difference between the left and the right wheel due to displacement in the center of gravity of the car. The plots show the tire-slip control in a scenario where braking commences on a high-friction surface ($\mu_H = 0.9$), then a spot with low friction is encountered ($\mu_H = 0.3$) and finally the braking finishes on a high-friction surface ($\mu_H = 0.6$). This is a scenario that would simulate braking on a dry surface with a wet or icy spot.

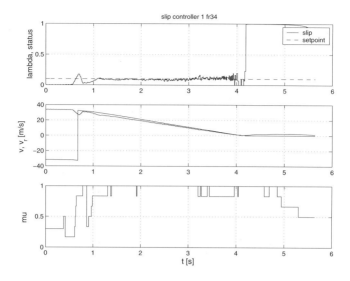

Figure 5.7 Experimental results for left front wheel

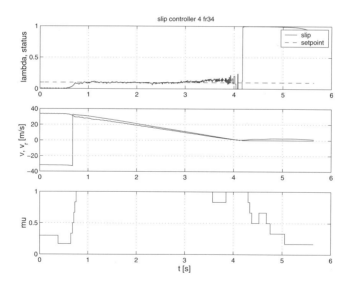

Figure 5.8 Experimental results for right bac -wheel

The third subplot shows the estimate of the maximum friction, (denoted μ) and the maximum friction used in the simulation. The friction curves used in simulation have no relative drop after reaching the maximum, *i.e.*, there is no unstable part in the dynamics. The controller has been tried out in a Mercedes E Class test vehicle provided by DaimlerChrysler, the results are presented in Figures 5.7-5.8. Here, braking on a dry surface with summer tires was tested. Summer tires present a prominent peak at the maximum friction coefficient, which will give rise to an unstable region for the slip dynamics.

For the tests as well as the simulation, a gain-scheduled controller has been used with two local controllers, one for the regions left from the maximum in the tire-friction curve and one for the region to the right of the maximum value in the tire-friction curve. Hence, seven parameters have been tuned.

5.5 Conclusion

A simple but powerful design model for ABS control has been presented. Fundamental limitations on the control performance have been pointed out. A gain-scheduled PI design approach has been used for the controller. Simulations and preliminary experiments in a test vehicle were performed with satisfactory results.

Acknowledgment

Financial support from the EU-ESPRIT project Heterogeneous Hybrid Control (H2C) is gratefully acknowledged.

5.6 References

[1] K. J. Åström. Limitations on control system performance. In *European Control Conference*, Brussels, Belgium, July, 1997.

[2] K. J. Åström and T. Hägglund. *PID Controllers: Theory, Design, and Tuning*. Instrument Society of America, Research Triangle Park, North Carolina, second edition, 1995.

[3] K. J. Åström, H. Panagopoulos, and T. Hägglund. Design of PI controllers based on non-convex optimization. *Automatica*, 34(5):585–601, May, 1998.

[4] S. Drakunov, Ü. Özgüner, P. Dix, and B. Ashrafi. ABS control using optimum search via sliding mode. In *IEEE Transactions on Control System Technology*, 3, 79–85, 1995.

[5] P. Hattwig. Synthesis of ABS hydraulic systems. *SAE, 930509*, 1993.

[6] J. Kalkkuhl, T. A. Johansen, and J. Lüdemann. Nonlinear adaptive backstepping with estimator resetting using multiple observers. Separate chapter of this book, 2002.

[7] U. Kiencke and L. Nielsen. *Automotive Control Systems*. Springer-Verlag, 2000.

[8] Yong Liu and Jing Sun. Target slip tracking using gain-scheduling for braking systems. In *Proc. American Control Conference*, 1178–1182, Seattle, WA, 1998.

[9] M. Maier and K. Müller. ABS5.3: The new and compact ABS5 unit for passenger cars. *SAE, 930757*, 1995.

[10] W. Maisch, W.-D. Jonner, R. Mergenthaler, and A. Sigi. ABS5 and ASR5: The new ABS/ASR family to optimize directional stability and traction. *SAE, 930505*, 1993.

[11] H. B. Pacejka, E. Bakker, and L. Lidner. A new tire model with application in vehicle dynamics studies. *SAE, 890087*, 1989.

[12] S. Solyom and A. Ingimundarson. A synthesis method of robust PID controllers for a class of uncertainties. Submitted for publication, 2001.

[13] P. E. Wellstead and N. B. O. L Pettit. Analysis and redesign of an antilock brake system controller. In *IEE Proceedings—Control Theory Appl.*, 144, 413–425, 1997.

6

Controller Design for Hybrid Systems using Simultaneous D-stabilisation and its Application to Anti-lock Braking Systems (ABS)

K. J. Hunt   Yongji Wang   M. Schinkel
T. Schmitt-Hartmann

Abstract

In recent years hybrid systems have been widely studied. Many controller design approaches are based on a state space plant model and use full state feedback to satisfy certain stability conditions, for example, the existence of a common Lyapunov function. Pole assignment and LMI-based controller design techniques have also been used. A new controller design method, called simultaneous D-stabilisation and strong simultaneous D-stabilisation, is developed, which can deal with the multiple plant requirement resulting from hybrid systems. Both the simultaneous stabilisation problem (SSP) and strong simultaneous stabilisation problem (SSSP) with the simplest Hurwitz stability requirement are open problems in the control community. In this paper, the new synthesis approach developed is extended to the general D-stability requirement rather than simply the Hurwitz-stability requirement. The advantages of this approach include 1) the desired D-stability region can be of any form and be connected or disjoint. This leads to a unified treatment of continuous and discrete systems, and consequently encompasses the Hurwitz and Schur stability regions as special cases; 2) the size of the family of plants may be finite and more than three, which is the upper bound with the available approaches; 3) both the SSPs and the SSSPs can be dealt with in a unified way, although an SSSP is more complicated than an SSP; 4) the traditional PID-controller tuning problem can be treated within this general framework.

The SSP approach is then applied to the synthesis of a robust controller for a typical application in the automotive industry, controller design for anti-lock braking systems (ABS). The controller robustly assigns the closed-loop poles for every plant from different tyre friction models in a prescribed D region. Simulation results with the DaimlerChrysler test vehicle for ABS will also be presented.

6.1 Introduction

We consider the problem of simultaneously stabilising, or co-stabilising, a family of single-input single-output (SISO) linear systems by one fixed controller of given order. This fundamental problem, recognised as one of the difficult open issues in linear system theory, arises in many applications such as reliable control, robust control and nonlinear control [5, 8], [9], [10]. The simultaneous stabilisation problem (SSP) can be formally stated as follows: Given an r-tuple $G_1(s)$, ..., $G_r(s)$, of proper distinct transfer functions for plants, find a compensator $C(s)$ (if it exists) such that all closed-loop systems $C(s)G_i(s)/(1+C(s)G_i(s))$, $(i = 1, 2, \ldots, r)$ are internally stable. If the compensator $C(s)$ is further restricted to be stable, *i.e.*, it has no unstable poles, then the SSP is referred to as a strong simultaneous stability problem (SSSP). The requirement for a stable compensator is due to Shaw who found that in certain cases the instability of the compensator appears to result in poor overall system sensitivity to variations in plant parameters [41]. The difficulty of this kind of problem arises from the fact that only output feedback rather than full state feedback is allowed.

The main motivation for solving the SSP (or SSSP) is fourfold. First, the problem can sometimes arise naturally in the synthesis of a reliable controller for a system with redundant sensors and actuators in order that failures in some components will not cause immediate threats to the safety of the overall system. Such design is particularly important in safety-critical systems, such as space missions, aircraft, nuclear or chemical reactors, *etc.* [3], [28], [31], [53]. Second, it may be used for the design of a controller for a nonlinear system whose dynamics may be represented by a finite number of linear time-invariant plant models. A possible application of such a design approach is vehicle speed control, where the automatic gear switching results in inherent nonlinear characteristics of the vehicle plant whose dynamics for each gear are normally described by seven to nine linear models [22], [26]. If a suitable fixed controller for such applications can be found, it will lead to a smooth transient response of the system and a saving in complexity. Third, it is related to the robust stabilisation problem for parametric uncertain systems, where the goal is to ensure that a fixed controller will maintain closed-loop stability in the presence of continuous perturbations corresponding to the uncertainties in modelling the nominal system. Motivated by Kharitonov's theorem [27], special attention has been paid to finding a synthesis approach to the controller, based on the extreme results for the robust stabilisation problem [5], [6]. It has been shown that a first-order compensator robustly stabilises an interval plant if and only if it stabilises sixteen plants that are generated using the Kharitonov polynomials associated with the numerator and denominator [5]. Furthermore, when additional *a priori* information about the compensator is specified (sign of the gain, and signs and relative magnitudes of the pole and zero), then in some cases it is necessary and sufficient to stabilise eight critical plants while in other cases it is necessary and sufficient to stabilise twelve critical plants [5]. Finally, it is related to stabilisation and control of hybrid systems. Therefore, the solution to the SSP and SSSP will be useful for the development of control theory.

Historically, Youla et al [52] first studied the so-called strong stabilisation problem for one plant: under what conditions can a linear plant be stabilised by a stable compensator? In their paper, it is shown that the necessary and sufficient conditions are that its real unstable poles and zeros have the parity interlacing property. Based on the result in [52], the necessary and sufficient conditions for simultaneous stabilisation of two single-input single output (SISO) systems are found in [38]. This result is further generalised for two multi-input multi-output (MIMO) systems in [45], where the remarkable conclusion that the simultaneous stabilisation problem of r plants is equivalent to that of $r - 1$ plants with a stable compensator was made. For $r = 2$, a computationally tractable condition is obtained, based on the result in [52]. Ghosh and Byrnes [18] obtained some significant results concerning the generic nature of the simultaneous stabilisation problem, based on which Minto et al presented a practical algorithm for the solution to the simultaneous stabilisation problem of two plants [30]. Following the work of Youla on the parity interlacing property [52], Wei first studied the SSSP for one linear plant via a stable compensator having no real unstable zeros. In [49], Wei investigated the transcendental problem with its application to the SSSP. He gave a necessary and sufficient condition under which two linear plants having the strong interlacing property can be simultaneously stabilised by a compensator having no real poles in the closed right half of the complex plane, and the compensator is constructed from the strong interlacing property via a computational procedure developed there. He conjectured that it is also a necessary and sufficient condition that any two linear plants can be simultaneously stabilised by a stable compensator. From Vidyasagar's aforementioned results, this conjecture seems to bring one some hope of solving the problem of simultaneous stabilisation of three plants. But it is soon denied by Blondel et al [8], where they proved that Wei's condition is not sufficient for the problem of simultaneous stabilisation of three plants. Furthermore, they gave a surprising fact that the problem of simultaneous stabilisation of three plants is a rationally undecidable problem, i.e., it is not possible to find a necessary and sufficient condition that involves only the arithmetical operations (additions, subtractions, multiplications and divisions), logical operations (and/or), and sign test operations (equal to, greater than, and greater than or equal to) of the coefficients of the three plants. Thus, the solution of simultaneous stabilisation of three or more plants possibly depends exclusively on some computational algorithm. This motivates the development of an approach which solves the simultaneous stabilisation of three plants [24]. The design procedure developed in [24] is achieved by two steps. First, design two compensators with common denominator or numerator, which renders simultaneous stabilisation of one and two plants, respectively. Then, choose a desirable controller from the convex combination of the two compensators obtained above.

A common feature of the aforementioned approaches is that they have attempted to develop some new existence and stability conditions from the given plants' coefficients for the SSP or SSSP, which proves to be very hard. This leads to the development of mathematical programming approaches based on Routh-Hurwitz stability or similar conditions in recent years. Wu et al proposed an algorithm for simultaneous stabilisation of single-input systems via dynamic feedback [51]. Based on a sufficient stability condition given in [32], the SSP is first formulated as a constrained nonlinear optimisation problem and then approximated as a set of linear constraints to get the solution. In [5], a graphic-based approach is developed which is based on the Routh-Hurwitz stability criterion. Because of its graphic feature, this approach can only be applied to the first order compensator having less than three design variables (note that a PI-controller has two design variables and thus it is possible to use this approach, but a PID-controller has three, and thus it is not). Due to the complexity of the problem, several researchers focus on devising simpler but approximate design methods. Obviously, this is done at the expense of conservatism. For instance, the simple linear programming (LP) approach proposed in [17] is based on stability conditions that are only necessary. In other words, the controller obtained through this method may actually not be able to stabilise all plants. Another technique recently described in [4] relies upon nonlinear programming (NLP). The SSP is framed as a mathematical programming problem constrained by nonlinear necessary and sufficient stability conditions straightforwardly given by the Routh-Hurwitz criterion.

Many other researchers have contributed to the simultaneously stabilising problem via full state feedback or dynamic feedback [44]. Ackermann proposed a computer aided design (CAD) approach for stabilising a family of single-input single-output linear plants using linear state feedback [2]. Peterson studied a special quadratic simultaneous stabilisation problem using nonlinear state feedback, and he obtained the necessary and sufficient conditions for the quadratic simultaneous stabilisation of a set of single-input systems [33]. In [53], a state feedback approach is proposed, which requires that 1) the state equation for the plant has the standard canonical forms; 2) a template uncertain characteristic polynomial is required based on the analytic results from an interval polynomial. Using state-space arguments, it is proved in [53] that the existence of a compensator feedback gain is equivalent to that of the solution of a set of coupled quadratic matrix inequalities. A heuristic design algorithm based on linear matrix inequalities (LMIs) is then proposed. Finally, we point out another approach to simultaneous stabilisation obtained in the framework of the polynomial approach to system control [21]. It is shown that the SSP actually amounts to solving a non-convex feasibility problem, where all the non-convexity is concentrated into a matrix constraint that has rank 1.

It is noted that all of the approaches mentioned above for the SSP or the SSSP consider only the stability region of the left half complex plane. However, it is desirable to develop a computationally tractable approach which can synthesise a simultaneously stabilising compensator which itself is stable and meanwhile can guarantee that the characteristic roots of the closed-loop systems for the r-tuple plants all lie in a desired D-stability region (a subset of the Hurwitz stability region) rather than simply the Hurwitz-stability region. The D-stability specification allows for the consideration of both stability and transient performance and will be treated later.

Given the preceding literature, it seems that there are four factors affecting the synthesis methods of the SSP or the SSSP; 1) the stability regions specified. To the best of our knowledge, nobody has discussed the SSP or SSSP with D-stability region specification; 2) the problem formulation as an SSP or an SSSP. As shown later, an SSSP is more complicated than an SSP; 3) the method of feedback used, *i.e.*, static output feedback, dynamic output feedback, or state feedback and finally; 4) the number of plants considered. There does not exist a method considering simultaneously all four of the factors. Hence, new synthesis methods have to be developed. For this purpose, we will attack the SSP and SSSP from a new angle in this paper and present a new numerical approach to address the generalised SSP and SSSP in a unified way. By generalised, we mean that the stability region is a general D-region. The main idea behind this paper is to represent the D-stability requirements for the closed-loop system as a set of equalities and inequalities, then the SSPs or SSSPs are reduced to finding a feasible vector solution to a set of equalities and inequalities for which efficient numerical techniques have been developed in the field normally called numerical analysis or mathematical programming [13], [16], [15], [34],[36]. Once the equalities and inequalities are satisfied, the SSP (SSSP) with D-stability is automatically met.

6.2 Constraints for SSP with D-stable Regions

Equality constraints arising from the unknown controller design parameters and the closed-loop characteristic roots and coefficients of the *i*-th polynomial

Consider a single-input single-output (SISO) unity feedback closed-loop control system consisting of a family of plants $G_i(s)$, which is connected in cascade with a compensator $C(s)$, as shown in Figure 6.1. The discussion to follow also applies for the alternative set-up with $C(s)$ in the feedback path. Suppose the family of plants $G_i(s)$ are expressed by r distinct, proper rational functions

$$G_i(s) = \frac{N_i(s)}{D_i(s)}, \quad i = 1, 2, \ldots, r \tag{6.1}$$

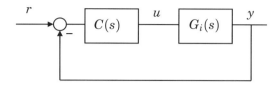

Figure 6.1 Feedback system

where the numerator $N_i(s)$ and the denominator $D_i(s)$ are given as

$$N_i(s) = a_{0,i} + a_{1,i}s + \ldots + a_{m_i-1,i}s^{m_i-1} + a_{m_i,i}s^{m_i} = A_i^T S_{m_i} \quad (6.2)$$
$$D_i(s) = b_{0,i} + b_{1,i}s + \ldots + b_{n_i-1,i}s^{n_i-1} + s^{n_i} = B_i^T S_{n_i} \quad (6.3)$$

where

$$\begin{aligned} A_i &= [a_{0,i}, a_{1,i}, \ldots, a_{m_i,i}]^T, \\ B_i &= [b_{0,i}, b_{1,i}, \ldots, b_{n_i-1,i}, 1]^T, \\ S_{m_i} &= [1, s, \ldots, s^{m_i}]^T \\ S_{n_i} &= [1, s, \ldots, s^{n_i}]^T \end{aligned}$$

Without loss of generality, it is assumed that the denominator is monic. Furthermore, it is assumed that the coefficients A_i and B_i are all real, and $n_i \geq m_i$.

Let $C(s)$ denote the compensator, which is also expressed by a proper rational function ($C(s)$ can also be another transfer function like a PID-controller, but this is not pursued here) and let

$$G(s) = \frac{N_c(s)}{D_c(s)} \quad (6.4)$$

where the numerator $N_c(s)$ and the denominator $D_c(s)$ are given as

$$N_c(s) = a_{0,c} + a_{1,c}s + \ldots + a_{m_c,c}s^{m_c} = A_c^T S_{m_c} \quad (6.5)$$
$$D_c(s) = b_{0,c} + b_{1,c}s + \ldots + b_{n_c-1,c}s^{n_c-1} + s^{n_c} = \begin{bmatrix} B_c \\ 1 \end{bmatrix}^T S_{n_c} \quad (6.6)$$

where

$$\begin{aligned} A_c &= [a_{0,c}, a_{1,c}, \ldots, a_{m_c,c}]^T, \quad B_c = [b_{0,c}, b_{1,c}, \ldots, b_{n_c-1,c}]^T, \\ S_{m_c} &= [1, s, \ldots, s^{m_c}]^T \quad S_{n_c} = [1, s, \ldots, s^{n_c}]^T \end{aligned}$$

Again, the coefficients A_c and B_c are all real, and $n_c \geq m_c$.

The i-th transfer function of the closed-loop system, denoted by $P_i(s)$ corresponding to the i-th plant $G_i(s)$ and the fixed compensator $C(s)$ is

$$\frac{y(s)}{r(s)} = P_i(s) = \frac{N_{p,i}(s)}{D_{p,i}(s)} = \frac{C(s)G_i(s)}{1 + C(s)G_i(s)} \quad (6.7)$$

Controller Design for Anti-lock Braking Systems (ABS)

where the denominator $D_{p,i}(s)$ (the characteristic polynomial of the closed system) is given as

$$D_{p,i}(s) = D_c(s)D_i(s) + N_c(s)N_i(s) \tag{6.8}$$

Since the orders of $D_i(s)$ and $D_c(s)$ are respectively n_i and n_c, $n_i \geq m_i$, and $n_c \geq m_c$, it is readily seen from Equations (6.2), (6.3), (6.5), (6.6) and (6.8), that $D_{p,i}(s)$ is a $(n_c + n_i)$ order polynomial. Suppose it is represented as

$$D_{p,i}(s) = c_{0,i} + c_{1,i}s + \ldots + c_{n-1,i}s^{n-1} + c_{n,i}s^n = C_i^T S_n \tag{6.9}$$

where $n = (n_c + n_i)$, $C_i = [c_{0,i}, c_{1,i}, \ldots, c_{n,i}]^T$ and $S_n = [1, s, \ldots, s^n]^T$. Substituting Equations (6.2), (6.3), (6.5), (6.6) into (6.8), and equating the coefficients of like power of s in both sides of equations (6.8) and (6.9) gives the expression of $c_{j,i}$ ($j = 0, 1, 2, \ldots, n$) in terms of A_c, B_c, A_i, and B_i. With the use of a symbolic manipulation this can be easily done. For the convenience of later discussion, suppose $c_{j,i}$ is expressed as

$$c_{j,i} = f_{j,i}(A_c, B_c, A_i, B_i), \quad i = 1, 2, \ldots, r, \text{ and } \quad j = 0, 1, 2, \ldots, n \tag{6.10}$$

We can observe that C_i are functions of the unknown variables (A_c, B_c) for the compensator and the given constants (A_i, B_i) for the i-th plant.

A conclusion drawn from algebra is that an n-th order polynomial $p(s)$ with real coefficients has, in the most general case, n complex characteristic roots (a real root is a special case with imaginary part being zero). The relationship between roots and coefficients of a real polynomial is well known in traditional algebra. However, the formulation presented in a traditional way is not suitable for our purpose because of two facts: first, fundamentally, traditional formulas include an imaginary unit in the formula which is not what we need and second, the traditional formulation is very lengthy and not suitable for symbolic manipulation. It suits only for the case that if the roots are given, then calculate the coefficients. In this section, we develop an efficient recursive formula which is required by the symbolic calculation because a point we should keep in mind is that the coefficients of the closed-loop system include uncertain parameters. The purpose of this section is to establish a set of equalities among the uncertain parameters and the corresponding uncertain roots as constraints.

Polynomial (6.9) has $n = (n_c + n_i)$ roots. Let

$$s_{kp,i} = \sigma_{kp,i} + j\omega_{kp,i}, (kp = 1, 2, \ldots, n) \tag{6.11}$$

denote the kp-th root for the i-th characteristic polynomial $D_{p,i}(s)$ expressed in Equation (6.9). Moreover, to simplify notation in the sequel, let $\Sigma_{j,i} = [\sigma_{1,i}, \sigma_{2,i}, \ldots, \sigma_{j,i}]^T$, $\Omega_{j,i} = [\omega_{1,i}, \omega_{2,i}, \ldots, \omega_{j,i}]^T$, $j = 1, 2, \ldots, n$. In this section, we will derive the relationship between $(\Sigma_{n,i}, \Omega_{n,i})$ and (A_c, B_c, A_i, B_i).

In the most general case, $D_{p,i}(s)$ can be written as the following complex polynomial

$$D_{p,i}(s) = \prod_{kp=1}^{n}(s - s_{kp,i}) = [\alpha_{0,i}(\Sigma_{n,i}, \Omega_{n,i}) + j\beta_{0,i}(\Sigma_{n,i}, \Omega_{n,i})] + \ldots$$
$$+ [\alpha_{n-1,i}(\Sigma_{n,i}, \Omega_{n,i}) + j\beta_{n-1,i}(\Sigma_{n,i}, \Omega_{n,i})]s^{n-1} + s^n \quad (6.12)$$

where the coefficients $\alpha_{jp,i}(\Sigma_{n,i}, \Omega_{n,i})$ and $\beta_{jp,i}(\Sigma_{n,i}, \Omega_{n,i})$, $jp = 0, 1, 2, \ldots, n-1$, mean that they are all functions of $\Sigma_{n,i}, \Omega_{n,i}$. To get expressions for $\alpha_{jp,i}$ and $\beta_{jp,i}$, one way is to directly substitute (6.11) into (6.12) and then expand $\prod_{kp=1}^{n}(s - s_{kp,i})$. However, as n becomes large (for example more than 4), manual derivation for $\alpha_{jp,i}$ and $\beta_{jp,i}$ becomes quite lengthy although it is quite straightforward. It can be observed that $\alpha_{jp,i}(\Sigma_{n,i}, \Omega_{n,i})$ and $\beta_{jp,i}(\Sigma_{n,i}, \Omega_{n,i})$ can be recursively represented by $\alpha_{jp,i}(\Sigma_{n-1,i}, \Omega_{n-1,i})$ and $\beta_{jp,i}(\Sigma_{n-1,i}, \Omega_{n-1,i})$, so in the following a recursive formula is derived. It gives a simple procedure for the calculation of the coefficients of an n-th order polynomial based on those of an $(n-1)$-th order polynomial. This is particularly suitable for symbolic manipulation.

Note that the polynomial in (6.12) can be rewritten as

$$D_{p,i}(s) = Re_{n,i}(s, \Sigma_{n,i}, \Omega_{n,i}) + jIm_{n,i}(s, \Sigma_{n,i}, \Omega_{n,i}) \quad (6.13)$$

Comparing (6.12) and (6.13), we can get

$$Re_{n,i}(s, \Sigma_{n,i}, \Omega_{n,i}) = \alpha_{0,i}(\Sigma_{n,i}, \Omega_{n,i}) + \alpha_{1,i}(\Sigma_{n,i}, \Omega_{n,i})s + \cdots$$
$$\cdots + \alpha_{n-1,i}(\Sigma_{n,i}, \Omega_{n,i})s^{n-1} + s^n \quad (6.14)$$
$$Im_{n,i}(s, \Sigma_{n,i}, \Omega_{n,i}) = \beta_{0,i} + \beta_{1,i}s + \ldots + \beta_{n-1,i}s^{n-1} \quad (6.15)$$

Since

$$\prod_{kp=1}^{n}(s - s_{kp,i}) = \prod_{kp=1}^{n-1}(s - s_{kp,i})(s - s_n)$$
$$= [Re_{n-1,i}(s, \Sigma_{n-1,i}, \Omega_{n-1,i})$$
$$+ jIm_{n-1,i}(s, \Sigma_{n-1,i}, \Omega_{n-1,i})](s - \sigma_{n,i} - j\omega_{n,i}) \quad (6.16)$$

where

$$Re_{n-1,i}(s, \Sigma_{n-1,i}, \Omega_{n-1,i}) = \alpha_{0,i}(\Sigma_{n-1,i}, \Omega_{n-1,i})$$
$$+ \alpha_{1,i}(\Sigma_{n-1,i}, \Omega_{n-1,i})s + \cdots + \alpha_{n-2,i}(\Sigma_{n-1,i}, \Omega_{n-1,i})s^{n-2} + s^{n-1}$$

$$Im_{n-1,i}(s, \Sigma_{n-1,i}, \Omega_{n-1,i}) = \beta_{0,i}(\Sigma_{n-1,i}, \Omega_{n-1,i})$$
$$+ \beta_{1,i}(\Sigma_{n-1,i}, \Omega_{n-1,i})s + \cdots + \beta_{n-2,i}(\Sigma_{n-1,i}, \Omega_{n-1,i})s^{n-2}$$

From (6.16), we have

$$Re_{n,i}(s, \Sigma_{n,i}, \Omega_{n,i}) = sRe_{n-1,i}(s, \Sigma_{n-1,i}, \Omega_{n-1,i})$$
$$- \sigma_{n,i} Re_{n-1,i}(s, \Sigma_{n-1,i}, \Omega_{n-1,i}) + \omega_{n,i} Im_{n-1,i}(s, \Sigma_{n-1,i}, \Omega_{n-1,i})$$

$$Im_{n,i}(s, \Sigma_{n,i}, \Omega_{n,i}) = sIm_{n-1,i}(s, \Sigma_{n-1,i}, \Omega_{n-1,i})$$
$$- \sigma_{n,i} Im_{n-1,i}(s, \Sigma_{n-1,i}, \Omega_{n-1,i}) - \omega_{n,i} Re_{n-1,i}(s, \Sigma_{n-1,i}, \Omega_{n-1,i})$$

Substituting (6.14), (6.15), (6.16) and equating the coefficients of like powers of s in both sides results in the following recursive formula. For $n = 1$,

$$\alpha_{0,i}(\Sigma_{1,i}, \Omega_{1,i}) = -\sigma_{1,i}$$
$$\beta_{0,i}(\Sigma_{1,i}, \Omega_{1,i}) = -\omega_{1,i} \qquad (6.17)$$

for $n > 1$,

$$\alpha_{0,i}(\Sigma_{n,i}, \Omega_{n,i}) = -\alpha_{0,i}(\Sigma_{n-1,i}, \Omega_{n-1,i})\sigma_{n,i} + \beta_{0,i}(\Sigma_{n-1,i}, \Omega_{n-1,i})\omega_{n,i}$$
$$\beta_{0,i}(\Sigma_{n,i}, \Omega_{n,i}) = -\beta_{0,i}(\Sigma_{n-1,i}, \Omega_{n-1,i})\sigma_{n,i} - \alpha_{0,i}(\Sigma_{n-1,i}, \Omega_{n-1,i})\omega_{n,i} \quad (6.18)$$
$$\alpha_{j,i}(\Sigma_{n,i}, \Omega_{n,i}) = \alpha_{j-1,i}(\Sigma_{n-1,i}, \Omega_{n-1,i}) - \alpha_{j,i}(\Sigma_{n-1,i}, \Omega_{n-1,i})\sigma_{n,i} \quad (6.19)$$
$$+ \beta_{j,i}(\Sigma_{n-1,i}, \Omega_{n-1,i})\omega_{n,i} \qquad (6.20)$$
$$\beta_{j,i}(\Sigma_{n,i}, \Omega_{n,i}) = \beta_{j-1,i}(\Sigma_{n-1,i}, \Omega_{n-1,i}) - \beta_{j,i}(\Sigma_{n-1,i}, \Omega_{n-1,i})\sigma_{n,i} \quad (6.21)$$
$$- \alpha_{j,i}(\Sigma_{n-1,i}, \Omega_{n-1,i})\omega_{n,i}, \quad j = 1, 2, \ldots, n-2$$

and

$$\alpha_{n-1,i}(\Sigma_{n,i}, \Omega_{n,i}) = \alpha_{n-2,i}(\Sigma_{n-1,i}, \Omega_{n-1,i}) - \sigma_{n,i}$$
$$\beta_{n-1,i}(\Sigma_{n,i}, \Omega_{n,i}) = \beta_{n-2,i}(\Sigma_{n-1,i}, \Omega_{n-1,i}) - \omega_{n,i} \qquad (6.22)$$

Recall that the i-th characteristic polynomial of the system is represented by (6.9), and note that (6.9) and (6.12) actually represent the same polynomial which has identical roots and coefficients, we thus have

$$c_{n,i}\alpha_{j,i}(\Sigma_{n,i}, \Omega_{n,i}) = c_{j,i}$$
$$\beta_{j,i}(\Sigma_{n,i}, \Omega_{n,i}) = 0, \quad j = 0, 1, 2, \ldots, n-1 \qquad (6.23)$$

Taking (6.10) into consideration yields

$$f_{n,i}\alpha_{j,i}(\Sigma_{n,i}, \Omega_{n,i}) = f_{j,i}$$
$$\beta_{j,i}(\Sigma_{n,i}, \Omega_{n,i}) = 0, \quad j = 0, 1, 2, \ldots, n-1 \qquad (6.24)$$

The number of equalities in either (6.23) and (6.24) is $2n$, the same as the number of unknown variables in the n complex roots for the i-th polynomial $D_{p,i}(s)$.

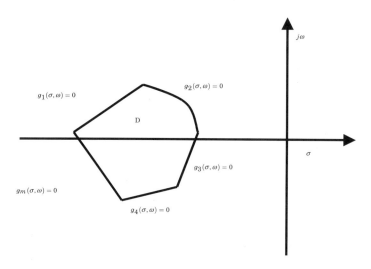

Figure 6.2 A general D-stable region bounded by a set of curves

General D-stable Region and Inequality Constraints

D-stable region. We first introduce some notation. A root-plane for both continuous and discrete systems is a complex plane. In this paper we use complex plane to represent the s-plane for the continuous case and the z-plane for the discrete case. Let D denote the interior of a region in the complex plane, ∂D denote its boundary. For a continuous time system, if a region is located in the left half of the complex plane this region is said to be a D-stable region, and its boundary is said to be a D-stable boundary. Similarly, for a discrete-time system if a region is within the open unit circle centred at the origin, this region is said to be a D-stable region, and the boundary is said to be a D-stable boundary. In control system design, all the closed-loop poles must be confined to a desired D-stable region or a set of disjoint desired D-stable regions. We shall say that the closed-loop system is D-stable if all the roots of its characteristic polynomial are located in the desired D-stable regions.

Inequality representation. The mathematical description of a desired D-stable region has to be discussed. We call an inequality of the form $g(x) < 0$ a strict one, and an inequality of the form $g(x) \leq 0$ a non-strict one. Although a region D may have different shapes in the complex plane, it is always bounded by an algebraic curve or a set of piecewise algebraic curves like straight lines, circles, ellipses, hyperbolas, parabolas and so on. Thus, the interior D of a region may be described by a set of strict inequalities, its boundary ∂D can be described by a set of equalities and $D \cup \partial D$ (the union of the interior and boundary) can be described by a set of non-strict inequalities. A general form for inequalities and equalities in the complex plane for the description of the desired pole regions is (see Figure 6.2):

$$D : g_l(\sigma, \omega) < 0, \quad l = 1, 2, \ldots, m \quad (6.25)$$
$$\partial D : g_l(\sigma, \omega) = 0, \quad l = 1, 2, \ldots, m \quad (6.26)$$
$$D \cup \partial D : g_l(\sigma, \omega) \leq 0, \quad l = 1, 2, \ldots, m \quad (6.27)$$

where the complex variable s is $s = \sigma + j\omega$. As a result, the D-stable requirement means that all the roots must satisfy (6.25) or (6.27), depending on whether the roots are allowed to be on the boundary. On the other hand, if a root satisfies (6.25) it must be on the D region. Later, we will show that if a feasible point or an optimal point is found, the D-stability requirement is automatically met.

In the following, we discuss the issue of transforming a strict equality into a non-strict one. This requirement only arises when a special case occurs, where the D-stability region is the open left-half complex plane and the location of the roots on the boundary is not allowed. In this case, $l = 1$ and the D-stability region can be described as

$$D : g_l(\sigma, \omega) = \sigma < 0 \quad (6.28)$$
$$\partial D : g_l(\sigma, \omega) = \sigma = 0 \quad (6.29)$$
$$D \cup \partial D : g_l(\sigma, \omega) = \sigma \leq 0 \quad (6.30)$$

The stability requirement in this case leads to the strict inequality (6.28). In any other case where the D-stability region is a subset of the left half plane, the stability requirement may be represented by a set of non-strict inequalities. The reason for discussing the transformation is that non-strict inequalities are exclusively required by certain numerical algorithms, for example, those adopted in nonlinear constraint optimisation to find an optimum solution. This requires the transformation of a strict inequality into a non-strict one. From a practical engineering point of view, this does not cause any difficulty because it is easy to transform a strict inequality into a non-strict one simply by introducing a small positive number δ, say 0.0001, such that the satisfaction of a strict inequality of the sort $g(x) < 0$ is guaranteed if a corresponding non-strict inequality $g(x) \leq -\delta$ is satisfied.

For simplicity, we will not distinguish the terms inequality from non-strict inequality in the sequel unless otherwise stated, and it is further assumed that D-stability requirement is equivalent to the satisfaction of the sort of inequalities given in (6.27).

Examples. Figures 6.3 and 6.4 show some widely-used D-stable regions and their boundaries for pole assignment, and they are represented respec-

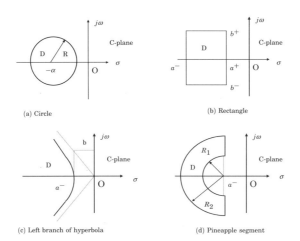

Figure 6.3 (a)-(d) Examples of widely-used D-stable regions

tively by the following inequalities:

a. $(\sigma + \alpha)^2 + \omega^2 \leq R^2$ (6.31)
b. $a^- \leq \sigma \leq a^+$ and $b^- \leq \omega \leq b^+$ (6.32)
c. $-(\sigma/a)^2 + (\omega/b)^2 \leq -1$ (6.33)
d. $\sigma + a \leq 0$ and $R_1^2 \leq (\sigma + a)^2 + \omega^2 \leq R_2^2$ (6.34)
e. $\sigma + a \leq 0, b\sigma + a\omega \leq 0$, and $b\sigma - a\omega \leq 0$ (6.35)
f. $\sigma + a \leq 0$ or $(\sigma + a_1)^2 + \omega^2 \leq R^2$ or
    $(\sigma + a_2)^2 + (\omega + b)^2 \leq R^2$ or $(\sigma + a_2)^2 + (\omega - b)^2 \leq R^2$ (6.36)

Inequality representation for root location requirements. Suppose that a desired D-stability region is given in the form of (6.27) and, as assumed previously, let $s_{kp,i} = \sigma_{kp,i} + j\omega_{kp,i}$, $kp = 1, 2, \ldots, n$, $i = 1, 2, \ldots, r$, denote the kp–th root of the i–th characteristic polynomial $D_{p,i}(s)$. Then $s_{kp,i}$ must satisfy (6.27) for an SSP system, thus

$$g_l(\sigma_{kp,i}, \omega_{kp,i}) \leq 0, \quad l = 1, 2, \ldots, m \qquad (6.37)$$

Let us introduce a vector design variable $x_{ssp} = [A_c, B_c, \Sigma_{n_c+n_1,1}, \Sigma_{n_c+n_2,2}, \ldots, \Sigma_{n_c+n_r,r}, \Omega_{n_c+n_1,1}, \Omega_{n_c+n_2,2}, \ldots, \Omega_{n_c+n_r,r}]$, where A_c and B_c are compensator coefficients and

$$\Sigma = [\Sigma_{n_c+n_1,1}, \Sigma_{n_c+n_2,2}, \ldots, \Sigma_{n_c+n_r,r}],$$
$$\Omega = [\Omega_{n_c+n_1,1}, \Omega_{n_c+n_2,2}, \ldots, \Omega_{n_c+n_r,r}]$$

are introduced previously. The total number of the elements (or the dimension) in vector x_{ssp} is $[m_c + (2r+1)n_c + 2\Sigma_{i=1}^r n_i]$, which is one of the two

Controller Design for Anti-lock Braking Systems (ABS) 109

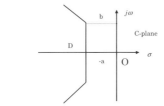

(e) D region determined by three straight lines

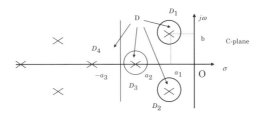

(f) Disjoint D region, $D = D_1 \cup D_2 \cup D_3 \cup D_4$

Figure 6.4 (e)-(f) Examples of widely-used D-stable regions

indicators for the problem size. The other indicator is the total number of the equalities and inequalities involved. From (6.24) and (6.37), we know that there are, respectively, $2[rn_c + \sum_{i=1}^{r} n_i]$ equalities and $m[rn_c + \sum_{i=1}^{r} n_i]$ inequalities. It is easily seen that reducing the number of m is very important. We stress here that in the above discussion, the curve number of the D-stability region for every root is m. However, it is allowed to have a more general case that the number for every D region is different. We will not discuss this any more.

Now, it is clear that if a vector design variable x_{ssp} as defined above can be found such that the equalities (6.24) and the inequalities (6.37) are simultaneously satisfied, then a controller for SSP is found. Finding x_{ssp} is a standard problem that will be treated below.

6.3 Constraints for SSSP with D-stable Regions

In addition to the satisfaction of the requirement for an SSP, an SSSP must satisfy an extra requirement that the compensator must also be stable, i.e. its poles must be located at the open left half complex plane. For a lower order compensator, this requirement can be simply expressed as a set of strict inequalities based on the Routh-Hurwitz-stability criterion [4]. For example, for a first order compensator having the following form

$$C(s) = \frac{N_c(s)}{D_c(s)} = \frac{a_{0,c} + a_{1,c}s}{b_{0,c} + s} \qquad (6.38)$$

the strong stabilising requirement simply means the following strict inequality is satisfied:

$$b_{0,c} > 0 \tag{6.39}$$

If a further requirement, say a lead first-order compensator, is raised, then two extra strict inequalities of the following forms must be satisfied:

$$a_{1,c}b_{0,c} > a_{0,c}, a_{1,c}b_{0,c} > 0 \tag{6.40}$$

Note that (6.39) and (6.40) are strict inequalities. However as mentioned previously, they can be transformed into non-strict ones by introducing a small positive number δ, say 0.0001, such that a strict inequality of the sort $g(x) < 0$ is satisfied if the corresponding non-strict inequality $g(x) \leq -\delta$ is satisfied.

A more general approach to representing the strong stability requirement as a set of equalities and inequalities is the one we introduced previously. Let $s_{kp,c} = \sigma_{kp,c} + j\omega_{kp,c}$, $(kp = 1, 2, \ldots, n_c)$ denote the kp-th root for the denominator $D_c(s)$ of the compensator $C(s)$, as expressed in Equation (6.6) and let $\Sigma_c = [\sigma_{1,c}, \sigma_{2,c}, \ldots, \sigma_{n_c,c}]^T$ and $\Omega_c = [\omega_{1,c}, \omega_{2,c}, \ldots, \omega_{n_c,c}]^T$. Then, $D_c(s)$ can be represented as

$$\begin{aligned} D_c(s) &= \prod_{kp=1}^{n_c}(s - s_{kp,c}) \\ &= [\alpha_{0,c}(\Sigma_c, \Omega_c) + j\beta_{0,c}(\Sigma_c, \Omega_c)] + [\alpha_{1,c}(\Sigma_c, \Omega_c) + j\beta_{1,c}(\Sigma_c, \Omega_c)]s \\ &\quad + \cdots + [\alpha_{n_c-1,c}(\Sigma_c, \Omega_c) + j\beta_{n_c-1,c}(\Sigma_c, \Omega_c)]s^{n_c-1} + s^{n_c} \end{aligned}$$

The coefficients $\alpha_{kp,c}, \beta_{kp,c} (kp = 0, 1, 2, \ldots, n_{c-1})$ can be calculated using the previous recursive formulas. From (6.6) and the above we have

$$\alpha_{j,c}(\Sigma_c, \Omega_c) = b_{j,c}, \beta_{j,c}(\Sigma_c, \Omega_c) = 0, \quad j = 0, 1, 2, \ldots, n_c - 1 \tag{6.41}$$

In addition, since $s_{kp,c}$ must also lie in the open left half plane, so we have

$$\sigma_{kp,c} < 0, \quad kp = 1, 2, \ldots, n_c \tag{6.42}$$

By introducing a new vector design variable $x_{sssp} = [x_{ssp}, \Sigma_c, \Omega_c]$, where $x_{ssp} = [A_c, B_c, \Sigma_{n_c+n_1,1}, \ldots, \Sigma_{n_c+n_r,r}, \Omega_{n_c+n_1,1}, \ldots, \Omega_{n_c+n_r,r}]$, we conclude that if a vector design variable x_{sssp} can be found such that the equalities (6.24), (6.41) and the inequalities (6.37), (6.42) are simultaneously satisfied, then an SSSP controller is found.

Notice that the requirement for the SSSP, compared with that for the SSP, leads to the extra constraints imposed: $2n_c$ equalities and n_c inequalities. Meanwhile, the number of design variables also increases from $[m_c + (2r+1)n_c + 2\sum_{i=1}^{r} n_i]$ to $[m_c + (2r+3)n_c + 2\sum_{i=1}^{r} n_i]$. This indicates that an SSSP is more complicated than an SSP.

6.4 Numerical Solution Techniques for the SSP and SSSP

Feasibility Problem (FP) and its solution

DEFINITION 6.1—FEASIBLE POINT
A vector point $x \in R^n$ satisfying a given set of inequalities and inequalities

$$g_i(x) = 0; \quad i = 1, 2, \ldots, m_1, g_j(x) \leq 0; \quad j = m_1+1, m_1+2, \ldots, m_2 \quad (6.43)$$

is said to be a feasible point of the given set. Finding a feasible point is said to be a feasibility problem (FP). □

Thus far, we have respectively formulated the requirements for both the SSP and SSSP as the problems of finding a feasible solution x_{ssp} or x_{sssp} for a set of equalities and inequalities. The focal point followed is how to find a solution to it. Fortunately, the feasibility problem (FP) is a standard problem in the field of numerical analysis and optimisation theory and has been widely studied. Generally speaking, there are two classes of techniques to attack the FP. The first are the direct techniques. A number of algorithms in this class have been developed. For example, Pshenichnyi [36], Robinson [37], and Daniel [14] extended Newton's method to systems of nonlinear equalities and inequalities. Polyak [34] used gradient methods for solving this problem. In [16], a trust-region approach was also presented.

The second class of techniques transfers the FP problem into finding the roots for a set of nonlinear equations. This is realised in the following way. The inequalities in (6.43) can be transformed to equality constraints by adding non-negative slack variables, $y_i^2, i = m_1+1, m_1+2, \ldots, m_2$, where the value of the slack variables are yet unknown. This reduces the FP to finding the solution of systems of nonlinear equations with new design variables (x, y), where $y = (y_{m_1+1}, \ldots, y_{m_2})^T$. Newton's method is a well-known and very powerful technique for solving systems of nonlinear equations, see for example, [13] and [15]. Another efficient numerical algorithm for solving systems of nonlinear equations is the LM method, which has been implemented in [29].

Due to restricted space, we will not give more details about the existing algorithms and their convergence analysis for the FP. As will be shown later, SSP and SSSP can also be formulated as a constrained optimisation problem, and in our experiment we run the optimisation algorithms (sequential quadratic programming) to solve the problem. For more details, and the relevant references on FP, the reader is referred to the latest reference [16]. The important point is that this formulation establishes the link between SSP (SSSP) and another well-studied mature field, mathematical programming, and allows us to use well-developed tools and algorithms.

Nonlinear Constrained Optimisation Problem (NLCOP)

Once a feasible solution is found using the algorithms presented above, then a controller satisfying the SSP and SSSP with D-stable requirement has been found. This represents one method for the design of a compensator. However, improvement of the system remains possible because the FP algorithm will stop once a feasible solution is found. In this subsection, we will formulate the SSP or SSSP as an NLCOP by introducing an objective function. The reason for that is obvious: we want to find an optimal solution rather than simply a feasible solution. This establishes a link between control system design and the well-developed field of nonlinear optimisation, to which a considerable amount of effort has been devoted since the 1960s [50]. The field is still active due to the development of interior-point algorithms [20].

A general NLCOP can be described as follows: find an optimal point $x^* \in R^n$ that minimises a scalar function

$$f(x) \tag{6.44}$$

subject to a set of equalities and inequalities with the following form:

$$g_i(x) = 0; \quad i = 1, 2, \ldots, m_1, \quad g_j(x) \leq 0; \quad j = m_1 + 1, m_1 + 2, \ldots, m_2 \tag{6.45}$$

where m_1 and m_2 are integers, $x = (x_1, x_2, \ldots, x_n)$, is an n-dimensional vector of unknowns, and f, g_i, $i = 1, 2, \ldots, m_1$ and g_j, $j = m_1 + 1, \ldots, m_2$, are real-valued functions of the variables $(x_1, x_2, \ldots, x_n)$. The function f is the objective function, and the equations and inequalities are constraints.

The difference between an FP and an NLCOP is obvious. An NLCOP is an FP with an objective function. The solution procedure of an NLCOP normally consists of two steps: first, find a feasible solution and then keep going without stop to find an optimal solution.

There are various ways to design the objective function. In the context of NLCOPs, this function must represent some meaning of the practical problem, for example, minimum time, minimum distance, minimum energy or minimum cost. For the SSP or SSSP, the objective function is naturally chosen as a quadratic function of the vector design variables x_{ssp} or x_{sssp} in order that the error signal is not amplified too much. This will reduce the risk of the input constraint violation for the actuator. From a computational point of view, the convex property of a quadratic function will also improve the efficiency of the numerical algorithms. Thus, we define the following weighted objective functions:

$$f(x_{ssp}) = x_{ssp}^T W_{ssp} x_{ssp} \quad \text{for SSP} \tag{6.46}$$

$$f(x_{sssp}) = x_{sssp}^T W_{sssp} x_{sssp} \quad \text{for SSSP} \tag{6.47}$$

Controller Design for Anti-lock Braking Systems (ABS) 113

where W_{ssp} and W_{sssp} are, respectively, $n_{ssp} \times n_{ssp}$ and $n_{sssp} \times n_{sssp}$ dimensional positive, diagonal matrices, given as:

$$W_{ssp} = \begin{bmatrix} w_1 & 0 & \ldots & 0 \\ 0 & w_2 & \ldots & 0 \\ \ldots & \ldots & \ldots & \ldots \\ 0 & 0 & \ldots & w_{n_{ssp}} \end{bmatrix} \quad W_{sssp} = \begin{bmatrix} w_1 & 0 & \ldots & 0 \\ 0 & w_2 & \ldots & 0 \\ \ldots & \ldots & \ldots & \ldots \\ 0 & 0 & \ldots & w_{n_{sssp}} \end{bmatrix}$$

where

$$n_{ssp} = [m_c + (2r+1)n_c + 2\sum_{i=1}^{r} n_i] \text{ and } n_{sssp} = [m_c + (2r+3)n_c + 2\sum_{i=1}^{r} n_i]$$

Numerical algorithms. For engineering applications, nonlinear programming problems are generally solved by two classes of approaches: a) reduced gradient methods [19]; b) iterative programming techniques: successive linear programming (SLP) and successive quadratic programming (SQP) [11], [19], [35], [39].

Of these approaches, SQP has emerged as the method of choice for the solution of nonlinear programming problems, especially because it can be adapted and tailored easily to a wide variety of specialised problem formulations. Schittowski, for instance, has implemented and tested a version which outperforms every other tested method in terms of efficiency, accuracy and percentage of successful solutions, over a large number of test problems. The SQP method for solving (6.44), (6.45) solves the QP subproblem (6.48) at the current point and iteration k to generate a search direction d_k:

$$\min \quad \nabla f(x_k)^T d_k + 0.5 d^T H_k d_k$$

$$\text{s.t.} \quad \begin{aligned} \nabla g_i(x_k)^T d_k + g_i(x_k) &= 0, & i &= 1, \ldots, m_1 \\ \nabla g_j(x_k)^T d_k + g_j(x_k) &\leq 0, & j &= m_1 + 1, \ldots, m_2 \end{aligned} \quad (6.48)$$

where H_k is the Hessian of the Lagrangian function or a positive-definite approximation to it. The solution d_k is used to form a new iterate:

$$x_{k+1} = x_k + \alpha_k d_k \tag{6.49}$$

The step length parameter α_k is determined by a one-dimensional line search procedure (for example the quadratic or cubic polynomial method) so that a sufficient decrease in a merit function is obtained. The SQP implementation consists of three main stages: a) updating of the Hessian matrix of the Lagrangian function; b) quadratic programming problem solution; c) linear search and merit function calculation. (See the section

"Update the Hessian Matrix" in [29] for more details about the Hessian update and line search techniques. The solution procedure for the QP can be found in [19]).

With the widespread use of SQP for process optimisation in design, operations, and control, a variety of different implementations and strategies have been developed for this algorithm. Most of the commonly used QP codes use active set strategies, *i.e.*, they iteratively determine the set of inequality constraints that are active at the solution of the problem, which is similar in principle to the simplex method for linear programming. However, it has been observed that an active set strategy can become combinatorially expensive for very large-scale problems. Consequently, popular alternatives to these active set methods are interior point methods, which have made inroads into the solution of NPLs in the mid-90s. A comprehensive discussion of active set *vs.* interior point algorithms can be found in Wright [50]. For the SSP and SSSP with low dimension, we still adopt the active set approach.

Finally, one thing must be clarified. The point one may argue about the NLCOP is its concave property, which does not offer any guarantee about a global solution. It is true that an NLCOP algorithm can only efficiently find a local minimum. However what we must stress here is that the local minima problem with a NLCOP is not a serious, but only a mild restriction for the SSP and SSSP in the sense that what we need is to find a stabilising compensator satisfying (6.45). A local minimum can also meet (6.45). On the other hand, although people expect a convex formulation of the SSP and SSSP, our experience shows that it is hard and this is still open because a convex formulation like LMI needs full state feedback that is not available for the SSP and SSSP.

6.5 Design Example and Application to ABS Control

In this section, the approach developed above is first illustrated with a simple example. By simple, we mean that the capability the approach can offer is higher. However, it is still much more complicated compared with the examples given in other papers, for example, [8], [9], [10]. From this example, we can see that the D-stable region parameters can be easily changed. Secondly, the approach is applied to a practical ABS controller design problem.

An Illustrative Example with Different D Regions

Given four plants of the following forms (note that G_3 and G_4 are unstable)

$$G_1(s) = \frac{s+4}{s+2}, \qquad G_2(s) = \frac{2(s+2)}{s+3} \qquad (6.50)$$

$$G_3(s) = \frac{s+2}{s^2 - 3s + 4}, \qquad G_4(s) = \frac{s+3}{s^2 - 2s + 1} \qquad (6.51)$$

Controller Design for Anti-lock Braking Systems (ABS) 115

find a first order simultaneous stabilisation compensator of the form

$$C(s) = \frac{a_{1,c}s + a_{0,c}}{s + b_{0,c}} \qquad (6.52)$$

such that all the roots are located in the - stability region described by a parabola of the following form

$$D \cup \partial D : g(\sigma, \omega) = \sigma + \omega^2 + 1 \leq 0 \qquad (6.53)$$

In the sequel, we will demonstrate the approach step by step. The four characteristic polynomials of the closed-loop system corresponding to the four plants are respectively given as

$$\begin{aligned}
D_{p,1}(s) &= D_c(s)D_1(s) + N_c(s)N_1(s) \qquad (6.54)\\
&= (a_{1,c}+1)s^2 + (a_{0,c} + 4a_{1,c} + b_{0,c} + 2)s + (4a_{0,c} + 2b_{0,c})\\
D_{p,2}(s) &= D_c(s)D_2(s) + N_c(s)N_2(s) \qquad (6.55)\\
&= (2a_{1,c}+1)s^2 + (2a_{0,c} + 4a_{1,c} + b_{0,c} + 3)s + (4a_{0,c} + 3b_{0,c})\\
D_{p,3}(s) &= D_c(s)D_3(s) + N_c(s)N_3(s) \qquad (6.56)\\
&= s^3 + (a_{1,c} + b_{0,c-3})s^2 + (a_{0,c} + 2a_{1,c} - 3b_{0,c} + 4)s\\
&\quad + (2a_{0,c} + 4b_{0,c})\\
D_{p,4}(s) &= D_c(s)D_4(s) + N_c(s)N_4(s) \qquad (6.57)\\
&= s^3 + (a_{1,c} + b_{0,c} - 2)s^2 + (a_{0,c} + 3a_{1,c} - 2b_{0,c} + 1)s\\
&\quad + (3a_{0,c} + b_{0,c})
\end{aligned}$$

In the following, we only derive the equalities and inequalities for plant 1, a similar procedure applies to other plants.

$D_{p,1}(s)$ is a second order polynomial, so let $s_{1,1} = \sigma_{1,1} + j\omega_{1,1}$ and $s_{2,1} = \sigma_{2,1} + j\omega_{2,1}$ denote its two roots. Therefore

$$\begin{aligned}
D_{p,1}(s) &= (s - s_{1,1})(s - s_{2,1}) = (s - \sigma_{1,1} - j\omega_{1,1})(s - \sigma_{2,1} - j\omega_{2,1})\\
&= (\sigma_{1,1}\sigma_{2,1} - \omega_{1,1}\omega_{2,1}) + j(\sigma_{1,1}\omega_{2,1} + \sigma_{2,1}\omega_{1,1})\\
&\quad + (-\sigma_{1,1} - \sigma_{2,1})s + j(-\omega_{1,1} - \omega 2, 1)s + s^2 \qquad (6.58)
\end{aligned}$$

From (6.54) and (6.58), we have

$$\begin{aligned}
(\sigma_{1,1}\sigma_{2,1} - \omega_{1,1}\omega_{2,1})(a_{1,c}+1) &= 4a_{0,c} + 2b_{0,c} \qquad (6.59)\\
(-\sigma_{1,1} - \sigma_{2,1})(a_{1,c}+1) &= a_{0,c} + 4a_{1,c} + b_{0,c} + 2 \qquad (6.60)\\
\sigma_{1,1}\omega_{2,1} + \sigma_{2,1}\omega_{1,1} &= 0 \qquad (6.61)\\
-\omega_{1,1} - \omega_{2,1} &= 0 \qquad (6.62)
\end{aligned}$$

It is interesting to note that (6.61) and (6.62) indicate that if the two roots are complex they must be a pair of conjugate roots ($\sigma_{1,1} = \sigma_{2,1}$ and $\omega_{1,1} = -\omega_{2,1} \neq 0$), and if the two roots are real they may not be the same. This verifies the well-known result in algebra that if a real polynomial has complex roots, the complex roots must be conjugate pairs. In addition, all the roots for $D_{p,1}(s)$ must belong to the $D \cup \partial D$ described by (6.53), so they must satisfy

$$\sigma_{1,1} + \omega_{1,1}^2 + 1 \leq 0 \tag{6.63}$$

$$\sigma_{2,1} + \omega_{2,1}^2 + 1 \leq 0 \tag{6.64}$$

where (6.59)-(6.64) are the equalities and inequalities for plant 1. Similar conditions can be obtained for other plants. If the SSS requirement is imposed, then we have

$$b_{0,c} > 0 \quad \text{leads to} \quad -b_{0,c} \leq -\delta, \quad \delta = 0.0001 \tag{6.65}$$

Simulation Results

SSP. Choose the objective function where $x_{ssp} = [a_{0,c}, a_{1,c}, b_{0,c}, \sigma_{1,1}, \sigma_{2,1}, \sigma_{1,2}, \sigma_{2,2}, \sigma_{1,3}, \sigma_{2,3}, \sigma_{3,3}, \sigma_{1,4}, \sigma_{2,4}, \sigma_{3,4}, \omega_{1,1}, \omega_{2,1}, \omega_{1,2}, \omega_{2,2}, \omega_{1,3}, \omega_{2,3}, \omega_{3,3}, \omega_{1,4}, \omega_{2,4}, \omega_{3,4}]$, the initial start point $x0_{ssp} = [1, 2, 3, -3, -3, -3, -3, -3, -3, -3, -3, -3, -3, 0, 0, 0, 0, 0, 0, 0.1, 0.1, 0.1, 0.1]$. Obviously the initial start point is not a feasible solution. However, after running the optimisation programming given in Matlab [29] with 1082 iterations (18 s), the programming succeeds in converging to the optimal point $x_{ssp} = [18.7222, 11.4998, 2.5161, -1.6397, -3.8994, -1.8534, -1.8534, -2.5436, -2.5436, -5.9287, -4.9937, -4.9937, -2.0284, 0.0000, 0.0000, 0.0000, 0.0000, 1.2424, -1.2424, 0.0000, 1.9984, -1.9984, 0.0000]$. The algorithm used in the implementation is the sequential quadratic programming (SQP). For a detailed description of this algorithm, interested readers are referred to the book [29] and the references therein.

The obtained coefficients of the optimal compensator are

$$a_{0,c} = 18.7222, \quad a_{1,c} = 11.4998, \quad b_{0,c} = 2.5161 \tag{6.66}$$

Consequently, the four closed-loop characteristic equations are respectively

$$D_{p,1}(s) = 12.4998s^2 + 69.2373s + 79.9208 \tag{6.67}$$
$$D_{p,2}(s) = 23.9995s^2 + 88.9594s + 82.4369 \tag{6.68}$$
$$D_{p,3}(s) = s^3 + 11.0158s^2 + 38.1734s + 47.5086 \tag{6.69}$$
$$D_{p,4}(s) = s^3 + 12.0158s^2 + 49.1893s + 58.6826 \tag{6.70}$$

and their roots are respectively

$$s_{1,1} = -3.8994, \quad s_{2,1} = -1.6397 \tag{6.71}$$
$$s_{1,2} = -1.8534, \quad s_{2,2} = -1.8534 \tag{6.72}$$
$$s_{1,3} = -2.54 + j1.24, \quad s_{2,3} = -2.54 - j1.24, \quad s_{3,3} = -5.93 \tag{6.73}$$
$$s_{1,4} = -4.99 + j1.99, \quad s_{2,4} = -4.99 - j1.99, \quad s_{3,4} = -2.03 \tag{6.74}$$

Controller Design for Anti-lock Braking Systems (ABS)

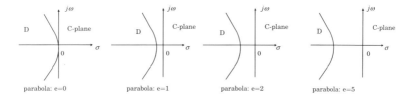

Figure 6.5 Four different D-stable regions

Obviously, the four closed systems are all stable and their roots are all in the specified D-stability region. The complex roots of the conjugate pairs $s_{1,3}, s_{2,3}$, and $s_{1,4}, s_{2,4}$ are located on the boundary of the specified D-stability region.

SSSP. After running the optimisation programming, the coefficients of the optimal compensator are

$$a_{0,c} = 18.7222, \quad a_{1,c} = 11.4998, \quad b_{0,c} = 2.5161 \tag{6.75}$$

They are exactly the same as those for the SSP. This result is exactly what we expect, because the compensator found from the SSP is stable. The D-stability region has an effect on the compensator. We have investigated the following situation that for the four plants given in (6.50), (6.51) and the compensator given in (6.52), the D-stability region is described by the following inequality (see Figure 6.5)

$$D \cup \partial D : g(\sigma, \omega) = \sigma + \omega^2 + e \leq 0 \tag{6.76}$$

where e varies from 0 to 5. The physical meaning for the change of e is that when e is bigger, the allowed region for the roots is smaller, and in this case it is more difficult to find a compensator. Table 6.1 shows the simulation results. It can be seen that when $e = 1$ and $e = 2$, respectively, both the SSP and SSSP converge to the same result. The convergence here means the D-stability is satisfied. In Table 6.1, it is marked as D-stable. When $e = 0.5$, both the SSP and SSSP converge, but the results are a little bit different due to the numerical error caused by the strong stability requirement when numerical difference is used in the SQP algorithm. When $e = 5$, the results for SSP and SSSP after 5000 iterations does not satisfy the D-stability region. However, the leftmost root among all the roots for the four plants is -2.0282. This means that the closed-loop systems for the four plants are still stable, so we mark stable rather than D-stable in Table 6.1.

Controller Design for ABS

Nonlinear model of the braking car. The nonlinear dynamic equa-

Table 6.1 Effect of D-stability region on the compensator

D region	$\sigma + \omega^2 \leq -0.5$	$\sigma + \omega^2 \leq -1$	$\sigma + \omega^2 \leq -2$	$\sigma + \omega^2 \leq -5$
$C_{ssp}(s)$	$\dfrac{9.89s + 4.12}{s + 2.41}$	$\dfrac{11.5s + 18.72}{s + 2.52}$	$\dfrac{13.52s + 27.05}{s + 2.39}$	$\dfrac{16.64s + 110.0}{s + 6.59}$
$C_{ssp}(s)$	$\dfrac{9.79s + 4.08}{s + 2.40}$	$\dfrac{11.5s + 18.72}{s + 2.52}$	$\dfrac{13.52s + 27.05}{s + 2.39}$	$\dfrac{17.39s + 92.2}{s + 5.33}$
$C_{ssp}(s)$	D-stable	D-stable	D-stable	Stable
$C_{ssp}(s)$	D-stable	D-stable	D-stable	Stable

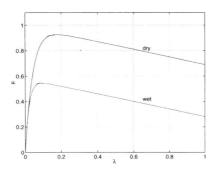

Figure 6.6 Tyre friction curves dependence on surface conditions

tions for describing the motion of a quarter-car can be given as:

$$\begin{aligned}\dot{\lambda} &= -\frac{1}{v}[\frac{1}{m}(1-\lambda) + \frac{r^2}{J}]F_z\mu(\lambda,\mu_H) + \frac{r}{vJ}T_b \\ \dot{v} &= -\frac{1}{m}F_z\mu(\lambda,\mu_H)\end{aligned} \qquad (6.77)$$

where v is the vehicle speed, m vehicle mass, J wheel inertia, r wheel radius, λ tyre slip, μ friction function between tyre and road, μ_H maximum friction coefficient between tyre and road, F_z vertical force (dynamic load), T_b brake torque.

The friction coefficient can vary in a very wide range, depending on factors like a) road surface conditions (dry or wet); b) tyre side slip angle; c) tyre brand (summer tyre, winter tyre). Its qualitative dependence on surface conditions is shown in Figure 6.6. The task of robust ABS controller design with SSP is to find a controller with the structure in Figure 6.1 such that linearised models at different road conditions are simultaneously stabilised with a desired D region.

Controller Design for Anti-lock Braking Systems (ABS)

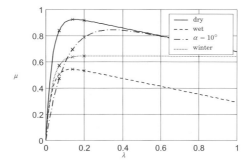

Figure 6.7 Tyre friction curves and 12 linearised points

Controller design. The first stage in designing an ABS controller using the SSP approach is to linearise the nonlinear model of a braking quarter car, as described in Equation (6.77). In order to cover all possible dynamic behaviours, the model is linearised at four different road conditions and at three different wheel slips, as depicted in Figure 6.7. Linearising (6.77) at the 12 operation points results in an array of linear transfer functions:

$$G_i(s) = \frac{0.32}{s + k_i}, \quad i = 1, 2, \ldots, 12 \tag{6.78}$$

The desired dynamic behaviour of the controlled ABS system is described by a D region as in Figure 6.4 (e), ($a = 0.75$, $b = 1.32$). Solving the set of equations and inequalities then yields to the following linear PI-controller:

$$C(s) = \frac{621s + 10248}{s} \tag{6.79}$$

Figure 6.8 shows the open- and closed-loop poles and the D-region of one of the 12 linear models. The simulation of the nonlinear model (6.77) and this controller (6.79) produces the results shown in Figure 6.9. In this example, the car first drives on a dry surface. After two seconds, road conditions change to a wet surface. The system is stable and relatively fast both on wet and dry surfaces.

6.6 Conclusions

In this paper, a new approach to the challenging open simultaneous stabilisation problem and the strong simultaneous stabilisation problem with extension to D-stable regions has been presented. The validation of this approach is based on the fundamental idea that the performance of a linear time-invariant system is determined by the distribution of its closed-loop characteristic roots. The main contributions of this paper include: a) the establishment of links between the open SSP(SSSP) and the well-developed

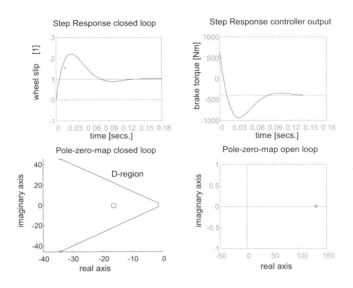

Figure 6.8 Dynamics of linear closed loop model

and mature fields in optimisation theory. This allows us to use the existing tools and algorithms; b) the stability region considered in this paper is extended from the previous left half complex plane to any subset of the left half complex plane. This extension enables us to deal with not only the stability but also the transient performance requirements; c) it also allows us to deal with both continuous systems and discrete systems in a unified way.

The main idea underlying this approach is the representation of the D stability requirement as a set of equalities and inequalities. We consider the problem from a new angle different from those presented by others [1], [9], [49]. This leads to a standard problem in optimisation theory. Unlike some conditions [42], these conditions possess the advantage that they result in a numerically tractable procedure. In fact from the viewpoint of representation of root location requirements by inequalities and equalities, this idea is simply a logical extension of the conventional pole assignment approach by full state feedback, where the relationship between the specified characteristic eigenvalues and the feedback gain is represented by a set of equalities, and then a numerical algorithm is used to find the solution to the equalities (equations). However, the approach presented here is more general in the sense that it includes inequalities and non-full state feedback. Because the full state feedback is not available, arbitrary pole assignment is not possible. However, the poles may be assigned into a region.

From the example given, it is shown that the desired D region can be changed. This is a distinguishing, useful feature of this approach. When the D region is the left half complex plane, the Routh-Hurwitz stability criteria

Controller Design for Anti-lock Braking Systems (ABS)

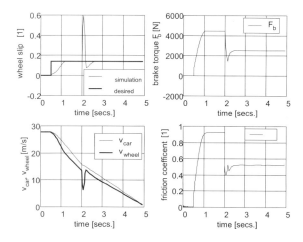

Figure 6.9 Nonlinear closed loop simulation

for continuous systems may be used to establish the inequality. However, for a general D region, there has not existed a similar condition yet. This is the reason why we directly establish the relationship between roots and the compensator coefficients. As mentioned in the introduction, the solution to the SSP and SSSP will find wider applications in reliable system control, robust control, and nonlinear control.

Acknowledgment

This work was supported by the European Commission through the Long Term Research Project 28104, H2C (Heterogeneous Hybrid Control).

6.7 References

[1] C. T. Abdallah, P. Dorato, and M. Bredemann. New sufficient conditions for strong simultaneous stabilisation. *Automatica*, 33:1193–1196, 1997.

[2] J. Ackermann. Parameter space design of robust control systems. *IEEE Trans. Automatic Control*, 25:1058–1082, 1980.

[3] J. Ackermann, A. Bartlett, D. Kaesbauer, W. Sienel, and R. Steinhauser. *Robust control: Systems with uncertain physical parameters*. Springer-Verlag, Berlin, 1993.

[4] A. B. Arehart and W. A. Wolovich. A nonlinear programming procedure and a necessary condition for the simultaneous stabilisation of 3 or more linear systems. *Proc. IEEE Conference on Decision and Control*, 2631–263, 1995.

[5] B. R. Barmish, C. V. Hollot, F. J. Kraus, and R. Tempo. Extreme point results for robust stabilisation of interval plants with first order compensators. *IEEE Trans. Automatic Control*, 37:707–714, 1992.

[6] B. R. Barmish and H. I. Kang. A survey of extreme point results for robustness of control systems. *Automatica*, 29:13–35, 1993.

[7] S. P. Bhattacharyya, H. Chapellat, and L. H. Keel. *Robust Control, the parametric approach*. Prentice Hall Int., London, 1995.

[8] V. Blondel and M. Gevers. Simultaneous stabilisation of three linear systems is rationally undecidable. *Mathematics of Control, Signals, and Systems*, 6:135–145, 1993.

[9] V. Blondel, M. Gevers, R. Mortini, and R. Rupp. Simultaneous stabilisation of three or more plants: Conditions on the positive real axis do not suffice. *SIAM J. Control and Optimisation*, 32:572–590, 1994.

[10] V. D. Blondel, E. D. Sontag, M. Vidyasagar, and J. C. Willems. *Open Problems in Mathematical Systems and Control Theory*. Springer-Verlag, London, 1998.

[11] P. T. Boggs and J. W. Tolle. Sequential quadratic programming. *Acta Numerica*, 1–51, 1995.

[12] G. Celentano and G. De Maria. New linear parameterisation of all stabilising controllers for single-input/single-output plants. *Proc. IEE, Part D, Control theory and applications*, 136:225–230, 1989.

[13] S. C. Chapra and R. P. Canale. *Numerical Methods for Engineers, Third Edition*. McGraw-Hill, London, 1998.

[14] J. W. Daniel. Newton's method for nonlinear inequalities. *Numer. Math.*, 40:381–387, 1973.

[15] J. E. Dennis and R. B. Schnabel. *Numerical methods for unconstrained optimisation and nonlinear equations*. Prentice Hall, Amsterdam, 1989.

[16] J. E. Dennis, M. El-alem, and K. Williamson. A trust-region approach to nonlinear systems of equalities and inequalities. *SIAM J. Optimisation*, 9:291–315, 1999.

[17] C. Fonte, H. Noura, and D. Sauter. Simultaneous compensation with non-switching controllers of low order. *Proc. American Control Conference*, 2365–2366, 1994.

[18] B. K. Ghosh and C. I. Byrnes. Simultaneous stabilisation and simultaneous pole placement by nonswitching dynamic compensation. *IEEE Trans. Automatic Control*, 28:735–741, 1983.

[19] P. E. Gill, W. Murray, and M. H. Wright. *Practical Optimisation*. Academic Press, London, 1981.

[20] V. Gopal and L. T. Biegler. Large scale inequality constrained optimisation and control. *IEEE Control System*, 59–68, 1998.

[21] D. Henrion, S. Tarbouriech, and V. Kucera. Rank-one LMI approach to simultaneous stabilisation of linear systems. *European Control Conference*, 116–123, 1999.

[22] K. J. Hunt, J. C. Kalkkuhl, H. Fritz, and T. A. Johansen. Constructive empirical modelling of longitudinal vehicle dynamics using local model networks. *Control Eng. Practice*, 4:167–178, 1996.

[23] S. T. Impram and N. Munro. Stability of nonlinear systems with complex disc and norm-bounded perturbations. *Proc. American Control Conference*, 3111–3115, 2000.

[24] Y. Jia and J. Ackermann. Some new results on simultaneous stabilisation of linear plants. *Proc. IFAC 14th Triennial world conference*, 219–224, 1999.

[25] Y. Jia, W. Gao, and M. Cheng. Robust strict positive real stabilisation and asymptotic hyperstability robustness. *Int. J. Control*, 59:1143–1157, 1994.

[26] T. A. Johansen, K. J. Hunt, P. J. Gawthrop, and H. Fritz. Off-equilibrium linearisation and design of gain-scheduled control with application to vehicle speed control. *Control Eng. Practice*, 6:167–180, 1998.

[27] V. L. Kharitonov. Asymptotic stability of an equilibrium position of a family of systems of linear differential equations. *Translation in Differential Equations*, 14:1483–1485, 1979.

[28] R. A. Luke. Simultaneous guaranteed-cost vector-optimal performance design for collections of systems. *J. of Guidance, Control, and Dynamics*, 22:96–102, 1999.

[29] MATLAB. *Optimisation Toolbox User's Guide*. The Math Works Inc. MATLAB, 1993.

[30] K. D. Minto and M. Vidyasagar. A state space approach to simultaneous stabilisation. *Control Theory and Advanced Technology*, 2:39–64, 1986.

[31] E. Muramatsu, M. Ikeda, and N. Hoshi. An interpolated controller for stabilisation of a plant with variable operating conditions. *IEEE Trans. Automatic Control*, 44:76–80, 1999.

[32] Y. Y. Nie. A new class of criterion for the stability of polynomials. *Acta Mechanica Sinica*, 15:110–116, 1976.

[33] P. I. Petersen. A procedure for simultaneously stabilising a collection of single input linear systems using nonlinear state feedback. *Automatica*, 23:33–40, 1987.

[34] B. T. Polyak. Gradient methods for solving equations and inequalities. *USSR Comput. Math.*, 4:17–32, 1964.

[35] M. J. D. Powell. A fast algorithm for non-linear constrained optimisation calculation. *Numerical analysis, ed. G. A. Watson, Lecture Notes in Mathematics*, 630, Springer-Verlag, 1978.

[36] B. N. Pshenichnyi. *Newton's method for the solution of equalities and inequalities*. Math. Notes Acad. Sci. USSR, 1970.

[37] S. M. Robinson. Extension of Newton's method to non-linear functions with values in a cone. *Numer. Math.*, 19:341–347, 1972.

[38] R. Saeks and J. Murray. Fractional representation, algebraic geometry, and the simultaneous stabilisation problem. *IEEE Trans. Automatic Control*, 27:895–903, 1982.

[39] K. Schittowski. NLQPL: A FORTRAN-subroutine solving constrained nonlinear programming problems. *Operations Research*, 5:485–500, 1985.

[40] W. E. Schmitendorf and C. V. Hollot. Simultaneous stabilisation via linear state feedback. *IEEE Trans. Automatic Control*, 34:1001–1005, 1989.

[41] L. Shaw. Pole placement: stability and sensitivity of dynamic compensators. *IEEE Trans. on Automatic Control*, 16:210–217, 1971.

[42] V. L. Syrmos, C. T. Abdallah, P. Dorato, and K. Grigoriadis. Static output feedback—A survey. *Automatica*, 33:125–137, 1997.

[43] O. Toker. On the order of simultaneous stabilisation compensators. *IEEE Trans. Automatic Control*, 41:430–433, 1996.

[44] C. C. Tsui. High-performance state feedback, robust, and output feedback stabilising control-a systematic design algorithm. *IEEE Trans. Automatic Control*, 44:560–563, 1999.

[45] M. Vidyasagar and N. Viswanadham. Algebraic design techniques for reliable stabilisation. *IEEE Trans. Automatic Control*, 27:1085–1095, 1982.

[46] Y. Wang and K. J. Hunt. The calculation of stability radius with d stability region and non-linear coefficients. *Proc. 3rd IFAC Symposium on Robust Control Design (ROCOND) 2000*, 240–246, 2000.

[47] Y. Wang, T. Schmitt-Hartmann, M. Schinkel, and K. J. Hunt. A new approach to simultaneous stabilisation with D stability and its application to control of antilock braking systems. *Proc. European Control Conf. (ECC-2001)*, 612–617, Portugal, 4-7 September, 2001.

[48] K. Wei. Stabilisation of a linear plant via a stable compensator having no real unstable zeros. *System Control Letter*, 15:259–264, 1990.

[49] K. Wei. The solution of a transcendental problem and its applications in simultaneous stabilisation problems. *IEEE Trans. Automatic Control*, 37:1305–1315, 1992.

[50] S. J. Wright. Optimisation strategies for process control. *Chem. Process Control-V*, 1997.

[51] D. Wu, W. Gao, and M. Cheng. Algorithm for simultaneous stabilisation of single-input systems via dynamic feedback. *Int. J. Control*, 51:631–642, 1990.

[52] D. C. Youla, J. J. Bongiorno, and C. N. Lu. Single loop feedback stabilisation of linear multivariable plants. *Automatica*, 10:159–173, 1974.

[53] Q. Zhao and J. Jiang. Reliable state feedback control system design against actuator failures. *Automatica*, 34:1267–1272, 1998.

7

Wheel Slip Control in ABS Brakes using Gain-scheduled Constrained LQR

I. Petersen   T.A. Johansen   J. Kalkkuhl   J. Lüdemann

Abstract

A wheel slip controller for ABS brakes is formulated using an explicit constrained LQR design. The controller gain matrices are designed and scheduled on the vehicle speed based on local linearisations. A Lyapunov function for the nonlinear control system is derived using the Riccati equation solution in order to prove stability and robustness with respect to uncertainty in the road/tyre friction characteristic. Experimental results from a test vehicle with electromechanical brake actuators and brake-by-wire show that high performance and robustness are achieved.

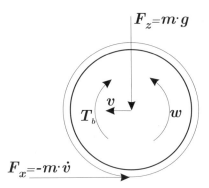

Figure 7.1 Quarter-car forces and torques.

7.1 Introduction

An anti-lock brake system (ABS) controls the slip of each wheel to prevent it from locking such that a high friction is achieved and steerability is maintained. ABS controllers are characterised by robust adaptive behaviour with respect to highly uncertain tyre characteristics and fast changing road surface properties [2].

The introduction of advanced functionality such as ESP (electronic stability program), drive-by-wire and more sophisticated actuators and sensors offers both new opportunities and requirements for higher performance in ABS brakes. The contribution of the present work is a study of a model-based design of ABS controllers, see also [7]. We consider electromechanical actuators [10] rather than hydraulic actuators, which allow continuous adjustment of the clamping force.

Here, we design the wheel slip controller based on a recently developed explicit LQR design method that takes into account input and state constraints [6], see also [1]. The wheel slip dynamics are highly nonlinear. Despite this, our control design relies on local linearisation and gain-scheduling. In order to analyse the effects of this simplification, we develop a Lyapunov-based stability and robustness analysis. Preliminary results from experiments using a test vehicle are also included. Other studies of model-based wheel slip control in ABS can be found in [5], [4], [3], [11].

7.2 Modelling

In this section, we review a mathematical model of the wheel slip dynamics, see also [2] and [4]. The problem of wheel slip control is best explained by looking at a quarter-car model as shown in Figure 7.1. The model consists of a single wheel attached to a mass m. As the wheel rotates, driven by inertia of mass m in the direction of the velocity v, a tyre reaction force F_x is generated by the friction between the tyre surface and the road surface.

The tyre reaction force will generate a torque that initiates a rolling motion of the wheel, causing an angular velocity ω. A brake torque applied to the wheel will act against the spinning of the wheel causing a negative angular acceleration. The equations of motion of the quarter-car are

$$m\dot{v} = -F_x \qquad (7.1)$$
$$J\dot{\omega} = r F_x - T_b \operatorname{sign}(\omega) \qquad (7.2)$$

where

v	horizontal speed at which the car travels
ω	angular speed of the wheel
F_z	vertical force
F_x	tyre friction force
T_b	brake torque
r	wheel radius
J	wheel inertia

The tyre friction force F_x is given by

$$F_x = F_z \cdot \mu(\lambda, \mu_H, \alpha) \qquad (7.3)$$

where the friction coefficient μ is a nonlinear function of

λ	tyre-slip
μ_H	friction coefficient between tyre and road
α	slip angle of the wheel

and the longitudinal slip λ, defined by

$$\lambda = \frac{v - \omega r}{v} \qquad (7.4)$$

The slip λ describes the normalised difference between horizontal speed v and the speed of the wheel perimeter ωr. The slip value of $\lambda = 0$ characterises the free motion of the wheel where no friction force F_x is exerted. If the slip attains the value $\lambda = 1$, then the wheel is locked, which means that it has come to a standstill.

The friction coefficient μ can span over a very wide range, but is differentiable with the property $\mu(0, \mu_H, \alpha) = 0$ and $\mu(\lambda, \mu_H, \alpha) > 0$ for $\lambda > 0$. Its typical dependence on slip λ is shown in Figure 7.2. The upper part shows how the friction coefficient μ increases with slip λ up to a value $\lambda_0 \approx 0.14\mu_H$, where it attains its maximum μ_H. For higher slip values, the friction coefficient will decrease to a minimum μ_G where the

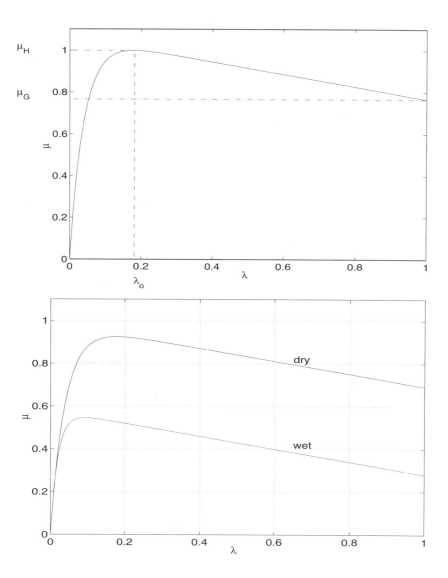

Figure 7.2 Tyre-slip/friction curves

wheel is locked and only the sliding friction will act on the wheel. The dependence of friction on the road condition is shown in the lower part of Figure 7.2. For wet or icy roads, the maximum friction μ_H is small and the right part of the curve is flatter. The tyre friction curve will also depend on the brand of the tyre. In particular, for winter tyres, the curve will cease to have a pronounced maximum.

If the motion of the wheel is extended to two dimensions, then the lateral slip of the tyre must also be considered. The slip angle α is the

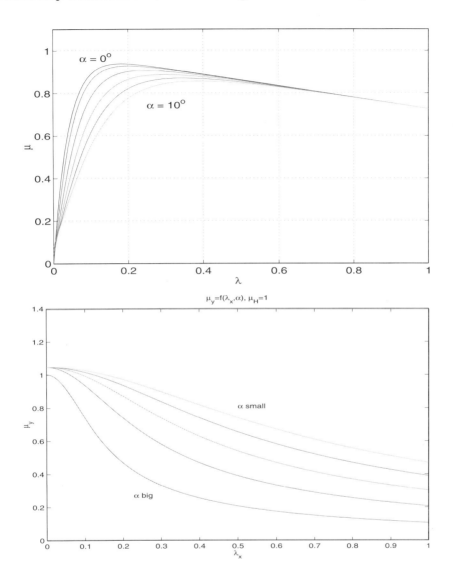

Figure 7.3 Tyre side slip/friction curves

angle between the wheel bearing and the velocity vector of the vehicle.

In this case, the longitudinal slip $\lambda_x = \frac{v_x - \omega r}{v}$ and the lateral slip $\lambda_y = \sin \alpha$ are distinguished as well as the corresponding friction coefficients μ_x and μ_y. The upper part of Figure 7.3 shows the dependence of the friction coefficient μ_x on the side slip angle α. The side force F_y depends greatly on the side slip angle α and is shown in the lower part of Figure 7.3. For large side slip angles, the force gets smaller. In the sequel, for simplification purposes unless otherwise stated, the side slip angle will be considered to be zero with $\mu_x = \mu$ and $v_x = v$.

Using Equations (7.1)–(7.4), we get for $v > 0$ and $\omega \geq 0$

$$\dot{\lambda} = -\frac{1}{v}\left(\frac{1}{m}(1-\lambda) + \frac{r^2}{J}\right) F_z \mu(\lambda, \mu_H, \alpha) + \frac{1}{v}\frac{r}{J}T_b \qquad (7.5)$$

$$\dot{v} = -\frac{1}{m} F_z \mu(\lambda, \mu_H, \alpha) \qquad (7.6)$$

Note that when $v \to 0$, the dynamics of the open-loop system become infinitely fast with infinite gain. This leads to a loss of controllability and the slip controller must be switched off for small v.

PROPOSITION 7.1
Consider the system (7.5)–(7.6) with $T_b(t) \geq 0$ for all $t \geq 0$. If $v(0) > 0$ and $\lambda(0) \in [0, 1]$, then $\lambda(t) \in [0, 1]$ and $\dot{v}(t) \leq 0$ for all $t \geq 0$ where $v(t) > 0$.

Proof. Note that $\lambda(t)$ is a continuous trajectory. Hence, the possible escape points are $\lambda = 0$ and $\lambda = 1$. Consider first $\lambda = 0$. Since $\mu(0) = 0$, it follows from (7.5) that $\dot{\lambda} = \frac{r}{vJ}T_b \geq 0$ due to $T_b \geq 0$. Hence, $\lambda(0) \geq 0$ implies $\lambda(t) \geq 0$ for all $t \geq 0$. Consider next $\lambda = 1$. Then, $\omega = 0$ and from (7.2) it follows that $\dot{\omega} \geq 0$ due to the discontinuity $sign(\omega)$ in (7.2). From (7.4), we conclude that $\dot{\lambda} \leq 0$, which implies $\lambda(t) \leq 1$ for all $t \geq 0$. Finally, note that $\dot{v} \leq 0$ from (7.1) because $F_x \geq 0$ for $\lambda \in [0, 1]$. $\square$

7.3 Control Problem

The control problem is essentially to control the value of the longitudinal slip λ to a given setpoint λ^* that is either constant or commanded from a higher-level control system such as ESP (electronic stability program). The controller must be robust with respect to uncertainties in the tyre characteristic and variations in the road surface conditions.

The actual control input is the clamping force F_b that is related to the brake torque as $T_b = k_b F_b$. There are limitations on the force that can be applied to the brake pads by the actuator during braking. The (small) minimum is to ensure that the pads are positioned close to the brake disc (no air gap). Maximum is what the actuator is capable of. There is also a rate limit at how fast the torque can be changed by the actuator.

Integral action or adaptation must be incorporated to remove errors due to model inaccuracies, in particular, the road/tyre friction coefficient μ_H. It is essential that the controller maintains a high performance and is robust w.r.t. to any road/tyre friction curve that can be encountered *cf.* Figure 7.2.

7.4 Gain-scheduled LQRC Controller Design and Analysis

The dynamics of the wheel and car body are given by (7.5) and (7.6), respectively. Due to large differences in inertia, the wheel dynamics and car dynamics will evolve on significantly different time scales. The speed v will change much more slowly than the slip λ and is therefore a natural candidate for gain-scheduling. Thus, for the control design, we consider only (7.5) and regard v as a slow time-varying parameter. A constrained LQR controller [6] is applied. This requires a nominal linearised model for design.

Linearised dynamics

Let $(\lambda^*, \hat{T}_b)$ be an equilibrium point for (7.5) defined by the constants α^*, F_z^*, and μ_H^*, where λ^* is the desired slip

$$\hat{T}_b = \left(\frac{J}{mr}(1-\lambda^*) + r\right) F_z^* \mu(\lambda^*, \mu_H^*, \alpha^*)$$

The linearised slip dynamics are given by

$$\dot{\lambda} = \frac{\alpha_1}{v}(\lambda - \lambda^*) + \frac{\beta_1}{v}(T_b - \hat{T}_b) + \text{higher-order terms} \qquad (7.7)$$

where α_1 and β_1 are linearisation constants given by

$$\alpha_1 = -F_z^* \left(\frac{1}{m}(1-\lambda^*) + \frac{r^2}{J}\right) \frac{\partial \mu}{\partial \lambda}(\lambda^*, \mu_H^*, \alpha^*) + \frac{F_z^*}{m}\mu(\lambda^*, \mu_H^*, \alpha^*) \qquad (7.8)$$

$$\beta_1 = \frac{r}{J} > 0 \qquad (7.9)$$

The nonlinear wheel slip dynamics (7.5) can be written in the form

$$\dot{x} = \frac{\phi(x)}{v} + \frac{\beta_1}{v}u \qquad (7.10)$$

where

$$\phi(x) = -\left(\frac{1}{m}(1-\lambda^*-x) + \frac{r^2}{J}\right) F_z \mu(x+\lambda^*, \mu_H, \alpha) + \frac{r}{J}T_b^* \qquad (7.11)$$

and $x = \lambda - \lambda^*$, $u = T_b - T_b^*$ and

$$T_b^* = \left(\frac{J}{mr}(1-\lambda^*) + r\right) F_z \mu(\lambda^*, \mu_H, \alpha) \qquad (7.12)$$

It can be seen that this system has an equilibrium point given by $x = u = 0$ and $\phi(0) = 0$, and the linearised slip model (7.7) with a perturbation term can be written as follows

$$\dot{x} = \frac{\alpha_1}{v}x + \frac{\beta_1}{v}u + \frac{\varepsilon_\mu(x)}{v} \qquad (7.13)$$

where $\varepsilon_\mu(x) = \phi(x) - \alpha_1 x$.

Wheel Slip Control without Integral Action

For simplicity, we first study the design without integral action and we define the infinite-horizon quadratic cost function

$$J(x(t), u[t, \infty)) = \int_t^\infty (x^2(\tau)Q(v) + u^2(\tau)R)d\tau \qquad (7.14)$$

with $R > 0$ and $Q(v) \geq 0$ for all $v > 0$. Assuming v is constant (due to the separation of time scales) and neglecting the nonlinearity $\varepsilon_\mu(x)$, the optimal control law is uniquely given by the gain-scheduled state feedback

$$\hat{u} = -R^{-1}\frac{\beta_1}{v}P(v)x = K(v)x, \text{ where } K(v) = -R^{-1}\frac{\beta_1}{v}P(v) \qquad (7.15)$$

The algebraic Riccati equation is

$$\frac{2P(v)\alpha_1}{v} - \left(\frac{P(v)\beta_1}{v}\right)^2 R^{-1} + Q(v) = 0 \qquad (7.16)$$

with the positive solution

$$P(v) = \frac{\alpha_1 + (\alpha_1^2 + \beta_1^2 R^{-1} Q(v))^{1/2}}{\beta_1^2 R^{-1}} v = P'(v)v \qquad (7.17)$$

Due to actuator constraints, we define the saturated control

$$u = \begin{cases} T_b^{\max} - T_b^*, & \hat{u} > T_b^{\max} - T_b^* \\ T_b^{\min} - T_b^*, & \hat{u} < T_b^{\min} - T_b^* \\ \hat{u}, & \text{otherwise} \end{cases} \qquad (7.18)$$

for some $T_b^{\max} > T_b^{\min} \geq 0$. Furthermore, we define the error (note that both u and $\hat{u}$ depend on v and x)

$$\varepsilon_s(x, v) = \beta_1(u - \hat{u}) \qquad (7.19)$$

With the definition $\varepsilon(x, v) = \varepsilon_s(x, v) + \varepsilon_\mu(x)$, the closed-loop dynamics can be written as

$$\dot{x} = \left(\frac{\alpha_1}{v} + \frac{\beta_1 K(v)}{v}\right)x + \frac{\varepsilon(x, v)}{v} \qquad (7.20)$$

It is easy to see that $\varepsilon(0, v) = 0$ for all $v > 0$.

PROPOSITION 7.2
Consider the system (7.5) with controller (7.18), where $R > 0$ and the smooth function Q satisfies $Q(v) \geq 0$ and $\frac{dQ(v)}{dv} \geq 0$ for all $v > 0$. Suppose for some $\delta \in (0,1)$

$$x\varepsilon(x,v) \leq (1-\delta)\frac{Q(v)}{2P'(v)}x^2 \tag{7.21}$$

for all $v > 0$ and $x \in [-\lambda^*, 1-\lambda^*]$. Then, for all $v(0) > 0$ and $\lambda(0) \in [0,1]$, the equilibrium $x = 0$ is uniformly exponentially stable.

Proof. Let a Lyapunov function candidate be defined as

$$V(x) = x^2 P(v) \tag{7.22}$$

Along trajectories of (7.20), we have

$$\dot{V} = \frac{d}{dt}V(x) = x^2\left(\frac{dP(v)}{dv}\frac{dv}{dt}\right) + 2\dot{x}xP(v) \tag{7.23}$$

Substituting for (7.13), (7.15) and (7.16) in (7.23) gives

$$\dot{V} = x^2\left(P'(v) + v\frac{dP'(v)}{dv}\right)\dot{v} + 2\varepsilon(x,v)P'(v)x - x^2 Q(v) \tag{7.24}$$

Note that $P'(v) \geq 0$ and $\frac{dP'(v)}{dv} \geq 0$ for all $v > 0$. It follows from Proposition 7.1 that

$$\dot{V} \leq 2\varepsilon(x,v)P'(v)x - x^2 Q(v) \leq -\delta Q(v)x^2 \tag{7.25}$$

and the equilibrium is uniformly exponentially stable, Corollary 4.2 [9]. The region of attraction follows from Proposition 7.1 since $[0,1]$ is positively invariant for λ. □

Essentially, (7.21) requires that the error weight $Q(v)$ must be chosen to be sufficiently large, which means that the gain $K(v)$ must be sufficiently large in order to stabilize the system. Note that the system is open-loop unstable when operating in a region where the friction curve has negative slope $\frac{\partial \mu}{\partial \lambda}$ (in this case, $\alpha_1 > 0$). The controller is gain-scheduled since K depends on v. From a practical point of view, it makes sense to choose $dQ(v)/dv > 0$ since this leads to $dK(v)/dv > 0$. Hence, the gain is reduced as $v \to 0$ and one avoids instability due to unmodelled dynamics that typically become dominant as $v \to 0$.

The above analysis shows that the heuristics of local linearisation and gain scheduling do not lead to instability when $Q(v)$ is chosen sufficiently large. Unfortunately, it is not possible to implement this controller without significant loss of performance since the steady state brake torque T_b^* is unknown; recall that $T_b = T_b^* + u$ and T_b^* depends on μ_H, which is highly uncertain and time-varying. Hence, all practical wheel slip controllers need some form of integral action or adaptation. We therefore proceed by extending our design and analysis with integral action without further discussion.

Wheel Slip Control with Integral Action

Let the system dynamics be augmented with a slip error integrator $\dot{x}_1 = \lambda - \lambda^* = x_2$

$$\begin{pmatrix} \dot{x}_1 \\ \dot{x}_2 \end{pmatrix} = A(v) \begin{pmatrix} x_1 \\ x_2 \end{pmatrix} + B(v)(u - T_b^*) + W(v)\varepsilon_\mu(x_2) \tag{7.26}$$

where

$$A(v) = \begin{pmatrix} 0 & 1 \\ 0 & \frac{\alpha_1}{v} \end{pmatrix}, B(v) = \begin{pmatrix} 0 \\ \frac{\beta_1}{v} \end{pmatrix}, W(v) = \begin{pmatrix} 0 \\ \frac{1}{v} \end{pmatrix} \tag{7.27}$$

Since T_b^* is assumed unknown, we define $u = T_b$, and the equilibrium point to be

$$x^* = \begin{pmatrix} x_1^* \\ 0 \end{pmatrix}, \quad u^* = T_b^* \tag{7.28}$$

where x_1^* is yet unspecified. This leads to

$$\dot{x} = A(v)(x - x^*) + B(v)(u - u^*) + W(v)\varepsilon_\mu(x_2) \tag{7.29}$$

Next, define the quadratic cost function for the purpose of local LQ design based on the nominal part of (7.29):

$$J(x(t), u[t, \infty)) = \int_t^\infty ((x(\tau) - x^*)^T Q(v)(x(\tau) - x^*) \tag{7.30}$$
$$+ (u(\tau) - u^*)^T R(u(\tau) - u^*))d\tau$$

Assuming constant v, the optimal control law is given by

$$\hat{u} = K(v)x \tag{7.31}$$

where the gain matrix $K(v) = -R^{-1}B^T(v)P(v)$ and we choose to neglect the unknown x^* and u^*, which will be accounted for due to the integral action. The symmetric matrix $P(v) > 0$ is defined by the solution to the algebraic Riccati equation for the design:

$$P(v)A(v) + A^T(v)P(v) - P(v)B(v)R^{-1}B^T(v)P(v) = -Q(v) \tag{7.32}$$

The elements of the matrix equation (7.32) are

$$\left(\frac{\beta_1}{v}\right)^2 \frac{P_{1,2}^2(v)}{R} = Q_{1,1}(v) \tag{7.33}$$

$$P_{1,1}(v) + P_{1,2}(v)\left(\frac{\alpha_1}{v} - \left(\frac{\beta_1}{v}\right)^2 \frac{P_{2,2}(v)}{R}\right) = 0 \tag{7.34}$$

$$2P_{1,2}(v) + P_{2,2}(v)\left(\frac{2\alpha_1}{v} - \left(\frac{\beta_1}{v}\right)^2 \frac{P_{2,2}(v)}{R}\right) = -Q_{2,2}(v) \tag{7.35}$$

This gives the following solution with $P(v) > 0$:

$$P_{1,1}(v) = \frac{\left(\alpha_1^2 + \beta_1^2 R^{-1}\left(Q_{2,2}(v) + \frac{2(Q_{1,1}(v)R)^{1/2}}{\beta_1}v\right)\right)^{1/2}}{(Q_{1,1}(v)R)^{-1/2}\beta_1} \tag{7.36}$$

$$P_{1,2}(v) = \frac{v}{\beta_1}(Q_{1,1}(v)R)^{1/2} \tag{7.37}$$

$$P_{2,2}(v) = \frac{\alpha_1 + \left(\alpha_1^2 + \beta_1^2 R^{-1}\left(Q_{2,2}(v) + \frac{2(Q_{1,1}(v)R)^{1/2}}{\beta_1}v\right)\right)^{1/2}}{\beta_1^2 R^{-1}} \cdot v \tag{7.38}$$

$$K_1(v) = -\left(Q_{1,1}(v)R^{-1}\right)^{1/2} \tag{7.39}$$

$$K_2(v) = -\frac{\alpha_1 + \left(\alpha_1^2 + \beta_1^2 R^{-1}\left(Q_{2,2}(v) + \frac{2(Q_{1,1}(v)R)^{1/2}}{\beta_1}v\right)\right)^{1/2}}{\beta_1} \tag{7.40}$$

We introduce the saturated control

$$T_b = u = \begin{cases} T_b^{\max}, & \hat{u} > T_b^{\max} \\ T_b^{\min}, & \hat{u} < T_b^{\min} \\ \hat{u}, & \text{otherwise} \end{cases} \tag{7.41}$$

Assuming $x_1^* = T_b^*/K_1(v)$ gives the closed-loop dynamics

$$\dot{\tilde{x}} = (A(v) + B(v)K(v))\tilde{x} + W(v)\varepsilon(x_2, v) \tag{7.42}$$

with $\tilde{x} = x - x^*$.

PROPOSITION 7.3
Consider the system (7.5) with controller (7.41). Assume $R > 0$ and the smooth matrix-valued function Q satisfies $Q_{1,2}(v) = Q_{2,1}(v) = 0$,

$Q_{1,1}(v) > 0$, $Q_{2,2}(v) > 0$, $dQ_{1,1}(v)/dv \geq 0$, $dQ_{2,2}(v)/dv \geq 0$ for all $v > 0$. Moreover, suppose (7.57) and

$$Q_{1,1}(v) > \frac{P_{2,1}(v)C}{v}(1-\delta) \tag{7.43}$$

$$Q_{2,2}(v)\tilde{x}_2^2 > \left(\frac{2}{v}\varepsilon(\tilde{x}_2)P_{2,2}(v)\tilde{x}_2 + \frac{P_{2,1}(v)\varepsilon^2(\tilde{x}_2)}{vC}\right)(1-\delta) \tag{7.44}$$

are satisfied for all $v > 0$, $\tilde{x}_2 \in [-\lambda^*, 1-\lambda^*]$ and some $C > 0$ and $\delta \in (0, 1)$. Then, the equilibrium $\tilde{x} = 0$ is uniformly exponentially stable.

Proof. Let a Lyapunov function candidate be

$$V(\tilde{x}) = \tilde{x}^T P(v)\tilde{x}$$

Its time-derivative along trajectories of (7.42)

$$\dot{V} = \frac{d}{dt}V(\tilde{x}(t)) = \tilde{x}^T\left(\frac{\partial P(v)}{\partial v}\dot{v}\right)\tilde{x} + \dot{\tilde{x}}^T P\tilde{x} + \tilde{x}^T P\dot{\tilde{x}} \tag{7.45}$$

is found by substituting for (7.29), (7.31) and (7.32) in (7.45):

$$\dot{V} = \tilde{x}^T \frac{\partial P(v)}{\partial v}\dot{v}\tilde{x} + \varepsilon(\tilde{x}_2)(W^T(v)P(v)\tilde{x} + \tilde{x}^T P(v)W(v)) - \tilde{x}^T Q(v)\tilde{x} \tag{7.46}$$

From Proposition 7.1, it is clear that $\dot{v} \leq 0$. Thus, the negativity of $\tilde{x}^T\left(\frac{\partial P(v)}{\partial v}\dot{v}\right)\tilde{x}$ requires $P'(v) = \frac{\partial P(v)}{\partial v} > 0$ for all $v > 0$. For $P'(v)$ to be positive definite, it is sufficient that $P'_{1,1}(v) > 0$ and $D(v) = P'_{1,1}(v)P'_{2,2}(v) - P'_{1,2}(v)P'_{2,1}(v) > 0$. Note that since $Q_{1,1}(v), Q_{2,2}(v), \beta_1 > 0$, it follows immediately that $P'_{1,1}(v) > 0$. In Appendix 7.A, it is shown that $D(v) > 0$ is satisfied due to (7.57). Then, (7.46) becomes:

$$\dot{V} \leq \varepsilon(\tilde{x}_2)(W^T P(v)\tilde{x} + \tilde{x}^T P(v)W(v)) - \tilde{x}^T Q(v)\tilde{x}$$
$$= -Q_{1,1}(v)\tilde{x}_1^2 - Q_{2,2}(v)\tilde{x}_2^2 + \frac{2}{v}\varepsilon(\tilde{x}_2)(P_{2,2}(v)\tilde{x}_2 + P_{2,1}(v)\tilde{x}_1) \tag{7.47}$$

To obtain all $\tilde{x}_1$-terms in (7.47) in a quadratic form, we apply Young's inequality $2ab \leq a^2/C + Cb^2$. Hence,

$$\dot{V} \leq -Q_{1,1}(v)\tilde{x}_1^2 + \frac{P_{2,1}(v)}{v}\tilde{x}_1^2 C - Q_{2,2}(v)\tilde{x}_2^2 + \frac{2}{v}\varepsilon(\tilde{x}_2)P_{2,2}(v)\tilde{x}_2$$
$$+ \frac{P_{2,1}(v)}{v}\frac{\varepsilon^2(\tilde{x}_2)}{C} \tag{7.48}$$

Due to (7.43) and (7.44), it follows that

$$\dot{V} \leq -\delta Q_{1,1}(v)\tilde{x}_1^2 - \delta Q_{2,2}(v)\tilde{x}_2^2$$

and we conclude that the equilibrium is uniformly exponentially stable using Corollary 4.2, [9]. □

Inequality (7.43) states that the error weight $Q_{1,1}(v)$ must be sufficiently large, leading to a sufficiently large controller gain. Note that $P_{2,1}(v)$ depends on $Q_{1,1}(v)$, but from (7.36), it is evident that $P_{2,1}(v)$ increases with $\sqrt{Q_{1,1}(v)}$ such that (7.43) will indeed hold for a sufficiently large $Q_{1,1}(v)$, except when $v \to 0$.

Inequality (7.44) states that the error weight $Q_{2,2}(v)$ must also be sufficiently large, leading to a sufficiently high gain to stabilize the system. From (7.38), it follows that $P_{2,2}(v)$ increases with $\sqrt{Q_{2,2}(v)}$ such that this is also possible, except for $v \to 0$.

Note that $Q_{2,2}(v)|\tilde{x}_2|$ is essentially chosen in (7.44) to dominate the perturbation $\varepsilon(\tilde{x}_2)$, which consists of two terms; one due to the linearisation and unknown friction curve (ε_μ) and one due to the control saturation (ε_s). Inequality (7.44) should be checked with respect to the perturbations ε_μ that are generated by *all* possible friction curves $\mu(\cdot)$ to ensure robust stability.

Inequality (7.57) can be seen to be non-restrictive since it will always be satisfied for α_1 close to zero. This corresponds to generating the nominal model by linearising near the peak of the friction curve. Experience shows that high performance is indeed achieved this way. Note that for $\alpha_1 = 0$, no information on the friction curves is actually utilised in the control design.

The constant $C > 0$ should be chosen to minimize conservativeness. However, the choice $Q_{1,2}(v) = Q_{2,1}(v) = 0$ and taking $P(v)$ from the solution of the Riccati equation are possibly conservative.

The controller gain $K(v)$ depends on the speed (gain-scheduling). From a practical point of view, a useful gain-schedule is achieved by letting $dQ_{1,1}(v)dv > 0$ and $dQ_{2,2}(v)dv > 0$, as this reduces the gain as $v \to 0$. As mentioned earlier, this is necessary to avoid instability due to the unmodelled (actuator) dynamics since these tend to dominate as $v \to 0$.

An idealised design example

We consider a design example with the following parameters $m = 450$ kg, $F_z = 4414$ N, $r = 0.32$ m, $J = 1.0$ kg·m² and the friction model in the upper part of Figure 7.2. Assuming $\lambda^* = 0.14$ and the nominal design $\mu_H^* = 0.8$ and $\alpha^* = 0$, we get $\alpha_1 = 10.2$ and $\beta_1 = 0.32$. We choose $R = 1$ and $Q(v) = \tilde{Q}v^{3/2}$ with $\tilde{Q}_{1,1} = 6 \cdot 10^9$ and $\tilde{Q}_{2,2} = 40 \cdot 10^6$. Note the scaling due to the different magnitudes of T_b and λ. The choice for $Q(v)$ leads to a gain-schedule with reduced gain as $v \to 0$, which is useful in avoiding instability due to unmodelled actuator dynamics as $v \to 0$.

Figure 7.4 shows that the robust stability requirement (7.44) is satisfied for all $\tilde{x}_2 \in [-\lambda^*, 1-\lambda^*]$ for all the friction curves in the lower part of Figure 7.2. Although curves are shown only for $v = 1$ m/s and $v = 32$ m/s, we have verified that (7.44) is fulfilled for intermediate values of v. The control design also satisfies the stability requirements (7.43) and (7.57), and we

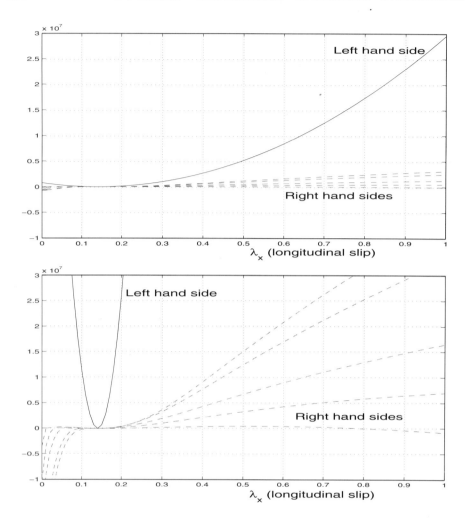

Figure 7.4 Illustration of the robust stability requirement, *i.e.*, the left and right hand sides of Equation (7.44). The upper part is for $v = 1\,\text{m/s}$ and the bottom part is for $v = 32\,\text{m/s}$

conclude robust uniform exponential stability of the equilibrium.

7.5 Implementation

The experimental vehicle is a Mercedes E220 equipped with electromechanical disc brake actuators and a brake-by-wire system. A previous version of this system is described in [10]. The system consists of four independent torque controllers, one for each actuator, and a central electronic control unit (ECU) where the four wheel slip controllers run.

Figure 7.5 illustrates the implemented wheel slip controller. With reference to the theoretical study presented in previous sections, we note the following:

1. The actual control implementation is based on the discretisation of the models and a discrete-time control design. The discrete-time model also takes into account communication delays on the brake-by-wire electronics. The sampling rate for the wheel slip control loop is 7 ms.

2. The electromechanical brake actuator is assumed to be linear,

$$\tau \dot{F}_b = -F_b + \tilde{F}_b \tag{7.49}$$

   where τ is the time constant and $\tilde{F}_b$ is the commanded clamping force. This is reasonable since it contains a local feedback loop. The model used for control design is augmented with a discretised version of (7.49) and the measured clamping force F_b is used for feedback.

3. The controller actually computes the change of the clamping force, rather than the clamping force itself, in order to enjoy the benefits of velocity-based gain-scheduling [8], as well as the ease with which it accounts for the actuator constraint (see [6])

$$-\dot{F}_b^{max} \leq \dot{F}_b \leq \dot{F}_b^{max} \tag{7.50}$$

   Anti-windup is implemented on the integrator state x_1.

4. Gain-scheduling is implemented by computing gain matrices for a finite number of operating points and then switching gain matrices. To achieve bumpless transfer, the integrator is reset at the switching instants. The gain matrices are scheduled on both speed and slip. The scheduling on slip only has an effect for very small slip values, and will improve the transient performance when the wheel slip controller is activated.

5. The slip λ and speed v are estimated online using an extended Kalman filter based on wheel speed (ω) and acceleration measurements.

6. The ABS system monitors the commands given by the driver using the brake pedal. Essentially, we set $T_b^{max} = k_b F_b^*$, where F_b^* is the clamping force commanded by the driver using the brake pedal. The slip controller is deactivated when the speed is below 1 m/s, and the controller state is reinitialised when the brake pedal is released.

7. The tuning of the implemented controller is similar to the tuning in the idealised set-up in Section 7.4. More specifically, the dominant pole of the nominal linear closed loop is almost the same in both cases (about 12.0 for $v = 12$ m/s).

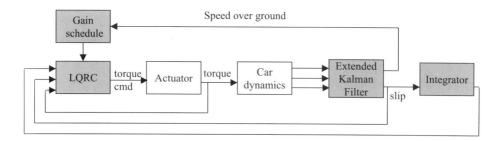

Figure 7.5 Wheel slip control—Block diagram

7.6 Experimental Results

The first test, Figure 7.6, is braking on dry asphalt, starting at $v(0) = 21.5$ m/s and without any steering manoeuvres. The slip setpoint is $\lambda^* = 0.09$ and we note that the regulation is highly accurate and satisfactory. When the speed approaches zero, significant variability in the slip emerges. Since the clamping force does not oscillate, we conclude that this is actually sensor noise that is known to increase as the speed goes to zero. We also note that the initial transient is not satisfactory, as the clamping force does not increase fast enough so that the slip is too low and the resulting friction force is too low in the interval $t \in (0.2, 0.7)$, leading to increased braking distance. This is due to the significant model error due to linearisation in the low-slip region, and redesign of the slip controller is necessary for this region.

The second test, Figure 7.7, is braking on snow, starting at $v(0) = 22.0$ m/s and without any steering maneuvers. The slip setpoint is $\lambda^* = 0.07$ and we note that the regulation is satisfactory. There are some small oscillations that we believe are due to unmodelled actuator nonlinearities that are expected to be more pronounced when operating at low clamping force levels due to low friction on snow.

The third test, Figure 7.8, is braking on a wet inhomogeneous surface (asphalt that is partially covered by a plastic coating and water) without any steering maneuvers. The initial speed is $v(0) = 22.0\,\text{m/s}$ and the slip setpoint is $\lambda^* = 0.09$. We note that in this case there are significant transients in the slip and clamping force. We still conclude that the regulation performance is highly satisfactory, since the surface is characterised by very large variations in the friction coefficient that cause large disturbances on the system.

7.7 Discussion and Conclusions

Using Lyapunov analysis and experimental verification, we have shown good performance and robustness of a model-based nonlinear wheel slip controller for ABS. In order to achieve the robustness, the approach does

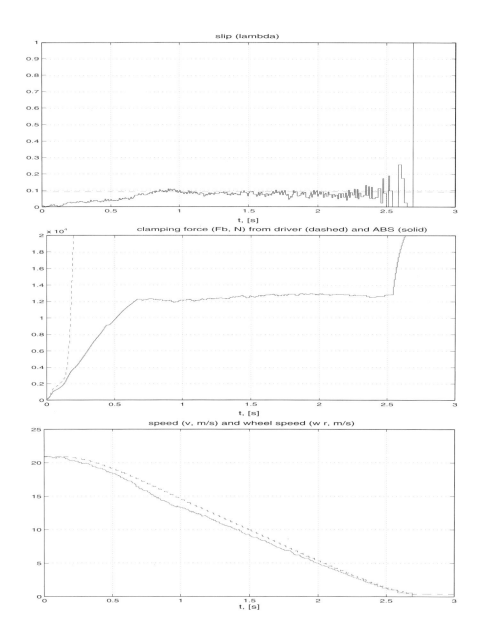

Figure 7.6 Experimental results with braking on dry asphalt.

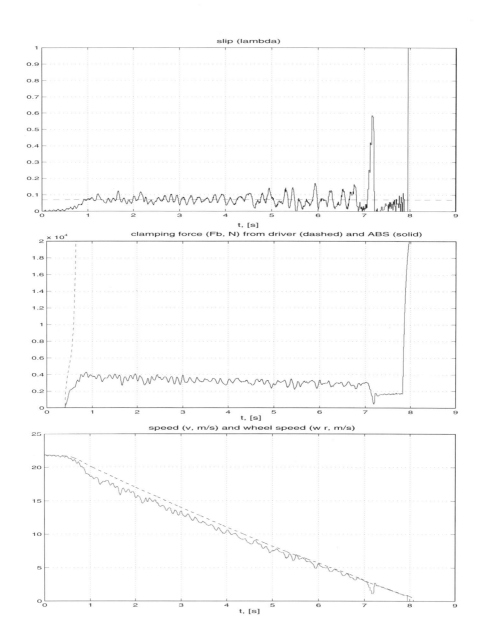

Figure 7.7 Experimental results with braking on snow

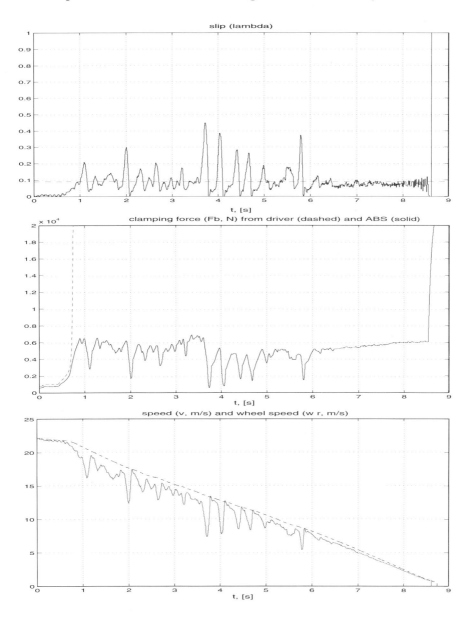

Figure 7.8 Experimental results with braking on a wet inhomogeneous surface

not rely on knowledge of the tyre/road friction curve. Static uncertainty (due to unknown μ_H) is eliminated using integral action, while dynamic uncertainty (due to unknown shape of $\mu(\cdot)$) is handled by a robust design with sufficient stability margin.

The results are preliminary, and can be possibly extended and improved.

In particular, the analysis might take into consideration some of the details that are present in the implementation, but omitted from the current analysis. Also, the use of the Riccati solution to define the Lyapunov function is not always the best choice. Finally, means for improving the transient performance of the controller (especially for low slip) are under investigation.

Although a detailed comparison with existing off-the-shelf ABS has not been conducted, the present results are encouraging, in particular when taking into account the modest time taken to design, tune and commission this model-based approach.

Acknowledgements

The work was sponsored by the European Commission under the ESPRIT LTR-project 28104 H_2C.

7.8 References

[1] A. Bemporad, M. Morari, V. Dua, and E. N. Pistikopoulos. The explicit solution of model predictive control via multiparametric quadratic programming. In *Proc. American Control Conference*, 872–876, Chicago, 2000.

[2] M. Burckhardt. *Fahrwerktechnik: Radschlupf-Regelsysteme*. Vogel Verlag, Würzburg, 1993.

[3] C. C. de Wit, R. Horowitz, and P. Tsiotras. Model-based observers for Tire/Road contact friction prediction. In H. Nijmeijer and T.I. Fossen, editors, *In New Directions in Nonlinear Observer Design*, 23–42. Springer-Verlag, 1999.

[4] S. Drakunov, Ü. Özgüner, P. Dix, and B. Ashrafi. ABS control using optimum search via sliding modes. *IEEE Trans. Control Systems Technology*, 3(1):79–85, March 1995.

[5] R. Freeman. Robust slip control for a single wheel. Technical Report CCEC 95-0403, University of California, Santa Barbara, 1995.

[6] T. A. Johansen, I. Petersen, and O. Slupphaug. Explicit suboptimal linear quadratic regulation with state and input constraints. In *IEEE Conference on Decision and Control, Sydney*, 2000.

[7] T. A. Johansen, J. Kalkkuhl, J. Lüdemann, and I. Petersen. Hybrid control strategies in ABS. In *American Control Conference*, Washington DC, 2001.

[8] I. Kaminer, A. M. Rascoal, P. P. Khargonekar, and E. E. Coleman. A velocity algorithm for the implementation of gain-scheduled controllers. *Automatica*, 31:1185–1191, 1995.

[9] H. K. Khalil. *Nonlinear Systems*. Prentice Hall, 1996.

[10] R. Schwarz. *Rekonstruktion der Bremskraft Bei Fahrzeugen mit Elektromechanisch Betätigten Radbremsen*. Ph.D. thesis, Institut für Automatisierungstechnik der TU Darmstadt, 1999.

[11] Jingang Yi, L. Alvarez, R. Horowitz, and C. Canudas-de-Wit. Adaptive emergency braking control using a dynamical Tire/Road friction model. In *IEEE Conference on Decision and Control*, Sydney, 2000.

7.A Appendix—Details of Proof

In this section, we prove that (7.57) implies $D(v) > 0$ for all $v > 0$.

$$\dot{P}_{1,1}(v) = \frac{\partial P_{1,1}(v)}{\partial v} \dot{v} \tag{7.51}$$

$$= \frac{\left(\alpha_1^2 + \beta_1^2 R^{-1}\left(Q_{2,2}(v) + \frac{2(Q_{1,1}(v)/R^{-1})^{1/2}}{\beta_1}v\right)\right)^{1/2} \frac{d}{dv}Q_{1,1}(v)}{2\beta_1 (Q_{1,1}(v)R^{-1})^{1/2}} \dot{v} \tag{7.52}$$

$$+ \frac{(Q_{1,1}(v)R^{-1})^{1/2} \beta_1 \left(\frac{d}{dv}Q_{2,2}(v) + \frac{2(Q_{1,1}(v)/R^{-1})^{1/2}}{\beta_1} + \frac{\frac{d}{dv}Q_{1,1}(v)}{\beta_1(Q_{1,1}(v)R^{-1})^{1/2}}v\right)}{2\left(\alpha_1^2 + \beta_1^2 R^{-1}\left(Q_{2,2}(v) + \frac{2(Q_{1,1}(v)/R^{-1})^{1/2}}{\beta_1}v\right)\right)^{1/2}} \dot{v}$$

$$\dot{P}_{1,2}(v) = \dot{P}_{2,1}(v) = \frac{\partial P_{1,2}(v)}{\partial v}\dot{v}$$

$$= \left(\frac{(Q_{1,1}(v)/R^{-1})^{1/2}}{\beta_1} + \frac{\frac{d}{dv}Q_{1,1}(v)}{2\beta_1(Q_{1,1}(v)R^{-1})^{1/2}}v\right)\dot{v} \tag{7.53}$$

$$\dot{P}_{2,2}(v) = \frac{\alpha_1 + \left(\alpha_1^2 + \beta_1^2 R^{-1}\left(Q_{2,2}(v) + \frac{2(Q_{1,1}(v)/R^{-1})^{1/2}}{\beta_1}v\right)\right)^{1/2}}{\beta_1^2 R^{-1}}\dot{v}$$

$$+ \frac{\left(\frac{d}{dv}Q_{2,2}(v) + \frac{2(Q_{1,1}(v)/R^{-1})^{1/2}}{\beta_1} + \frac{\frac{d}{dv}Q_{1,1}(v)}{\beta_1(Q_{1,1}(v)R^{-1})^{1/2}}v\right)v}{2\left(\alpha_1^2 + \beta_1^2 R^{-1}\left(Q_{2,2}(v) + \frac{2(Q_{1,1}(v)/R^{-1})^{1/2}}{\beta_1}v\right)\right)^{1/2}}\dot{v} \tag{7.54}$$

Define the following positive variables:

$$A = \alpha_1^2 + D\left(Q_{2,2}(v) + \frac{2C}{D}v\right) = \alpha_1^2 + DQ_{2,2}(v) + 2Cv$$

$$B = \frac{d}{dv}Q_{2,2}(v) + \frac{2C}{D} + \frac{\frac{d}{dv}Q_{1,1}(v)}{C}v$$

$$C = (Q_{1,1}(v)R^{-1})^{1/2}\beta_1$$

$$D = \beta_1^2 R^{-1}$$

Now

$$D(v) = \frac{\left(\alpha_1^2 + \beta_1^2 R^{-1}\left(Q_{2,2}(v) + \frac{2(Q_{1,1}(v)/R^{-1})^{1/2}}{\beta_1}v\right)\right)^{1/2} \frac{d}{dv}Q_{1,1}(v)\alpha_1}{2\beta_1(Q_{1,1}(v)R^{-1})^{1/2} \beta_1^2 R^{-1}}$$

$$+ \frac{\left(\alpha_1^2 + \beta_1^2 R^{-1}\left(Q_{2,2}(v) + \frac{2(Q_{1,1}(v)/R^{-1})^{1/2}}{\beta_1}v\right)\right) \frac{d}{dv}Q_{1,1}(v)}{2\beta_1(Q_{1,1}(v)R^{-1})^{1/2}\beta_1^2 R^{-1}}$$

$$+ \frac{\frac{d}{dv}Q_{1,1}(v)\left(\frac{d}{dv}Q_{2,2}(v) + \frac{2(Q_{1,1}(v)/R^{-1})^{1/2}}{\beta_1} + \frac{\frac{d}{dv}Q_{1,1}(v)}{\beta_1(Q_{1,1}(v)R^{-1})^{1/2}}v\right)v}{4\beta_1(Q_{1,1}(v)R^{-1})^{1/2}}$$

$$+ \frac{(Q_{1,1}(v)R^{-1})^{1/2}\beta_1 \left(\frac{d}{dv}Q_{2,2}(v) + \frac{2(Q_{1,1}(v)/R^{-1})^{1/2}}{\beta_1} + \frac{\frac{d}{dv}Q_{1,1}(v)}{\beta_1(Q_{1,1}(v)R^{-1})^{1/2}}v\right)}{2\left(\alpha_1^2 + \beta_1^2 R^{-1}\left(Q_{2,2}(v) + \frac{2(Q_{1,1}(v)/R^{-1})^{1/2}}{\beta_1}v\right)\right)^{1/2}} \frac{\alpha_1}{\beta_1^2 R^{-1}}$$

$$+ \frac{(Q_{1,1}(v)R^{-1})^{1/2}\beta_1 \left(\frac{d}{dv}Q_{2,2}(v) + \frac{2(Q_{1,1}(v)/R^{-1})^{1/2}}{\beta_1} + \frac{\frac{d}{dv}Q_{1,1}(v)}{\beta_1(Q_{1,1}(v)R^{-1})^{1/2}}v\right)}{2\beta_1^2 R^{-1}}$$

$$+ \frac{(Q_{1,1}(v)R^{-1})^{1/2}\beta_1 \left(\frac{d}{dv}Q_{2,2}(v) + \frac{2(Q_{1,1}(v)/R^{-1})^{1/2}}{\beta_1} + \frac{\frac{d}{dv}Q_{1,1}(v)}{\beta_1(Q_{1,1}(v)R^{-1})^{1/2}}v\right)^2}{4\left(\alpha_1^2 + \beta_1^2 R^{-1}\left(Q_{2,2}(v) + \frac{2(Q_{1,1}(v)/R^{-1})^{1/2}}{\beta_1}v\right)\right)}v$$

$$- \left(\frac{(Q_{1,1}(v)/R^{-1})^{1/2}}{\beta_1} + \frac{\frac{d}{dv}Q_{1,1}(v)}{2\beta_1(Q_{1,1}(v)R^{-1})^{1/2}}v\right)^2$$

The above expression can then be rewritten as

$$D(v) = \frac{A^{1/2}\frac{d}{dv}Q_{1,1}(v)\alpha_1}{2CD} + \frac{A\frac{d}{dv}Q_{1,1}(v)}{2CD} + \frac{\frac{d}{dv}Q_{1,1}(v)Bv}{4C}$$
$$+ \frac{\alpha_1 BC}{2A^{1/2}D} + \frac{CB}{2D} + \frac{CB^2}{4A}v - \left(\frac{C}{D} + \frac{\frac{d}{dv}Q_{1,1}(v)}{2C}v\right)^2 \quad (7.56)$$

In (7.56), there are two possible negative factors involved: α_1 and the last quadratic term with a negative sign $(-P'_{1,2}(v)P'_{2,1}(v))$. In order to cancel the term $P'_{1,2}(v)P'_{2,1}(v)$, parts from the third and fifth term in (7.56) are used which gives

$$\frac{\frac{d}{dv}Q_{1,1}(v)Bv}{4C} + \frac{CB}{2D} - \left(\frac{C}{D} + \frac{\frac{d}{dv}Q_{1,1}(v)}{2C}v\right)^2$$
$$= \frac{\frac{d}{dv}Q_{2,2}(v)\frac{d}{dv}Q_{1,1}(v)}{4C}v + \frac{d}{dv}Q_{2,2}(v)$$

The following inequality thus ensures $D(v) > 0$:

$$D(v) = \frac{A^{1/2}\frac{d}{dv}Q_{1,1}(v)\alpha_1}{2CD} + \frac{A\frac{d}{dv}Q_{1,1}(v)}{2CD} + \frac{\frac{d}{dv}Q_{2,2}(v)\frac{d}{dv}Q_{1,1}(v)}{4C}v$$
$$+ \frac{\alpha_1 BC}{2A^{1/2}D} + \frac{d}{dv}Q_{2,2}(v) + \frac{CB^2}{4A} > 0 \quad (7.57)$$

where only α_1 may have a negative value.

8

Friction Tire/Road Modeling, Estimation and Optimal Braking Control

C. Canudas-de-Wit   P. Tsiotras   X. Claeys   J. Yi
R. Horowitz

Abstract

The modeling part of the paper discusses a new dynamic friction force model for the longitudinal road/tire interaction for wheeled ground vehicles. The model is based on a dynamic friction model developed previously for contact-point friction problems, called the LuGre model [13]. By assuming a contact patch between the tire and the ground, the model is given by a PDE. It is, however, possible to approximate this PDE by an ODE (the lumped model). This ODE for the friction force is developed based on the patch boundary conditions and the normal force distribution along the contact patch. We also discuss possible extension of longitudinal models to high-dimensional models including lateral force and self-alignment torque. Finally, we discuss how these models may be affected by other external factors, such as road conditions (roughness), road/tire interfaces (wet road), temperature variation, *etc*.

In the estimation section, we discuss the problem of tire-road friction estimation using only angular wheel velocity, which cannot always be computed from actual sensors. Tire forces information is relevant to problems like optimization of anti-lock brake systems (ABS), traction systems, diagnostic of the road friction conditions, *etc.* These results may suggest alternative traction control methodologies other than the current ones based on the use of tracking of the "optimal" slip coefficient using, for example, sliding mode control. These aspects are presented in the last two sections of the paper. First, we investigate optimal braking strategies for a simplified model of a wheeled vehicle. For the case of static friction force models, we derive the *exact* optimality conditions for the minimum braking distance problem by solving the associated optimal control problem. It is shown that the optimal control is singular and can be written in a state feedback form. Using a singular perturbation approach, we also derive the optimal strategy for the case of the dynamic LuGre model. Then, we present a control scheme for emergency braking of vehicles. The controller utilizes estimated state feedback control to achieve a near-maximum deceleration.

8.1 Introduction

The problem of predicting the friction force between the tire and the ground for wheeled vehicles is of enormous importance to the automotive industry. Since friction is the major mechanism for generating forces on the vehicle, it is extremely important to have an accurate characterization of the magnitude (and direction) of the friction force generated at the ground/tire interface. However, accurate tire/ground friction models are difficult to obtain analytically. Subsequently, in the past several years, the problem of modeling and predicting tire friction has become an area of intense research in the automotive community. In particular, ABS and traction control systems rely on knowledge of the friction characteristics. Such systems have enhanced safety and maneuverability to such an extent that they have become almost mandatory for all current passenger vehicles.

Traction control systems reduce or eliminate excessive slipping or sliding during vehicle acceleration and thus enhance the controllability and maneuverability of the vehicle. Proper traction control design has a paramount effect on safety and handling qualities for passenger vehicles. Traction control aims to achieve maximum torque transfer from the wheel axle to forward acceleration. Similarly, anti-lock braking systems (ABS) prohibit wheel-lock and skidding during braking by regulating the pressure applied on the brakes, thus increasing lateral stability and steerability, especially during wet and icy road conditions. As with the case of traction control, the main difficulty in designing ABS is the nonlinearity and uncertainty of the tire/road models. In either case, the friction force at the tire/road interface is the main mechanism for converting wheel angular acceleration or deceleration (due to the motor torque or braking) to forward acceleration of deceleration (longitudinal force). Therefore, the study of the friction force characteristics at the road/tire interface is of paramount importance for the design of ABS and/or traction control systems. Moreover, tire friction models are also indispensable for accurately reproducing friction forces for simulation purposes. Active control mechanisms, such as ESP, TCS, ABS, steering control, active suspension, *etc.* may be tested and optimized using vehicle mechanical 3D simulators with suitable tire/road friction models.

A common assumption in most tire friction models is that the normalized tire friction μ

$$\mu = \frac{F}{F_n} = \frac{\text{Friction force}}{\text{Normal force}}$$

is a nonlinear function of the normalized relative velocity between the road and the tire (slip coefficient s) with a distinct maximum; see Fig. 8.1. In addition, it is understood that μ also depends on the velocity of the vehicle and road surface conditions, among other factors (see [10] and [27]). The curves shown in Figure 8.1 illustrate how these factors influence the shape of μ.

The curves shown in Figure 8.1 are derived empirically, based solely on steady state (*i.e.*, constant linear and angular velocity) experimental

Friction Tire/Road Modeling,...and Optimal Braking Control

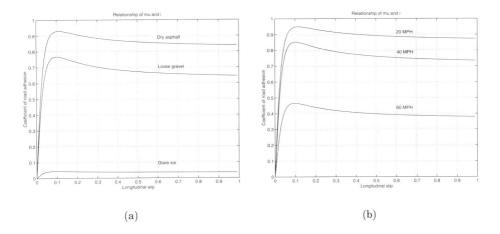

Figure 8.1 Typical variations of the tire/road friction profiles for different road surface conditions (a), and different vehicle velocities (b). Curves given by Harned *et al* [27].

data [27], [6] in a highly controlled laboratory environment or using specially designed test vehicles. Under such steady-state conditions, experimental data seem to support the force vs. slip curves of Figure 8.1. In reality, the linear and angular velocities can never be controlled independently and hence, such idealized steady-state conditions are not reached except during the rather uninteresting case of cruising with constant speed. The development of the friction force at the tire/road interface is very much a dynamic phenomenon. In other words, the friction force does not reach its steady-state value shown in Figure 8.1 instantaneously, but rather exhibits transient behavior, which may differ significantly from its steady-state value. Experiments performed in commercial vehicles have shown that the tire/road forces do not necessarily vary along the curves shown Figure 8.1, but rather "jump" from one value to another when these forces are displayed in the μ–s plane [50]. In addition, in realistic situations, these variations are most likely to exhibit hysteresis loops, clearly indicating the dynamic nature of friction.

In this paper, we present a series of results concerning the problem of modeling, and estimation of road/tire contact friction. We also present some aspects related to the problem of optimal braking control.

In the first part of the paper, we discuss a new dynamic friction force model for the longitudinal road/tire interaction for wheeled ground vehicles. The model is based on a dynamic friction model developed previously for contact-point friction problems, called the LuGre model [13]. By assuming a contact patch between the tire and the ground the model is given by a PDE. It is, however, possible to approximate this PDE by an ODE (the lumped model) based on the patch boundary conditions and the normal force distribution along the contact patch. The solution to this ODE provides the friction force developed at the tire. We also discuss possible extensions of longitudinal models to high dimensional models including lateral force and self-alignment torque. Finally, we discuss how these models may be affected by other external factors such as road conditions (roughness), road/tire interfaces (wet road), temperature variation, *etc.*

In the estimation section we discuss the problem of tire-road friction estimation using only angular wheel velocity measurements which can always be computed from actual sensors. Tire forces information is relevant to problems like optimization of ABS, traction systems, diagnostic of the road friction conditions, *etc.*

These results suggest alternative traction control methodologies, other than the current ones based on the use of tracking of the "optimal" slip coefficient using, for example, sliding mode control. These aspects are presented in the last two sections of the paper. First, we investigate optimal braking strategies for a simplified model of a wheeled vehicle. For the case of static force/slip curves, we derive the *exact* optimality conditions for the minimum braking distance problem by solving the associated optimal control problem. It is shown that the optimal control is singular and can be written in a state feedback form. We extend these results for the case of the LuGre dynamic friction model using a singular perturbation approach. Next, we present a control scheme for emergency braking of vehicles. The controller utilizes estimated state feedback control to achieve a near-maximum deceleration.

8.2 Road/Tire Contact Friction Models

In this section we consider the simplified motion dynamics of a quarter-vehicle model. The system is then of the form

$$m\dot{v} = F \qquad (8.1)$$
$$J\dot{\omega} = -rF + u, \qquad (8.2)$$

where m is 1/4 of the vehicle mass and J, r are the inertia and radius of the wheel, respectively. v is the linear velocity of the vehicle, ω is the angular velocity of the wheel, u is the accelerating (or braking) torque, and F is the tire/road friction force. For the sake of simplicity, only longitudinal motion will be considered in this section. The lateral motion as well as combined longitudinal/lateral dynamics are discussed in subsequent sections. The dynamics of the braking and driving actuators, suspension dynamics are also neglected.

Static Slip/Force Models

The most common tire friction models used in the literature are those of algebraic slip/force relationships. They are defined as one-to-one (memoryless) maps between the friction F, and the longitudinal slip rate s, which

Friction Tire/Road Modeling,...and Optimal Braking Control

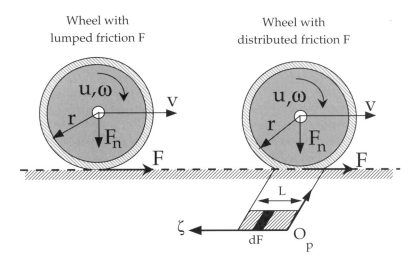

Figure 8.2 One-wheel system with lumped friction (left), and distributed friction (right)

is defined as

$$s = \begin{cases} s_b = \dfrac{r\omega}{v} - 1 & \text{if } v > r\omega, \ v \neq 0 \quad \text{for braking} \\ s_d = 1 - \dfrac{v}{r\omega} & \text{if } v < r\omega, \ \omega \neq 0 \quad \text{for driving} \end{cases} \quad (8.3)$$

The slip rate results from the reduction of the effective circumference of the tire as a consequence of the tread deformation due to the elasticity of the tire rubber [36]. This, in turn, implies that the ground velocity v will not be equal to $r\omega$. The absolute value of the slip rate is defined in the interval $[0, 1]$. When $s = 0$, there is no sliding (pure rolling), whereas $|s| = 1$ indicates full sliding/skidding. It should be pointed out that in this paper we always define the relative velocity as $v_r = r\omega - v$. As a result, the slip coefficient in (8.3) is positive for driving and negative for braking. This is somewhat different than what is done normally in the literature, where the relative velocity (and hence the slip s) is always kept positive by redefining $v_r = v - r\omega$ in the case of braking. Since we wish our results to hold for both driving and braking, we feel that it is more natural to keep the same definition for the relative velocity for both cases. This also avoids any inconsistencies and allows for easy comparison between the braking and driving regimes.

The slip/force models aim at describing the shapes shown in Figure 8.1 via static maps $F(s) : s \mapsto F$. They may also depend on the vehicle velocity v, i.e., $F(s,v)$, and vary when the road characteristics change.

One of the most well-known models of this type is Pacejka's model (see, Pacejka and Sharp [41]), also known as the "magic formula". This model has been shown to suitably match experimental data, obtained under particular conditions of constant linear and angular velocity. The Pacejka model has the form[1]

$$F(s) = c_1 \sin(c_2 \arctan(c_3 s - c_4(c_3 s - \arctan(c_3 s))))$$

where the c_is are the parameters characterizing this model. These parameters can be identified by matching experimental data, as shown in Bakker et al [6]. Another static model is the one proposed by Burckhardt [10]. The tire/road friction characteristic is of the form

$$F(s, v) = (c_1(1 - e^{-c_2 s}) - c_3 s) e^{-c_4 v} \tag{8.4}$$

where $c_1, \ldots, c_4$ are some constants. This model has a velocity dependency, seeking to match variations like the ones shown in Figure 8.1(b).

Kiencke and Daiß [30] neglect the velocity-dependent term in equation (8.4) and, after approximating the exponential function in (8.4), they obtain the following expression for the friction/slip curve

$$F(s) = k_s \frac{s}{c_1 s^2 + c_2 s + 1} \tag{8.5}$$

where k_s is the slope of the $F(s)$ vs. s curve at $s = 0$, and c_1 and c_2 are properly chosen parameters.

Alternatively, Burckhardt [9] proposes a simple, velocity-independent three-parameter model as follows:

$$F(s) = c_1(1 - e^{-c_2 s}) - c_3 s \tag{8.6}$$

All the previous friction models are highly nonlinear in the unknown parameters and, thus, they are not well adapted to be used for on-line identification. For this reason, simplified models like

$$F(s) = c_1 \sqrt{s} - c_2 s \tag{8.7}$$

have been proposed in the literature.

It is also well understood that the "constant" c_is in the above models, are not really invariant, but they may strongly depend on the tire characteristics (e.g., compound, tread type, tread depth, inflation pressure, temperature), on the road conditions (e.g., type of surface, texture, drainage, capacity, temperature, lubricant, etc.), and on the vehicle operational conditions (velocity, load); see, for instance, the discussion in [40].

---

[1] In the formulas that follow, it is assumed that $s \in [0, 1]$. Hence, these formulas give the magnitude of the friction force. The sign of F is then determined from the sign of $v_r = r\omega - v$.

Lumped Dynamic Friction Models

The static friction models of the previous section are appropriate when we have steady-state conditions for the linear and angular velocities. In fact, the experimental data used to validate the friction/slip curves are obtained using specialized equipment that allows independent linear and angular velocity modulation so as to transverse the whole slip range. This steady-state point of view is rarely true in reality, especially when the vehicle goes through continuous successive phases between acceleration and braking.

As an alternative to the static $F(s)$ maps, different forms of dynamic models can be adopted. The so-called "dynamic friction models" attempt to capture the transient behavior of the tire-road contact forces under time-varying velocity conditions. Generally speaking, dynamic models can be formulated either as lumped or as distributed models, as shown in Figure 8.2. A *lumped* friction model assumes a point tire-road friction contact. As a result, the mathematical model describing such a model is an ordinary differential equations that can be easily solved by time integration. *Distributed* friction models, on the other hand, assume the existence of a *contact patch* between the tire and the ground with an associated normal pressure distribution. This formulation results in a partial differential equation that needs to be solved both in time and space.

A number of dynamic models have been proposed in the literature that can be classified under the term "dynamic friction models." One such model, for example, has been proposed by Bliman *et al.* in [7]. In that reference, the friction is calculated by solving a differential equation of the following form:

$$\begin{aligned} \dot{z} &= |v_r|Az + Bv_r \\ F(z, v_r) &= Cz + \text{sgn}(v_r)D \end{aligned} \qquad (8.8)$$

The matrix A is required to be Hurwitz of dimension either one or two, with the latter case being more accurate. Another lumped dynamic model that can be used to accurately predict the friction forces during transients is the LuGre friction model [14]. The derivation of this model is discussed in great detail [15], and summarized in Section 8.2. Before doing that, we next present an example of a frequently used dynamic friction model, the so-called "kinematic model". This will allow us to motivate our introduction in Section 8.2 of a new distributed (and later, an average lumped) friction model. In addition, it will be shown that these two commonly used lumped models are not able to reproduce the steady-state characteristics, similar to those of Pacejka's "magic formula".

Kinematic-based models. These models are derived from the idealization of a contact point deformation and from kinematic considerations (velocity relations between the points that concern the tire deformations). Their derivation follows semi-empirical considerations and assumes that the contact forces result from the product of the tire deformation and the tire stiffness.

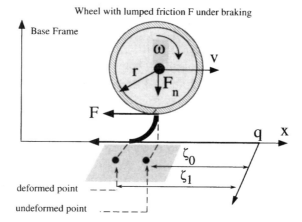

Figure 8.3 View of the contact area with the position of the undeformed contact point ζ_0 and the point ζ_1 that deforms under longitudinal shear forces during a brake phase

An example of a two-dimensional model characterizing the lateral force and the aligning moment can be found in [34]. A brush model for the longitudinal tire dynamics has been derived in [20], [5].

For simplicity, next we only discuss the breaking case. The traction case follows a similar development, with an appropriate definition of z. In that case, the relaxation length may be defined as the wheel arc length required to build friction.

A model for the longitudinal dynamics in [20] is derived by defining the *normalized* longitudinal slip z as

$$z = \frac{\zeta_1 - \zeta_0}{\zeta_0} \tag{8.9}$$

where ζ_1 locates a hypothetical element that follows the road, and it defines the distance from the *deformed* wheel point to a forward point q. ζ_0 locates a hypothetical element that in undeformed under longitudinal shear forces, and defines the distance from the *undeformed* wheel point (*i.e.*, center of the wheel's rotational axis) to a forward point q as shown in Figure 8.3. The undeformed point ζ_0 and the forward point q are moving at the same ground speed v, thus the distance ζ_0 is constant with respect to the moving point q. The length ζ_0 is known a s the *longitudinal relaxation length*.

Differentiating z in (8.9) with respect to time, and noticing that $\dot{\zeta}_0 = v$, $\dot{\zeta}_1 = r\omega$, we get[2]

$$\frac{1}{\sigma}\frac{dz}{dt} = v_r \text{sgn}(v) - |v|z \tag{8.10}$$

$$F = h(z) \tag{8.11}$$

---

[2]In this expression, both positive and negative v are considered.

where v is the linear velocity, $v_r = r\omega - v$ is the relative velocity, and the friction force F is defined by the function $h(z)$ that describes the stationary slip characteristics. In the simplest case, $h(z)$ is given by a linear relationship between the longitudinal slip and the tire (linear) stiffness k:

$$h(z) = kz$$

The constant $1/\sigma = \zeta_0$ is called the *relaxation length*, and can be defined as the distance required to reach the steady-state value of F

$$F_{ss} = h(z_{ss}) = kz_{ss} = k\frac{r\omega - v}{v} = ks$$

after a step change of the slip longitudinal velocity, $s = s_b = v_r/v = (r\omega - v)/v$. The role of the relaxation length $1/\sigma$ in equation (8.10), can be better understood by rewriting this equation in terms of the spatial coordinate η,

$$\eta(t) = \int_0^t |v(\tau)| d\tau$$

rather than as a time-differential equation, *i.e.*,

$$\frac{1}{\sigma}\frac{dz}{dt} = \frac{1}{\sigma}\frac{dz}{d\eta}\frac{d\eta}{dt} = v_r \mathrm{sgn}(v) - |v|z \qquad (8.12)$$

$$\frac{1}{\sigma}\frac{dz}{d\eta} = -z + \frac{v_r}{v} = -z + s \qquad (8.13)$$

Equation (8.13) can thus be seen as a first order spatial equation with the sliding velocity s as its input. It thus becomes clear that σ represents the *spatial* constant of this equation.

As pointed out in [20], this model works well for high speeds, but it generates lightly-damped oscillations at low speeds. The reason for this is that at quasi-steady-state regimes, z is close to its steady-state value ($z \approx s$), and the friction F is dominated by its spring-like behavior ($F \approx kz$), resulting in a lightly-damped mechanical system. Additional considerations are necessary to make this model consistent for all possible changes in the velocity sign. This restricts the usefulness of this model for tire friction analysis and control development. Reference [20] actually provides a twelve-step algorithm for implementing this model in simulations.

Note that the kinematic model does not exhibit a maximum for values $|s| < 1$, as suggested by experimental data and also by Pacejka's formula. However, the kinematic model can be modified by redefining the function $h(z)$ in (8.11) so as to produce a steady-state behavior similar to the one predicted by the "magic formula" [5].

The Lumped LuGre Model The LuGre model is an extension of the Dahl model that includes the Stribeck effect (see, [13]). This model will be used for further developments and will provide the basis for the final model proposed in this paper. The lumped, LuGre model as proposed in [14], and [12] is given as,

$$\dot{z} = v_r - \frac{\sigma_0 |v_r|}{g(v_r)} z \quad (8.14)$$

$$F = (\sigma_0 z + \sigma_1 \dot{z} + \sigma_2 v_r) F_n \quad (8.15)$$

with

$$g(v_r) = \mu_c + (\mu_s - \mu_c) e^{-|v_r/v_s|^\alpha} \quad (8.16)$$

where σ_0 is the rubber longitudinal lumped stiffness, σ_1 the rubber longitudinal lumped damping, σ_2 the viscous relative damping, μ_c the normalized Coulomb friction, μ_s the normalized static friction, ($\mu_c \leq \mu_s$), v_s the Stribeck relative velocity, F_n the normal force, $v_r = r\omega - v$ the relative velocity, and z the internal friction state. The constant parameter α is used to capture the steady-steady friction/slip characteristic[3].

In contrast to the Dahl model, the lumped LuGre model does exhibit a maximum friction for $|s| \leq [0, 1]$, but it still displays a discontinuous steady-state characteristic, at zero relative velocity. Nevertheless, by letting the internal bristle deflection z depend on both the time and the contact position ζ along the contact patch, it is possible to show that this model will have the appropriate steady-state properties. This is discussed next.

The LuGre Distributed Model

Distributed models assume the existence of an area of contact (or patch) between the tire and the road, as shown in Figure 8.2. This patch represents the projection of the part of the tire that is in contact with the road. With the contact patch is associated a frame O_p, with the ζ-axis along the length of the patch in the direction of the tire rotation. The patch length is L.

Brief review of existing models. Distributed dynamical models, have been studied previously, for example, in the works of Bliman *et al* [7]. In these models, the contact patch area is discretized to a series of elements, and the microscopic deformation effects are studied in detail. In particular, Bliman *et al* [7] characterize the elastic and Coulomb friction forces at each point of the contact patch, and they give the aggregate effect of these distributed forces by integrating over the whole patch area. They propose a second order rate-independent model (similar to Dahl's model), by applying the point friction model (8.8) to a rubber element situated at ζ at time t.

---

[3]The model in (8.15) differs from the point-contact LuGre model in [13] in the way that the function $g(v)$ is defined. Here we propose to use $\alpha = 1/2$ instead of $\alpha = 2$ as in the LuGre point-contact model in order to better match the pseudo-stationary characteristic of this model (map $s \mapsto F(s)$) with the shape of the Pacejka model. Other values of α in the range $[1/2, 2]$ can be used.

By letting $z(\zeta,t)$ denote the corresponding friction state, they obtain the partial differential equation

$$\frac{\partial z}{\partial t} + r\omega \frac{\partial z}{\partial \zeta} = |v_r|Az + Bv_r, \qquad z(\zeta,0) = z(0,t) = 0 \qquad (8.17)$$

$$F(t) = \frac{F_n}{L} \int_0^L Cz(\zeta,t)\, d\zeta \qquad (8.18)$$

They also show that, under constant v and ω, there exists a choice of parameters A, B and C that closely match a curve similar to the one characterizing the "magic formula".

In [50], van Zanten et al use a distributed brush model. The contact patch is described by a brush-type model where the displacement of each bristle is characterized by the state z_i, $i = 1,\ldots,N$. The discretized version of this model, in its simplest form, is given by

$$\frac{dz_i}{dt} = v - \omega r - \frac{z_i - z_{i-1}}{L/N}\omega r \qquad (8.19)$$

$$F = \sum_{i=1}^N c_i z_i \qquad (8.20)$$

where N is the number of discrete elements (bristles) and c_i is the stiffness of the bristle. The imposed boundary condition $dz_1/dt = 0$ implies that the bristle at the beginning of the contact area has no displacement.

Distributed LuGre model. One can also extend the point friction model (8.14)-(8.15) to a distributed friction model along the patch by letting $z(\zeta,t)$ denote the friction state (deflection) of the bristle/patch element located at the point ζ along the patch at a certain time t. At every time instant $z(\zeta,t)$ provides the deflection distribution along the contact patch. The model (8.14)-(8.15) can now be written as

$$\frac{dz}{dt}(\zeta,t) = v_r - \frac{\sigma_0|v_r|}{g(v_r)} z \qquad (8.21)$$

$$F = \int_0^L dF(\zeta,t) \qquad (8.22)$$

with $g(v_r)$ defined as in (8.16) and where

$$dF(\zeta,t) = \left(\sigma_0 z(\zeta,t) + \sigma_1 \frac{\partial z}{\partial t}(\zeta,t) + \sigma_2 v_r\right) dF_n(\zeta,t)$$

where $dF(\zeta,t)$ is the differential friction force developed in the element $d\zeta$ and $dF_n(\zeta,t)$ is the differential normal force applied in the element $d\zeta$ at time t. This model assumes that the contact velocity of each differential state element is equal to v_r.

Assuming a steady-state normal force distribution $dF_n(\zeta,t) = dF_n(\zeta)$ and introducing a normal force density function $f_n(\zeta)$ (force per unit length) along the patch, i.e.

$$dF_n(\zeta) = f_n(\zeta)d\zeta$$

one obtains the total friction force as

$$F(t) = \int_0^L (\sigma_0 z(\zeta,t) + \sigma_1 \frac{\partial z}{\partial t}(\zeta,t) + \sigma_2 v_r) f_n(\zeta) d\zeta \qquad (8.23)$$

Noting that[4] $\dot{\zeta} = |r\omega|$, and that

$$\frac{dz}{dt}(\zeta,t) = \frac{\partial z}{\partial \zeta}\frac{\partial \zeta}{\partial t} + \frac{\partial z}{\partial t},$$

we have that Equation (8.21) describes a partial differential equation, *i.e.*

$$\frac{\partial z}{\partial \zeta}(\zeta,t)|r\omega| + \frac{\partial z}{\partial t}(\zeta,t) = v_r - \frac{\sigma_0 |v_r|}{g(v_r)} z(\zeta,t) \qquad (8.24)$$

that should be solved in both in time and space. More details on the derivation of this distributed friction model are given in [15].

Steady-state characteristics. The time steady-state characteristics of the model (8.21)–(8.22) are obtained by setting $\frac{\partial z}{\partial t}(\zeta,t) \equiv 0$ and by imposing that the velocities v and ω are constant. Enforcing these conditions in (8.24) results in

$$\frac{\partial z(\zeta,t)}{\partial \zeta} = \frac{1}{|\omega r|}\left(v_r - \frac{\sigma_0 |v_r|}{g(v_r)} z(\zeta,t)\right) \qquad (8.25)$$

At steady-state, v, ω (and hence v_r) are constant, and (8.25) can be integrated along the patch with the boundary condition $z(0,t) = 0$. A simple calculation shows that

$$z_{ss}(\zeta) = \text{sgn}(v_r)\frac{g(v_r)}{\sigma_0}\left(1 - e^{-\frac{\sigma_0}{g(v_r)}\left|\frac{v_r}{\omega r}\right|\zeta}\right) = c_2(1 - e^{c_1 \zeta}) \qquad (8.26)$$

where

$$c_1 = -\frac{\sigma_0}{g(v_r)}\left|\frac{v_r}{\omega r}\right|, \qquad c_2 = \text{sgn}(v_r)\frac{g(v_r)}{\sigma_0} \qquad (8.27)$$

Notice that when $\omega = 0$ (locked wheel case) the distributed model, and hence the steady-state expression (8.26) collapses into the one predicted by the standard point-contact LuGre model. This agrees with the expectation that for a locked wheel, the friction force is only due to pure sliding.

The steady-state value of the total friction force is calculated from (8.23)

$$F_{ss} = \int_0^L (\sigma_0 z_{ss}(\zeta) + \sigma_2 v_r) f_n(\zeta) d\zeta \qquad (8.28)$$

To proceed with the calculation of F_{ss}, we need to postulate a distribution for the normal force $f_n(\zeta)$. The typical form of the normal force distribution

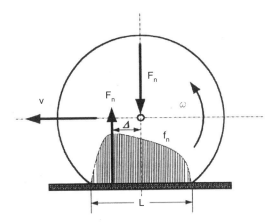

Figure 8.4 Typical normal load distribution along the patch; from [25].

reported in the literature [51], [36], [25], [24], is shown in Figure 8.4. However, for the sake of simplicity, other forms can be adopted. Some examples are given next.

- *constant norm distribution*. A simple result can be derived if we assume uniform load distribution, as done in [14] and [22]. For uniform normal load

$$f_n(\zeta) = \frac{F_n}{L}, \quad 0 \leq \zeta \leq L \quad (8.29)$$

and one obtains,

$$F_{ss} = \left(\text{sgn}(v_r)g(v_r)\left[1 - \frac{Z}{L}(1 - e^{-L/Z})\right] + \sigma_2 v_r \right) F_n \quad (8.30)$$

where

$$Z = \left|\frac{\omega r}{v_r}\right| \frac{g(v_r)}{\sigma_0} \quad (8.31)$$

- *exponentially decreasing distribution*. In this case, the decrease of the normal load along the patch shown in Figure 8.4 is approximated with an exponentially decreasing function

$$f_n(\zeta) = e^{-\lambda(\frac{\zeta}{L})} f_{n0}, \quad 0 \leq \lambda, \ 0 \leq \zeta \leq L \quad (8.32)$$

where $f_n(0) = f_{n0}$ denotes the distributed normal load at $\zeta = 0$. This particular choice will become clear later on, when we reduce the

---

[4]It is assumed here that the origin of the ζ-frame changes location when the wheel velocity reverses direction, such that $\dot{\zeta} = r\omega$, for $\omega > 0$, and $\dot{\zeta} = -r\omega$, for $\omega < 0$.

infinite dimension distributed model to a simple lumped one having only one state variable. Moreover, for $\lambda > 0$, we have a strictly decreasing function of f_n. With the choice (8.32), one obtains

$$F_{ss} = \sigma_0 c_2 k_1 \left(1 - e^{-\lambda} + k_2 e^{(-\lambda + c_1 L)} + k_2\right) + \sigma_2 v_r k_1 (1 - e^{-\lambda}) \quad (8.33)$$

where c_1 and c_2 as in (8.27), and

$$k_1 = \frac{f_{n0} L}{\lambda} \quad \text{and} \quad k_2 = \frac{\lambda}{c_1 L - \lambda}$$

The details of these calculations are given [15]. The value of f_{n0} can be computed from λ, L and the total normal load F_n acting on the wheel shaft. That is,

$$\begin{aligned}
F_n &= \int_0^L f_n(\zeta)\, d\zeta = f_{n0} \int_0^L e^{-\frac{\lambda}{L}\zeta}\, d\zeta \\
&= -\frac{L}{\lambda} f_{n0} \left[e^{-\frac{\lambda}{L}\zeta}\right]_0^L = -\frac{L}{\lambda} f_{n0} \left(e^{-\lambda} - 1\right) \\
&= \frac{L}{\lambda} f_{n0} \left(1 - e^{-\lambda}\right)
\end{aligned} \quad (8.34)$$

which yields

$$f_{n0} = F_n \frac{\lambda}{(1 - e^{-\lambda}) L} \quad (8.35)$$

- *distributions with zero boundary conditions.* As shown in Figure 8.4, a realistic force distribution has, by continuity, zero values for the normal load at the boundaries of the patch. Several forms satisfy this constraint. Some possible examples proposed herein are given below:

$$f_n(\zeta) = \frac{3 F_n}{2L} \left[1 - \left(\frac{\zeta - L/2}{L/2}\right)^2\right] \quad \text{parabolic} \quad (8.36)$$

or,

$$f_n(\zeta) = \frac{\pi F_n}{2L} \sin(\pi \zeta / L) \quad \text{sinusoidal} \quad (8.37)$$

or,

$$f_n(\zeta) = \frac{\gamma^2 L^2 + \pi^2}{\pi L (e^{-\gamma L} + 1)} \exp^{-\gamma \zeta} \sin(\pi \zeta / L) \quad \text{sinusoidal/exponential} \quad (8.38)$$

where F_n denotes the total normal load.

Relation with the magic formula. The previously derived steady-state expressions, depend on both v and ω. They can also be expressed as a function of s and either v or ω. For example, for the constant distribution case, we have that $F_{ss}(s)$ can be rewritten as:

- *driving case.* In this case $v < r\omega$, see also (8.3), and the force at steady state is given by

$$F_d(s) = \text{sgn}(v_r) F_n g(s) \left(1 + \frac{g(s)}{\sigma_0 L |s|} (e^{-\frac{\sigma_0 L |s|}{g(s)}} - 1) \right) + F_n \sigma_2 r\omega s \quad (8.39)$$

with $g(s) = \mu_c + (\mu_s - \mu_c) e^{-|r\omega s/v_s|^\alpha}$, for some constant ω, and $s = s_d$.

- *braking case.* Noticing that the following relations hold between the braking s_b and the driving s_d slip definitions,

$$r\omega s_d = v s_b, \qquad s_d = \frac{s_b}{s_b + 1}$$

the steady-state friction force for the braking case can be written as

$$F_b(s) = \text{sgn}(v_r) F_n g(s) \left(1 + \frac{g(s)|1+s|}{\sigma_0 L |s|} (e^{-\frac{\sigma_0 L |s|}{g(s)|1+s|}} - 1) \right) + F_n \sigma_2 v s \quad (8.40)$$

where $g(s) = \mu_c + (\mu_s - \mu_c) e^{-|vs/v_s|^\alpha}$ for constant v, and $s = s_b$; see also (8.3).

Remark. Note that the above expressions depend not only on the slip s, but also on either the vehicle velocity v or the wheel velocity ω, depending on the case considered (driving or braking). Therefore, static plots of F vs. s can only be obtained for a specified (constant) velocity. This dependence of the steady-state force/slip curves on vehicle velocity is evident in experimental data found in the literature. Nonetheless, it should be stressed here that it is impossible to reproduce such curves from experimental data obtained from standard vehicles during normal driving conditions, since v and ω cannot be independently controlled. For that, specially designed equipment is needed. Figure 8.5(a) shows the steady-state dependence on the vehicle velocity for the braking case.

Dependency on road conditions. The level of tire/road adhesion can be modeled by introducing a multiplicative parameter θ in the function $g(v_r)$. To this aim, we substitute $g(v_r)$ by

$$\tilde{g}(v_r) = \theta g(v_r)$$

where $g(v_r)$ is the nominal known function given in (8.16). Computation of the function $F(s, \theta)$, from equation (8.40) as a function of θ, gives the

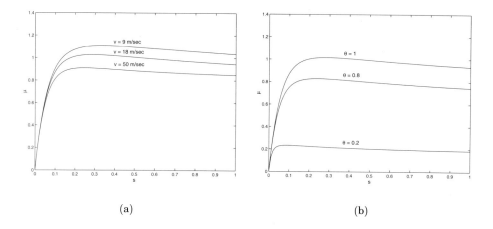

Figure 8.5 Static view of the distributed LuGre model with uniform force distribution (braking case) under: (a) different values for v, (b) different values for θ with $v = 20\,\mathrm{m/s} = 72\,\mathrm{Km/h}$. These curves show the normalized friction $\mu = F(s)/F_n$, as a function of the slip rate s.

curves shown in Figure 8.5 (b). These curves match reasonably well the experimental data shown in Figure 8.1 (a) for several coefficients of road adhesion.

We also note that the steady-state representation of Equations (8.39) or (8.40) can be used to identify most of the model parameters by fitting this model to experimental data. These parameters can also be used in a simple one-dimensional lumped model, which can be shown to suitably approximate the (average) solution of the partial differential equation (8.21) and (8.22). This approximation is discussed next.

Average Lumped Model

It is clear that the distributed model captures reality better than the lumped, point-contact model. It is also clear that in order to use the distributed model for control purposes, it is necessary to choose a discrete number of states to describe the dynamics for each tire. This has the disadvantage that a possibly large number of states is required to describe the friction generated at each tire. Alternatively, one could define a *mean friction state* $\bar{z}$ for each tire and then derive an *ordinary differential equation* for $\bar{z}$. This will simplify the analysis and can also lead to much simpler control design synthesis procedures for tire friction problems.

To this end, let us define

$$\bar{z}(t) \equiv \frac{1}{F_n} \int_0^L z(\zeta, t) f_n(\zeta) d\zeta \qquad (8.41)$$

where F_n is the total normal force, given by

$$F_n = \int_0^L f_n(\zeta)\,d\zeta$$

Thus,

$$\dot{\bar{z}}(t) = \frac{1}{F_n}\int_0^L \frac{\partial z}{\partial t}(\zeta,t) f_n(\zeta)\,d\zeta \tag{8.42}$$

Using (8.24), we get

$$\begin{aligned}\dot{\bar{z}}(t) &= \frac{1}{F_n}\int_0^L \left(v_r - \frac{\sigma_0|v_r|}{g}z(\zeta,t) - \frac{\partial z(\zeta,t)}{\partial \zeta}|\omega r|\right) f_n(\zeta)\,d\zeta \\ &= v_r - \frac{\sigma_0|v_r|}{g}\bar{z}(t) - \frac{|\omega r|}{F_n}\int_0^L \frac{\partial z(\zeta,t)}{\partial \zeta} f_n(\zeta)\,d\zeta \\ &= v_r - \frac{\sigma_0|v_r|}{g}\bar{z}(t) - \frac{|\omega r|}{F_n}\Big[z(\zeta,t)f_n(\zeta)\Big]_0^L + \frac{|\omega r|}{F_n}\int_0^L z(\zeta,t)\frac{\partial f_n(\zeta)}{\partial \zeta}\,d\zeta\end{aligned}$$

The term in the square brackets describes the influence of the boundary conditions, whereas the integral term accounts for the particular form of the force distribution.

From (8.23), the friction force is

$$\begin{aligned}F(t) &= \int_0^L \left(\sigma_0 z(\zeta,t) + \sigma_1 \frac{\partial z}{\partial t}(\zeta,t) + \sigma_2 v_r\right) f_n(\zeta)\,d\zeta \\ &= (\sigma_0 \bar{z}(t) + \sigma_1 \dot{\bar{z}}(t) + \sigma_2 v_r)\, F_n\end{aligned}$$

As a general goal, one wishes to introduce normal force distributions, which leads to the following form for the lumped LuGre model:

$$\dot{\bar{z}}(t) = v_r - \frac{\sigma_0|v_r|}{g}\bar{z}(t) - \kappa(\cdot)|\omega r|\bar{z}(t) \tag{8.43}$$
$$F(t) = (\sigma_0 \bar{z}(t) + \sigma_1 \dot{\bar{z}}(t) + \sigma_2 v_r)\, F_n \tag{8.44}$$

where $\kappa(\cdot)$ is defined as:

$$\kappa(\cdot)\bar{z} = \frac{1}{F_n}\left\{\Big[z(\zeta,t)f_n(\zeta)\Big]_0^L - \int_0^L z(\zeta,t)\frac{\partial f_n(\zeta)}{\partial \zeta}\,d\zeta\right\} \tag{8.45}$$

and F_n as above. When comparing this model with the point contact LuGre model (8.14)–(8.15), it is clear that κ captures the distributed nature of the former model. It is also expected that $\kappa > 0$, so that the map $v_r(t) \mapsto F(t)$ preserves the passivity properties of the point-contact LuGre model [13], [4].

Influence of the Force Distribution on $\kappa(\cdot)$

Depending on the postulated normal force distribution density function, several expressions for the average lumped model can be developed. For instance, κ may be a constant, an explicit or an implicit function of the mean friction state $\bar{z}$. We study some of these forms next.

Exponentially decreasing distribution. Assuming (8.32) along with $z(0,t) = 0$ one obtains

$$\begin{aligned}\kappa(\cdot)\bar{z} &= \frac{1}{F_n}\left[z(\zeta,t)e^{-\lambda(\zeta/L)}f_{n0}\right]_0^L + \frac{1}{F_n}\int_0^L z(\zeta,t)\frac{\lambda}{L}e^{-\lambda(\zeta/L)}f_{n0}\,d\zeta \\ &= \frac{1}{F_n}z(L,t)e^{-\lambda}f_{n0} + \frac{\lambda}{L}\bar{z}(t) \end{aligned} \quad (8.46)$$

Next, recall that we require $\lambda \geq 0$. For large values of λ, it is possible to ignore the term containing $z(L,t)$ in the equation above, and approximate $\kappa(\cdot)$ by a constant

$$\kappa = \frac{\lambda}{L}, \quad \text{with} \quad 0 \leq \lambda \quad (8.47)$$

Uniform normal distribution. The case of the uniform normal distribution can be viewed as a special case of (8.32) with $\lambda = 0$. In this case, $f_n(\zeta) = f_{n0} = F_n/L$ and we obtain the following expression

$$\kappa(\cdot)\bar{z} = \frac{1}{F_n}z(L,t)f_{n0} = \frac{1}{L}z(L,t) \quad (8.48)$$

Deur [22] proposed that the boundary condition for the last element $z(L,t)$ be approximated by a linear expression of the average deflection $\bar{z}$

$$z(L,t) \approx \kappa_0(\cdot)\bar{z} \quad (8.49)$$

resulting in the relation

$$\kappa(\cdot) = \frac{\kappa_0(\cdot)}{L} \quad (8.50)$$

The function $\kappa_0(\cdot)$ in (8.49) is chosen in [22] so that the steady state solutions of the total friction force for the average-lumped model in (8.43)-(8.44), and the one of the distributed model (8.30) are the same. This approximation results in the following expression for κ_0

$$\kappa_0 = \kappa_0(Z) = \frac{1 - e^{-L/Z}}{1 - \frac{Z}{L}(1 - e^{-L/Z})} \quad (8.51)$$

In [22], it is also shown that, κ_0 belongs to the range $1 \leq \kappa_0 \leq 2$. Often, a constant value for $\kappa_0 \in [1,2]$ can be chosen without significantly changing the steady states of the distributed and lumped models.

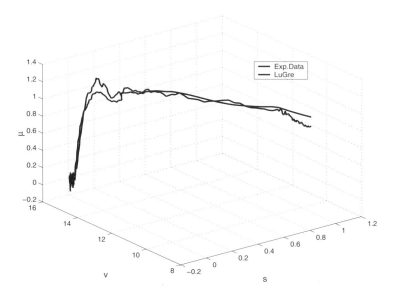

Figure 8.6 3D Plots of the corresponding (μ,s,v) curves for the collected data and the estimated predicted steady-state LuGre average lumped model, with $\alpha = 2$

Measurements of the longitudinal slip s, friction coefficient μ, and linear velocity v collected during brakings of a specially equipped test vehicle, have been used to identify the parameters of the average-lumped LuGre tire friction model [15]. These parameters can be used as a basis to validate the dynamic friction model by comparing the time histories of the friction force predicted by our model with the friction force from the three experiments.

The 3D (μ, v, s) steady-state solution of the distributed dynamical LuGre model at the mean velocity, is compared to the friction coefficient μ given by the experiments, as shown in Figure 8.6. This allows, via an optimization procedure, to identify the parameters $(\sigma_0, \sigma_2, \mu_s, \mu_c, \text{and } v_s)$. By comparing the time histories of the friction force given by our the model with the ones given by the experiments, we can determine the rest of the parameters (e.g., σ_1). An example of such comparison is shown in Figure 8.7, for time-varying κ.

8.3 Higher-dimensional Models

In this section, we recall the definitions of slip that are commonly used in tire/road friction modeling. We use $V = [V_x, V_y]$ to denote the translational longitudinal velocity at O_p, the extremity of the contact patch (see Figure 8.8) in the wheel plane XO_pY. The angular velocity of the wheel is denoted by ω, V_c as the equivalent wheel translational velocity at the point O_p, and $V_c = r\omega$ where r is the free radius of the tire. When ω is positive (the vehicle is moving forward), the wheel equivalent longitudinal velocity V_c

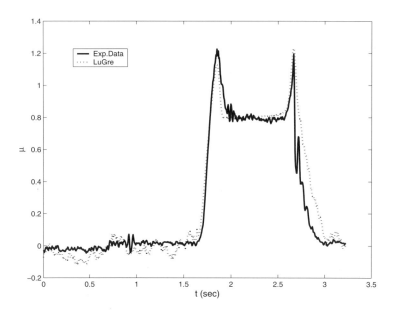

Figure 8.7 Case (ii): varying κ

is positive in the coordinate system R_0, where R_0 is defined as a moving frame with velocity V_x with the origin O_p in the extremity of the contact patch.

We define the slip velocity of the point O_p as $V_s = [V_{sx}, V_{sy}]$ in the wheel plane XO_pY. The slip angle is denoted by α. The slip ratios used to parameterize the friction model are defined as S_s and S_α, for longitudinal and lateral directions, respectively. Two conventions will be used to separate the braking and traction cases (see Fig 8.8). The pseudo-static braking curves are performed under constant velocity whereas pseudo-static traction curves are given for constant angular velocity. These conventions also prevent the slip from becoming undefined when either the wheel speed or the longitudinal speed reach zero.

1. In the braking case, longitudinal slip S_s and lateral slip S_α are given by

$$S_s = \frac{V_c - V_x}{V_x} = \frac{V_{sx}}{V_x} \tag{8.52}$$

$$S_\alpha = \frac{V_{sy}}{V_x} = |\tan \alpha| \tag{8.53}$$

In braking, $V_x - V_c < 0$, $V_x \neq 0$ then $-1 \leq S_s < 0$.

2. In the traction case, longitudinal slip S_s and lateral slip S_α are given

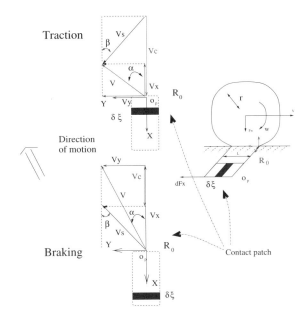

Figure 8.8 Convention for the slip definition for both braking and traction cases

by

$$S_s = \frac{V_c - V_x}{V_c} = \frac{V_{sx}}{V_c} \qquad (8.54)$$

$$S_\alpha = \frac{V_{sy}}{V_c} = (1 - S_s)|\tan \alpha| \qquad (8.55)$$

In traction, if $V_x - V_c < 0$, $\omega \neq 0$ then $1 \geq S_s > 0$.

Other conventions, like those used for the "magic formula" (see [6]), can easily be employed and do not change the final results. In this particular case, a specific definition needs to be considered when the longitudinal speed V_x or V_c tend to zero, in order to prevent a singularity in the definition of the slip.

In this section, we consider both longitudinal and lateral motions, and investigate the resulting forces and torque at the center of the contact patch. Some of these ideas can be found in [47].

A Two-dimensional Model

Let $\delta\xi$ represent a small slice of the deformed belt crossing the contact patch at position ξ in coordinate R_0 (see Figure 8.8). The slice $\delta\xi$ is moving at the speed $V_{\delta\xi} = [V_x, V_y + \xi\dot\varphi]$ with $\dot\varphi$ the yaw speed of the rim [47]. We can model the dry friction occurring in each slice using the LuGre dynamic friction model. The contact between the two surfaces can be represented by microscopic bristle deflections with the coordinates $z(\xi,t) = [z_x(\xi,t), z_y(\xi,t)]$

and the relative velocity of each slice at ξ with respect to O_p is given as $V_r(\xi,t) = [V_{rx}(\xi,t), V_{ry}(\xi,t)] = [V_{sx}(t), V_{sy}(t) + \xi\dot{\varphi}]$ (the direction of the total force is opposite to the slip vector). However, in this paper we consider the tire belt as a rigid body. The model might be extended in order to include dynamic properties of the rubber belt. In this case, the longitudinal velocity of each slide $V_{\delta\xi}$ of the contact patch would have a more complex expression that might include camber angle dependencies or other factors. These notions have already been explained in the literature [47], [26].

For the rigid tire belt model, the extended two-dimensional distributed tire/road friction model is given by

$$\dot{z}_x(\xi,t) = V_{rx} - \frac{\sigma_{0_x}}{g_x(V_{rx})} z_x(\xi,t)|V_{rx}| \tag{8.56a}$$

$$\dot{z}_y(\xi,t) = V_{ry} - \frac{\sigma_{0_y}}{g_y(V_{ry})} z_y(\xi,t)|V_{ry}| \tag{8.56b}$$

and the friction forces

$$\delta F_x = \{\sigma_{0_x} z_x(\xi,t) + \sigma_{1_x}\dot{z}_x(\xi,t) + \sigma_{2_x} V_{rx}\} \delta F_n \tag{8.57a}$$

$$\delta F_y = \{\sigma_{0_y} z_y(\xi,t) + \sigma_{1_y}\dot{z}_y(\xi,t) + \sigma_{2_y} V_{ry}\} \delta F_n \tag{8.57b}$$

with σ_{j_i} as the dynamic coefficients of the LuGre friction model for lateral and longitudinal direction ($i = x, y$; $j = 1, 2, 3$) known as the normalize rubber stiffness (σ_{0_i}), the normalized rubber damping (σ_{1_i}), and the normalized viscous relative damping (σ_{2_i}). The normal load δF_n is considered uniformly distributed over the patch along the ξ direction, thus $\delta F_n = F_n/L$, and

$$g_x(V_{rx}) = \mu_{c_x} + (\mu_{s_x} - \mu_{c_x})e^{-\left|\frac{V_{rx}}{v_{s_x}}\right|^{1/2}} \tag{8.58a}$$

$$g_y(V_{ry}) = \mu_{c_y} + (\mu_{s_y} - \mu_{c_y})e^{-\left|\frac{V_{ry}}{v_{s_y}}\right|^{1/2}} \tag{8.58b}$$

are two functions that characterize the steady state properties of the friction, where $\mu_{c_i}, \mu_{s_i}, v_{s_i}$ are, respectively, the Coulomb friction coefficient, the normalized static friction and the Stribeck relative velocity. The two LuGre models for longitudinal and lateral motions are using different parameters since the friction properties of the contact tire/road are different in longitudinal and lateral directions. The fact that the tire has non-isotropic properties is well known in this area and has already been introduced in most of the current models (see [6], [9]).

The system given in the Equations (8.56) and (8.57) is both time- and space-dependent, as a consequence, the derivative of $z_i(\xi,t)$, $i = x, y$, for longitudinal and lateral directions is a full derivative given by

$$\dot{z}_i(\xi,t) = \frac{\partial z_i(\xi,t)}{\partial \xi}\frac{\partial \xi}{\partial t} + \frac{\partial z_i(\xi,t)}{\partial t} \tag{8.59}$$

Self-aligning Torque

The self-aligning torque is an important part of the tire model because the reaction force applied to the vehicle is strongly dependent upon it (steering

Friction Tire/Road Modeling,...and Optimal Braking Control

wheel feedback force). The self-aligning torque consists of two important effects, the yaw motion of the tire that creates a friction torque $\delta M_{z_1}(\xi,t)$ and the transfer of the forces in the center of wheel frame where forces apply. Both the two effects generate the torque at the center of the patch, known as the self-aligning torque.

A self-aligning torque results from a non-symmetry of the contact patch deformation $z(\xi,t)$, or forces $\delta F_y(\xi,t)$, over the contact patch length L. The equivalent forces and torque produced by a slice $\delta\xi$ at position ξ with respect to O_p is given by $[\delta F_x, \delta F_y, \delta M_z = \xi \delta F_y]$ and the total tire/road interaction in the patch center is now expressed by two forces F_x, F_y and the self-aligning torque M_z:

$$F_x = \int_0^L \delta F_x(\xi,t)d\xi \tag{8.60a}$$

$$F_y = \int_0^L \delta F_y(\xi,t)d\xi \tag{8.60b}$$

$$M_z = \int_0^L \delta M_{z_2}(\xi,t)d\xi = \int_0^L (\frac{L}{2}-\xi)F_y(\xi,t)d\xi \tag{8.60c}$$

Stationary Properties

The stationary characteristics of the tire are widely used in this research area. To produce these characteristics, a complex experiment set-up is usually utilized. These conditions are hardly obtainable on a real vehicle since the maneuvers would then be very severe for the passengers. Each point on the stationary curve is given for a constant slip and a constant wheel velocity or wheel angular velocity, therefore V_c and V remain constant and a defined slip angle α is obtained. The yaw motion of the rim is also not considered i.e., $\dot\varphi = 0$. Therefore, during the stationary conditions, ξ and t are no longer independent because of constant velocity, we have $z_i(\xi,t) = z_i(t)$ if we like to find a time-varying solution, or $z_i(\xi,t) = z_i(\xi)$ if we look for the spatial solution. We choose to calculate the spatial solution in the frame R_0 defined previously. Notice that if $\dot\xi = V_c$ is constant during stationary conditions, we have

$$\frac{d}{dt}z_i(\xi,t) = \frac{dz_i(\xi)}{d\xi}\frac{d\xi}{dt} = \frac{dz_i(\xi)}{d\xi}V_c \tag{8.61}$$

with spatial coordinates. As a consequence, the stationary bristle model using spatial coordinates becomes similar to Equation 8.25.

Remark. $z_i(\xi)$ does not represent the behaviour of a slide z_i at a given time instant t and a given position ξ, both t and ξ are varying since $\dot\xi = V_c$. The stationary solution gives the history of one slide $z_i(\xi,t)$ during the interval $\Delta t = \Delta\xi/V_c$.

Integrating the forces and torque along the contact patch for a constant load distribution, we obtain three components for the stationary tire model

F_x, F_y, M_z for traction case. As discussed in the previous remark, these expressions do not represent the instantaneous forces and torques at the contact interval (like those given in Equations (8.60)), but the average force produce by one slide during the time interval $\Delta t = \frac{L}{V_c}$ necessary to cross the contact patch L.

$$F_x = \left(\text{sgn}(V_{sx})g_x(V_{sx})\left[1 - \frac{Z_x}{L}(1 - e^{-L/Z_x})\right] + \sigma_{2_x} V_{sx}\right) F_n \quad (8.62a)$$

$$F_y = \left(\text{sgn}(V_{sy})g_y(V_{sy})\left[1 - \frac{Z_y}{L}(1 - e^{-L/Z_y})\right] + \sigma_{2_y} V_{sy}\right) F_n \quad (8.62b)$$

$$M_z = \text{sgn}(V_{sy})g_i(V_{sy})\left(\frac{L}{2} + Z_y e^{-\frac{L}{Z_y}} + \frac{1}{L}Z_y^2\left(e^{-\frac{L}{Z_y}} - 1\right)\right) F_n \quad (8.62c)$$

with

$$Z_i = \left|\frac{\omega r}{V_{si}}\right| \frac{g_i(V_{si})}{\sigma_{0_i}} \quad (8.63)$$

σ_{i_j} are the dynamic coefficients of the LuGre friction model for lateral and longitudinal direction ($i = x, y$; $j = 1, 2, 3$). Parameter identification, and comparison with "magic formula" will be presented later. For the braking case, we can find the similar formula for F_x, F_y, and M_z.

Lumped Dynamic Tire/Road Friction Model

Distributed model are difficult to use for estimation and control purposes. Thus, we will now develop a simplified lumped parametric representation. In this paper, we obtained the lumped model by defining lumped variables $\bar{z}_i$ as follows:

$$\bar{z}_i(t) = \frac{1}{F_n}\int_0^L z_i(\xi, t) f_n(\xi) d\xi, \quad i = x, y \quad (8.64)$$

where L is defined as an "elementary surface length", which could be a *tread block element* or the *full contact patch length* between the tire and the road. Let us define κ_1^i ($i = [x, y]$) and κ_2 as

$$\kappa_1^i(\cdot)\bar{z}_i = \frac{1}{F_n}\int_0^L \frac{\partial z_i(\xi, t)}{\partial \xi} f_n(\xi) d\xi \quad (8.65)$$

$$= \frac{1}{F_n}\left\{[z_i(\xi, t) f_n(\xi)]_0^L - \frac{1}{F_n}\int_0^L z_i(\xi, t)\frac{\partial f_n(\xi)}{\partial \xi} d\xi\right\}$$

and

$$\kappa_2(\cdot)\bar{z}_y = \frac{1}{F_n}\int_0^L \xi\frac{\partial z_y(\xi, t)}{\partial \xi} f_n(\xi) d\xi \quad (8.66)$$

$$= \frac{1}{F_n}\left\{[\xi z_i(\xi, t) f_n(\xi)]_0^L - \int_0^L f_n(\xi) z_y(\xi, t) d\xi - \int_0^L \xi\frac{\partial f_n(\xi)}{\partial \xi} z_y(\xi, t) d\xi\right\}$$

Neglecting the yaw motion of the rim, *i.e.*, $\dot{\varphi} = 0$, the distributed friction model changes into ($i = x, y$)

- internal states

$$\dot{\bar{z}}_i = -V_{si} - \frac{\sigma_{0_i}}{g(V_{si})}\bar{z}_i|V_{si}| - \kappa_1^i(\cdot)|wr|\bar{z}_i \qquad (8.67)$$

with $\bar{z}_i(0) = 0$.
- lumped forces

$$F_i = \{\sigma_{0_i}\bar{z}_i + \sigma_{1_i}\dot{\bar{z}}_i + \sigma_{2_i}V_{si}\}F_n \qquad (8.68)$$

- lumped self-aligning torque

$$\dot{M}_z = V_c F_y + \left(\frac{\dot{\gamma}_y(V_{sy})}{\gamma_y(V_{sy})} - \frac{|r\omega|}{Z_y}\right)M_z - \sigma_{0_y}\gamma_y(V_{s_y})|rw|\kappa_2(\cdot)\bar{z}_y F_n \qquad (8.69)$$

The function γ_i defined as $\gamma_i(V_{s_i}) = 1 - \frac{\sigma_{1_i}|V_{s_i}|}{g_i(V_{s_i})}$, for $i = x, y$ (see [18]).

The calculation of κ_1^i and κ_2 is similar to the one-dimensional case.

Influence of External Factors

Friction characteristics can change dramatically with external conditions like humidity or different type of road, *etc*. In this subsection, we provide some insight into the way the friction model is affected by typical perturbations. We first discuss some variations that can be parameterized using the actual model parameters (elementary effects), then we discuss external perturbation that may affect the accuracy of the used model (*i.e.*, road conditions, temperature variation, *etc.*).

Elementary Effects

- dry friction characteristic is represented by the function g_i of Equations (8.58) and several dynamic coefficients σ_i. In the following, we will consider that a change in friction will affect primarily the static characteristics of the model (*i.e.*, the Coulomb friction and Stribeck effect) in a uniform manner. Hence, we may use a parameter θ_1 to scale the function g_i.;

- normal load variations can easily be introduced in the model using a second parameter θ_2 to scale F_N in Equations (8.57). Finally, as reported in [13], normal load may affect the static parameters of the LuGre model. This can be captured by changing the parameter θ_1;

- variations of the contact patch length could be introduced using a third coefficient θ_3 that scales the dynamic parameter σ_{0_i}. As shown in Equations (8.62), a variation of L can be captured by such a parameter.

And we propose the following structure for the tire friction model:

$$\begin{aligned}
\dot{\bar{z}}_i &= -V_{si} - \theta_1 \frac{\sigma_{0_i}\theta_3}{g(V_{si})}\bar{z}_i|V_{si}| - \kappa_1^i(\cdot)|r\omega|\bar{z}_i \\
\dot{M}_z &= V_c F_y + \left(\frac{\dot{\gamma}_y(V_{sy})}{\gamma_y(V_{sy})} - \frac{\theta_3|r\omega|}{Z_y}\right) M_z - \sigma_{0_y}\gamma_y(V_{s_y})|r\omega|\kappa_2(\cdot)\bar{z}_y F_n \\
F_i &= \theta_2 \left\{\theta_3 \sigma_{0_i}\bar{z}_i + \sigma_{1_i}\dot{\bar{z}}_i + \sigma_{2_i}V_{ri}\right\} F_n
\end{aligned} \qquad (8.70)$$

with $\bar{z}_i(0) = 0$ and $i = x, y$.

External Perturbations

1. different road surfaces result in variations of the dry friction characteristic of the model. This effect could easily be introduced using the parameter θ_1 described in the previous section;

2. contamination of the interface has a tremendous effect on the tire/road contact behavior. The depth of water on the road greatly influences tire/road friction. These phenomena are complex and related to the different aspects of the contact patch between the tire and the road. Previous analysis made use of different water level categories such as thin or thick film of water on the road [8], [44], [45]. Even if the distinction between these two situations sounds artificial, it helps in understanding the contamination of the tire/road contact patch by water. In the limiting situation of a thin water layer, the contact between tire and road is completely lost due to full contamination of the interface[5]. This problem is called *viscous hydroplaning* [42], [37]. On the other hand, for a thick water layer, an extra force is generated in front of the tire due to the accumulation of water (hydrodynamic forces). In this situation, the contact between the two surfaces could be lost without necessarily full contamination of the interface but mostly because of the hydrodynamic force. This phenomenon is named *dynamic hydroplaning*. The last case is more general than viscous hydroplaning due to the fact that other hydrodynamical effects take place during the contact process.

Consequently, we can summarize these remarks as follows: for every set of experimental conditions in wet weather, there exists a definite velocity called the *hydroplaning limit*, beyond which hydroplaning occurs (see [37]). At this velocity, the wheel is, in general, supported by (see Figure 8.9):

- A hydrodynamic upward thrust, just ahead of the contact area of the tire. The magnitude of this force is determined by the water layer depth h_∞ (for thick films) and vehicle velocity v. An expression for the upward thrust, F_H, can be found in [37]. As the hydrodynamic upward thrust results in a decrease of the

---

[5]When water has invaded the complete contact area between the tire and the road

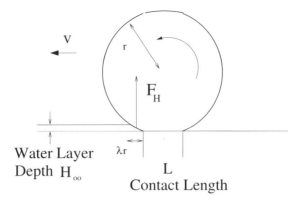

Figure 8.9 Schematic representation of tire under wet road

effective normal load, we propose in this paper to include this effect using the parameter θ_2 scaling the load F_N;

– A squeezed film and wiping effect in the contact area. The squeezed film effect is due to the inability of the tread elements traversing the contact area to remove the last few tenths of a millimeter of water in the available time. The wiping effect is due to the longitudinal and lateral relative motion of the tire elements in the contact area.

Remark. F_H approaches the load of the wheel at high velocities (over 200 km/h, which is not realistic) depending of the amount of water h_∞ on the road. Thus, as stated before, the hydrodynamic thrust F_H is not the only factor contributing to hydroplaning. This discrepancy is due to variations in the contact area due to a thin film of water squeezed at the tire/road interface: the thin water layer supports a fraction of the load of the wheel, and hence the effective contact area and the effective normal load decreases. The inclusion of all of these effects in the tire model is possible if we can develop dynamic expressions for the contact patch and the effective load of the tire, based on physic laws. Determining the contact patch is a difficult problem, and several empirical or semi-empirical approaches have been proposed in the literature [28], [42]. In these references, the road/tire interface (contact area) is divided into three areas where the major effects are different: dry contact, micro-water layer region, and macro-water layer region. A simple model can then be obtained to scale the effective contact area, given certain drainage characteristics of the ground and the tire. A more sophisticated interpretation of the contact patch dynamics based on fluid mechanics has also been proposed in [38], [37], [19]. These discontinuous interpretations are not satisfactory for simulation or control purposes, because they require a significant amount of characterizations and experimental measurements.

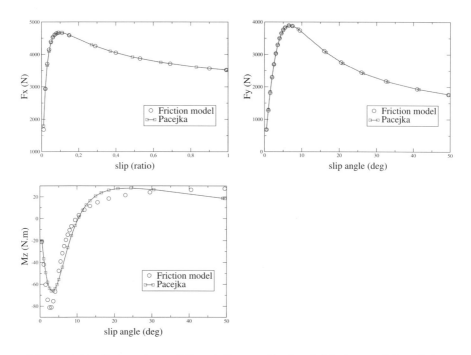

Figure 8.10 Comparison of the stationary tire/road friction model and the "magic formula" (constant velocity during braking $v = 15\,\text{m/s}$)

To simplify the problem, variations of the contact patch length could be included using a third parameter θ_3 scaling σ_{0_i} as mentioned previously;

3. pressure and temperature variations can also be decomposed in terms of the elementary effects given in Section 8.3. Hence, for example temperature variation changes the dry friction characteristic, therefore θ_1 and pressure variations affect the contact patch length, which therefore affects θ_3. However, these two perturbations may be more complex to include in the model since they may modify more than one parameter. Further investigation will be necessary into this problem.

A simple theory is discussed to include different road condition. Effects of perturbations could be investigated in terms of three varying coefficients of the friction model. This method facilitates the design of a new scheme for estimation and control using the dynamic friction model (*i.e.*, braking control, friction estimator). Since a single perturbation may have a complex effect on the friction model, a multi-parameter analysis might be necessary and experimental verification has to be done to validate this theory.

Figure 8.10 shows a comparison of the identified stationary tire/road model, with the curves obtained from the "magic formula" calibrated on dry surface conditions for a typical tire model 165-65R14, from Renault SA.

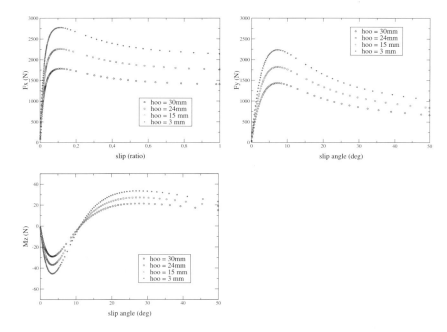

Figure 8.11 Effect of water depth h_∞ on a stationary tire model ($v = 25\,\text{m/s}$ and triangular texture of amplitude $\varepsilon = 0.06\,\text{mm}$)

The influence of different water films and various texture amplitudes are illustrated in Figures 8.11 and 8.12. Figure 8.11 illustrates the hydrodynamic properties of the tire under wet road conditions. As predicted, increasing the thickness of the water on the road h_∞ for a given travel velocity v dramatically reduces the resulting friction. Furthermore, Figure 8.12 shows the effect of varying the texture amplitude ε on the friction characteristics using the relation given in [19]. In this case, the water layer h_∞ is given and the road pattern is specified as triangular. As ε is varied, the setting time is slightly affected. We observe that friction increases as texture amplitude increases.

8.4 Road/Tire Friction Observers

This section is devoted to the problem of tire-road friction estimation using only angular wheel velocity, which can always been computed from actual sensors. Information about the tire forces is relevant to problems like optimization of anti-lock brake systems (ABS), traction systems, diagnostic of the road friction conditions, *etc*.

In applications like ABS, the linear velocity cannot be computed from existing sensors. It is therefore important to design observers that also estimate the linear wheel (or car) velocity. In this section, we present a nonlinear observer that estimates on-line both the tire/road characteristics

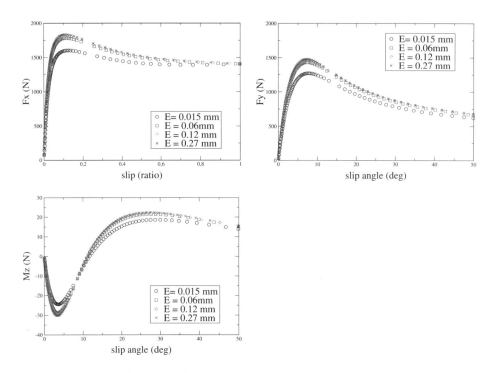

Figure 8.12 Effect of road texture amplitude E on a stationary tire model (texture: triangular motif, $v = 25 m/s$ and $h_\infty = 30\,\text{mm}$)

and the slip ratio using only rotational wheel velocity, see [12] and [11].

The Estimation Problem

We consider the one-wheel model with the lumped tire/road friction model [6], i.e.

$$m\dot{v} = F_n(\sigma_0 z + \sigma_1 \dot{z}) + F_n \sigma_2 v_r \qquad (8.71)$$

$$J\dot{\omega} = -rF_n(\sigma_0 z + \sigma_1 \dot{z}) - \sigma_\omega \omega + u_\tau \qquad (8.72)$$

$$\dot{z} = v_r - \theta \frac{\sigma_0 |v_r|}{g(v_r)} z \qquad (8.73)$$

with,

$$g(v_r) = \mu_C + (\mu_S - \mu_C)e^{-|v_r/v_s|^{1/2}}$$

where we have neglected the term σ_2 in Equation (8.72), σ_ω is the viscous rotational friction, and θ is the parameter related to changes in the road conditions, that is to be estimated. The observation problem can be now formulated as follows.

---

[6] We do not include the variation term κ in the lumped LuGre model in this section for simplicity. The same approach can be used if κ is included.

Friction Tire/Road Modeling,...and Optimal Braking Control

Problem formulation. Assume that the lumped friction parameters with $\theta = 1$, has been identified off-line, and assume that only the variable ω is measurable from some sensor. The problem is to design an on-line observer for θ that allows the controller to monitor the eventual changes in the road conditions by using only the measure of the rotational wheel velocity ω.

8.5 General Observer Design

Consider the following system:

$$\dot{x} = Ax + B\left[\theta\varphi(y,u,x)\right] + Ru + Ey \quad (8.74)$$
$$\dot{\theta} = 0 \quad (8.75)$$
$$y = C^T x \quad (8.76)$$

with $y, \theta, \varphi(y,u,x) \in R, x \in R^n$, and $u \in R^m$.
We assume that system states are bounded, and that the following hold:

A1) (A, C^T) is an observable pair.

A2) There exists a known function $\infty > \rho_0 \geq \rho(y,u) \geq 0$, and a known upper bound f_{max} such that either

(i) $|\varphi(y,u,x_1) - \varphi(y,u,x_2)| \leq \rho(y,u)\|x_1 - x_2\|, \quad \forall x_1, x_2$

(ii) $|\varphi(y,u,x)| \leq f(\|x\|) \leq f_{max}, \quad \forall |x| < \infty, \forall |y| < \infty, |u| < \infty$ holds.

A3) The map $G(s) : \psi \mapsto \tilde{y}$ of the system

$$\dot{\tilde{x}} = [A - KC^T]\tilde{x} + B\psi \quad (8.77)$$
$$\tilde{y} = C^T \tilde{x} \quad (8.78)$$

with $[A - KC^T]$ Hurwitz, is strictly passive, i.e. $\exists Q > 0$, $P = P^T > 0$, and K, such that

$$P[A - KC^T] + [A - KC^T]^T P = -Q \quad (8.79)$$
$$PB = C \quad (8.80)$$

A4) The trajectories of the system $(y(t), u(t), x(t))$ satisfy

$$\lim_{t \to \infty} \varphi(y(t), u(t), x(t)) \neq 0$$

A5) The following condition holds:

$$\lambda_{min} Q - 2\|C^T\|\theta_{max}\rho_0 = q_0 > 0$$

with $\theta_{max} \geq |\theta|$.

Under these hypotheses, we propose the following observer structure:

$$\dot{\hat{x}} = A\hat{x} + B\left[\hat{\theta}\varphi(y,u,\hat{x})\right] + Ru + Ey + K(y-\hat{y}) + Bv_1 \quad (8.81)$$
$$\dot{\hat{\theta}} = v_2 \quad (8.82)$$
$$\hat{y} = C^T\hat{x} \quad (8.83)$$

where v_1 and v_2 are design variables, which will be defined subsequently.
Introducing the error variables:

$$\tilde{x} = x - \hat{x} \quad (8.84)$$
$$\tilde{\theta} = \theta - \hat{\theta} \quad (8.85)$$
$$\tilde{y} = y - \hat{y} = C^T\tilde{x} \quad (8.86)$$

The error equation becomes

$$\dot{\tilde{x}} = [A - KC^T]\tilde{x} + B\left[\theta\varphi(y,u,x) - \hat{\theta}\varphi(y,u,\hat{x})\right] - Bv_1 \quad (8.87)$$
$$\dot{\tilde{\theta}} = -v_2 \quad (8.88)$$
$$\tilde{y} = C^T\tilde{x} \quad (8.89)$$

where

$$\theta\varphi(y,u,x) - \hat{\theta}\varphi(y,u,\hat{x}) = \tilde{\theta}\varphi(y,u,\hat{x}) + \theta\left[\varphi(y,u,x) - \varphi(y,u,\hat{x})\right]$$

Now, defining the Lyapunov function

$$V = \tilde{x}^T P\tilde{x} + \frac{1}{\gamma}\tilde{\theta}^2$$

and using properties A1 and A3 we have

$$\dot{V} = -\tilde{x}^T Q\tilde{x} + 2\tilde{\theta}\left[\tilde{y}\varphi(y,u,\hat{x}) - \gamma^{-1}v_2\right]$$
$$+ 2\tilde{y}\theta\left[\varphi(y,u,x) - \varphi(y,u,\hat{x})\right] - \tilde{y}v_1 \quad (8.90)$$

Defining the adaptation law v_2 as

$$v_2 = \gamma\varphi(y,u,\hat{x})\tilde{y} \quad (8.91)$$

we obtain

$$\dot{V} \leq -\tilde{x}^T Q\tilde{x} + 2|\tilde{y}||\theta|\,|\varphi(y,u,x) - \varphi(y,u,\hat{x})| - \tilde{y}v_1$$

If A2–(i) holds, then we have

$$\dot{V} \leq -\tilde{x}^T Q\tilde{x} + 2|\tilde{y}||\theta|\rho(y,u)||\tilde{x}|| - \tilde{y}v_1 \quad (8.92)$$
$$\leq -q||\tilde{x}||^2 + 2||C^T||\,|\theta|\rho(y,u)||\tilde{x}||^2 - \tilde{y}v_1 \quad (8.93)$$
$$\leq -||\tilde{x}||^2(\lambda_{min}Q - 2||C^T||\,|\theta|\rho_0) - \tilde{y}v_1 \quad (8.94)$$
$$\leq -q_0||\tilde{x}||^2 - \tilde{y}v_1 \quad (8.95)$$

where the last inequality follows from Assumption A4. In this case, we can simple set $v_1 = 0$ to get $\dot{V} \leq -q_0||\tilde{x}||^2$.

In the second case, when only A2–(ii) holds, we have

$$\dot{V} \leq -\tilde{x}^T Q \tilde{x} + 2|\tilde{y}||\theta|(f(\|x\|) + f(\|\hat{x}\|)) - \tilde{y}v_1$$
$$\leq -\tilde{x}^T Q \tilde{x} - |\tilde{y}|[-2\theta_{max}(f_{max} + f(\|\hat{x}\|)) + \text{sgn}(\tilde{y})v_1]$$

which suggests that v_1 should be defined to have a high-gain component, i.e.

$$v_1 = 2\theta_{max}(f_{max} + f(\|\hat{x}\|))\text{sgn}(\tilde{y}),$$

where θ_{max} is an upper bound of the parameter θ. With this choice of v_1, we have that

$$\dot{V} \leq -q\|\tilde{x}\|^2$$

Thus, in both of the cases considered by Assumption A2, $\tilde{x}$, and $\tilde{\theta}$ are bounded, and $\tilde{x} \to 0$. Finally, from the error equation (8.87), we have that

$$\lim_{t \to \infty} \{B\tilde{\theta}\varphi(y,u,x)\} = 0$$

which together with Assumption A4 leads us to conclude that

$$\lim_{t \to \infty} \hat{\theta} = \theta$$

We have thus proved the following theorem:

THEOREM 8.1
Consider the following system

$$\dot{x} = Ax + B[\theta\varphi(y,u,x)] + Ru + Ey$$
$$\dot{\theta} = 0$$
$$y = C^T x$$

under Assumptions A1) − A5), with $y, \theta, \varphi(y,u,x) \in R$, $x \in R^n$, and $u \in R^m$. Then, the following observer

$$\dot{\hat{x}} = A\hat{x} + B[\hat{\theta}\varphi(y,u,\hat{x})] + Ru + Ey + K(y - \hat{y}) + Bv_1$$
$$\dot{\hat{\theta}} = \gamma\varphi(y,u,\hat{x})\tilde{y}$$
$$\hat{y} = C^T \hat{x}$$

with

$$v_1 = \begin{cases} 0 & \text{if } A2\text{-(i) holds} \\ 2\theta_{max}(f_{max} + f(\|\hat{x}\|))\text{sgn}(\tilde{y}) & \text{if } A2\text{-(ii) holds} \end{cases}$$

ensures—under verification of Assumption A4—that

$$\lim_{t \to \infty} \hat{\theta} = \theta$$

□

Application to One-wheel Model

We consider the one-wheel model with lumped friction as described by the equations (8.71)–(8.73). We assume that only ω is measurable. To set our system in the same framework as the structure (8.74)–(8.76), we introduce the following change of coordinates:

$$\eta = rmv + J\omega \tag{8.96}$$
$$\chi = J\omega + rF_n\sigma_1 z \tag{8.97}$$

from which we get:

$$\dot{\eta} = -\frac{F_n\sigma_2}{m}\eta + (\frac{JF_n\sigma_2}{m} + r^2 F_n\sigma_2 - \sigma_\omega)\omega + u_\tau \tag{8.98}$$
$$\dot{\chi} = -\frac{\sigma_0}{\sigma_1}\chi + (J\frac{\sigma_0}{\sigma_1} - \sigma_\omega)\omega + u_\tau \tag{8.99}$$
$$\dot{z} = (r\omega - v) - \theta\sigma_0\frac{|r\omega - v|}{g(v_r)}z \tag{8.100}$$
$$y = \frac{1}{J}(\chi - rF_n\sigma_1 z) = \omega \tag{8.101}$$

Defining x, u, and y, respectively as

$$x = \begin{bmatrix} \eta \\ \chi \\ z \end{bmatrix}, \quad u = u_\tau, \quad y = \omega$$

we can rewrite the above system as

$$\dot{x} = \begin{bmatrix} -\frac{F_n\sigma_2}{m} & 0 & 0 \\ 0 & -\frac{\sigma_0}{\sigma_1} & 0 \\ -\frac{1}{rm} & 0 & 0 \end{bmatrix} x + \begin{bmatrix} 0 \\ 0 \\ -1 \end{bmatrix} \theta\varphi(y, u, x)$$
$$+ \begin{bmatrix} (\frac{JF_n\sigma_2}{m} + r^2 F_n\sigma_2 - \sigma_\omega) \\ (J\frac{\sigma_0}{\sigma_1} - \sigma_\omega) \\ r \end{bmatrix} y + \begin{bmatrix} 1 \\ 1 \\ 0 \end{bmatrix} u$$

where $\varphi(y, u, x)$, is defined as

$$\varphi(y, x) = \frac{\sigma_0|ry - v|}{g(ry - v)}z \quad v = (\eta - J\omega)/rm$$

With this representation, we shall now verify conditions under which Assumptions A1—A3 hold. Assumption A4 depends on the operational conditions, in particular, on the applied torque u_τ, and Assumption A5 depends on the level of friction involved.

Assumption A1 (linear observability). With A and C defined as above:

$$A = \begin{bmatrix} -\frac{F_n \sigma_2}{m} & 0 & 0 \\ 0 & -\frac{\sigma_0}{\sigma_1} & 0 \\ -\frac{1}{rm} & 0 & 0 \end{bmatrix}, \quad C = \begin{bmatrix} 0 \\ \frac{1}{J} \\ \frac{-rF_n \sigma_1}{J} \end{bmatrix}$$

Then

$$\text{rank } O = \text{rank} \begin{bmatrix} 0 & \frac{F_n \sigma_1}{m} & \frac{F_n^2 \sigma_1 \sigma_2}{m^2} \\ 1 & -\frac{\sigma_0}{\sigma_1} & (\frac{\sigma_0}{\sigma_1})^2 \\ -rF_n \sigma_1 & 0 & 0 \end{bmatrix} = 3, \text{ if and only if } \frac{F_n \sigma_2}{m} \neq \frac{\sigma_0}{\sigma_1}$$

with $O = [C, A^T C, (AA)^T C]$. Hence, Assumption A1 holds. This rank condition clearly shows that the existence of a non-zero normal force F_n is necessary to design the friction observer.

Assumption A2-ii (bounds). With

$$\varphi(y, x) = \frac{\sigma_0 |ry - v|}{g(ry - v)} z$$

we have that

$$|\varphi(y, x)| \leq \frac{\sigma_0 |ry - v|}{g(ry - v)} |z| \leq \frac{\sigma_0}{\mu_C} |ry - v||z|$$

$$\leq \frac{\sigma_0}{\mu_C} (|ry| + |v|)|z|$$

$$\leq \frac{\sigma_0}{\mu_C} ((r + \frac{J}{rm}) y_{sup} + \frac{|\eta|}{rm}) |z|$$

$$\leq \frac{\sigma_0}{\mu_C} ((r + \frac{J}{rm}) y_{sup} + \frac{|\eta|}{rm}) |z| = f(\|x\|)$$

and that

$$|\varphi(y, \hat{x})| \leq \frac{\sigma_0}{\mu_C} ((r + \frac{J}{rm}) y_{sup} + \frac{|\hat{\eta}|}{rm}) |\hat{z}| = f(\|\hat{x}\|)$$

Assumption A3 (passivity). Finding a vector K, so that the map $\psi \mapsto \tilde{y}$ of the system description (8.77)–(8.78) is strictly passive, is equivalent to searching for a vector $K = [k_1, k_2, k_3]^T$ such that the I/O-map $G(s)$, defined as

$$G(s) = C^T [Is - A + KC^T]^{-1} B \tag{8.102}$$

is strictly positive real (SPR), i.e., $\text{Re}\{G(jZ)\} > 0, \forall Z \in [0, \infty]$. Computation of $G(s)$ with the corresponding values for A, B, C, and with $k_2 = 0$, gives the map

$$G(s) = k_G \frac{(s + \beta)(s + \rho)}{(s^2 + \alpha_1 s + \alpha_2)(s + \rho)} \tag{8.103}$$

with

$$k_G = \frac{rF_n\sigma_1}{J}$$
$$\rho = \frac{\sigma_0}{\sigma_1}$$
$$\beta = \frac{F_n\sigma_2}{m}$$
$$\alpha_1 = \beta - k_3 r F_n \sigma_1$$
$$\alpha_2 = -k_3 \beta r F_n \sigma_1 + k_1 \frac{F_n \sigma_1}{m}$$

Since ρ is positive, a zero/pole cancellation can be done in the map $G(s)$. A sufficient condition for this function be SPR is thus that $\alpha_1 > \beta$, and that all the coefficients $\beta, \alpha_1, \alpha_2$ be positive. This condition can be achieved if

$$k_1 > 0$$
$$k_3 < 0$$

From the Kalman-Yakubovich-Popov lemma, we thus ensure with this choice of K that there exist a $P > 0$ satisfying the Lyapunov equation with $PB = C$.

Assumption A4 (persistence of excitation). To ensure parameter convergence, we need to guarantee that

$$\lim_{t\to\infty} \varphi(y(t), u(t), x(t)) = \lim_{t\to\infty} \frac{\sigma_0 |ry(t) - v(t)|}{g(ry - v)} z(t) \neq 0$$

This implies that the relative velocity should not tend to zero in order for the estimated parameter to converge. This in turn implies that the internal friction state $z(t)$ will not asymptotically converge to zero.

Assumption A5 is not required for this application. Summarizing, we have,

THEOREM 8.2
Consider the one-wheel model with lumped dynamic friction (8.99)–(8.101), then the following observer:

$$\dot{\hat{\eta}} = -\frac{F_n\sigma_2}{m}\hat{\eta} + (\frac{F_n\sigma_2}{m} + rF_n\sigma_2 - \sigma_\omega)\omega + u_\tau + k_1(\omega - \hat{y})$$
$$\dot{\hat{\chi}} = -\frac{\sigma_0}{\sigma_1}\hat{\chi} + (J\frac{\sigma_0}{\sigma_1} - \sigma_\omega)\omega + u_\tau$$
$$\dot{\hat{z}} = \hat{v}_r - \hat{\theta}\frac{\sigma_0|\hat{v}_r|}{g(\hat{v}_r)}\hat{z} - k_3(\omega - \hat{y})$$
$$\quad - 2\theta_{max}(f_{max} + f(||\hat{x}||))\text{sgn}(\tilde{y})$$
$$\dot{\hat{\theta}} = \gamma\frac{\sigma_0|\hat{v}_r|}{g(\hat{v}_r)}\hat{z}(\omega - \hat{y})$$
$$\hat{y} = \frac{1}{J}(\hat{\chi} - rF_n\sigma_1\hat{z})$$

Friction Tire/Road Modeling,...and Optimal Braking Control

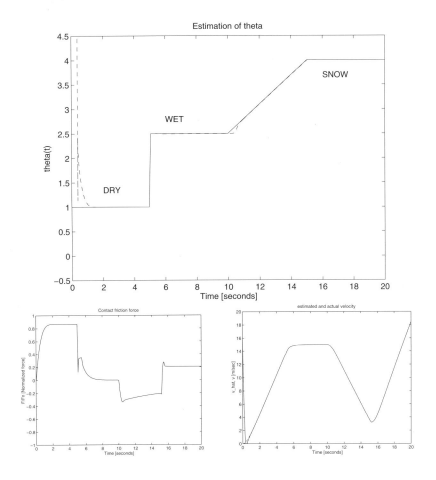

Figure 8.13 (a) Estimated parameter $\hat{\theta}(t)$, and evolution of θ (upper figure); (b) contact torque friction (normalized) $F(t)/F_n$ (low left); (c) actual and estimated wheel linear velocity $v(t)$ (low right)

where $\hat{v}_r = r\omega - \hat{v}$ with positive non-zero k_1, k_3 and γ ensures that all the estimated states are bounded, and that:

$$\lim_{t\to\infty} \hat{\eta} = \eta$$
$$\lim_{t\to\infty} \hat{\chi} = \chi$$
$$\lim_{t\to\infty} \hat{z} = z$$

If, in addition, the relative contact velocity does not vanish, then we also have that

$$\lim_{t\to\infty} \hat{\theta} = \theta$$

□

Simulations have been performed with the one-wheel system and the lumped LuGre model. The simulation has first an acceleration phase, and then a breaking phase. Figure 8.13 (b) shows the time-profile of the contact friction force. The estimation of linear velocity v is shown in Figure 8.13 (c).

Several simulations were done under variations of the parameter θ, which represent the road variation conditions. Figure 8.13 (a) shows in bold lines the value of θ, which evolves among four different conditions: the first quarter of the simulation corresponds to dry asphalt conditions. The second quarter corresponds to a sudden change from dry to wet. During the third quarter, there is a smooth variation from wet to snow. The last quarter is kept constant at the snow conditions. In the dotted lines, we can see the evolution of the estimate $\hat{\theta}(t)$. As we can observe, a good parameter tracking is obtained as long as the relative contact velocity is different from zero.

8.6 Optimal Braking

In this section, we investigate optimal braking strategies for a simplified model of a wheeled vehicle. We derive the optimality conditions for the minimum braking distance problem by solving the associated optimal control problem. For the case of a static tire friction model, it is shown that the optimal control is singular and can be written in a state feedback form. In the most general case, the optimal solution involves both subarcs of singular and bang-bang controls. In accordance with classical results, the bang-bang subarcs are included in order to satisfy the initial and final boundary conditions. In addition, we also provide a formal proof of the optimality of the "maximum friction" strategy used in standard ABS control system design. Thus, under the hypothesis that the friction force is an algebraic (memoryless) function of the slip coefficient, it is shown that the optimal singular control is the one that keeps the slip rate constant at the maximum friction value. Such static friction models may not be realistic, however. Dynamic friction models are more accurate in representing the actual friction generation mechanism. We thus also provide the corresponding optimality conditions for the case of the LuGre tire dynamic friction model.

State Equations

Consider Equations (8.71)–(8.72) and let $\rho = mr^2/J$, $u = ru_r/J$, and $f(s) = F_r(v_r, v)/m$, where F_r is the friction force, assumed to be a static map of the slip coefficient s. Here, $v_r = \omega r - v$ is the relative velocity. For braking, recall that $s = r\omega/v - 1 = v_r/v$. Let now the new variables

$$z_1 = v_r + (1+\rho)v, \qquad z_2 = v \tag{8.104}$$

Then, the state equations (8.71)–(8.72) can be written in the form

$$\dot{z}_1 = u \tag{8.105a}$$
$$\dot{z}_2 = f(s) \tag{8.105b}$$

where $s = (z_1 - (1+\rho)z_2)/z_2 = z_1/z_2 - (1+\rho)$.

Given Equations (8.105), the objective is to minimize the performance index

$$J = \int_0^{t_f} z_2(\tau)\,d\tau \qquad (8.106)$$

subject to the following initial and final conditions:

$$z_1(0) = z_{10}, \qquad z_2(0) = z_{20}, \qquad z_1(t_f) = z_2(t_f) = 0 \qquad (8.107)$$

and the final time t_f is free. To solve the previous optimization problem we introduce the associated Hamiltonian

$$\mathcal{H} = z_2 + \lambda_1 u + \lambda_2 f(z) \qquad (8.108)$$

The adjoint system is then given by

$$\dot{\lambda}_1 = -\lambda_2 \frac{\partial f}{\partial z_1} \qquad (8.109a)$$

$$\dot{\lambda}_2 = -\lambda_2 \frac{\partial f}{\partial z_2} \qquad (8.109b)$$

Since the final time is not specified, the transversality condition gives

$$\mathcal{H}(t_f) = 0 \qquad (8.110)$$

Along with the fact that the Hamiltonian is not an explicitly function of time, the last equation implies that $\mathcal{H}(t) = 0$ for all $t \in [0, t_f]$ along the optimal trajectory. From Pontryagin's Maximum Principle, the optimal control is given by

$$u_{\text{opt}} = \arg\min \mathcal{H}(z, \lambda, u) \qquad (8.111)$$

It is assumed here that the control input u is bounded, that is, any allowable control must satisfy the constraint

$$u_{\min} \leq u \leq 0 \qquad (8.112)$$

From (8.108), the switching function is

$$\mathcal{H}_1 = \lambda_1 \qquad (8.113)$$

and using Equations (8.111)–(8.112), one obtains the following optimal control strategy:

$$u_{\text{opt}} = \begin{cases} u_{\min} & \text{for } \mathcal{H}_1 > 0 \\ 0 & \text{for } \mathcal{H}_1 < 0 \\ u_{\text{sing}} & \text{for } \mathcal{H}_1 \equiv 0 \end{cases} \qquad (8.114)$$

Note that the singular control u_{sing} is used when the switching function remains zero over a *finite* time interval, i.e., $\mathcal{H}_1(t) = 0$ for $t \in [t_1, t_2] \subset [0, t_f]$. The singular control is given by [48]

$$u_{\text{sing}} = \frac{\alpha(z)}{\beta(z)} \tag{8.115}$$

where

$$\alpha(z) = -\frac{\partial^2 f}{\partial z_1 \partial z_2} f + \frac{\partial f}{\partial z_1} \frac{\partial f}{\partial z_2} \tag{8.116a}$$

$$\beta(z) = \frac{\partial^2 f}{\partial z_1^2} \tag{8.116b}$$

For optimality, the control in Equation (8.115) must also satisfy a local second-order convexity condition, known as the *generalized Legendre-Clebsch condition* or the Kelley-Contensou test [29]

$$\frac{\partial}{\partial u} \left(\frac{d^{2q} \mathcal{H}_1}{dt^{2q}} \right) \geq 0 \tag{8.117}$$

where q is the order of the singular arc. In our case, $q = 1$ and (8.117) reduces to testing whether

$$\beta(z) \leq 0 \tag{8.118}$$

along the singular part of the trajectory.

Computation of Singular Control

The computation of the singular control from Equation (8.115) requires the calculation of the partial derivatives in (8.116). To proceed with the calculation of the singular control, we need to consider the explicit expression of the friction force as a function of z_1 and z_2. Given the assumption that f is a function of only the slip coefficient $s = z_1/z_2 - (1+\rho)$, one obtains for the partial derivatives in (8.116)

$$\frac{\partial f}{\partial z_1} = \frac{\partial f}{\partial s} \frac{\partial s}{\partial z_1}, \quad \frac{\partial f}{\partial z_2} = \frac{\partial f}{\partial s} \frac{\partial s}{\partial z_2} \tag{8.119}$$

and hence

$$\frac{\partial^2 f}{\partial z_1 \partial z_2} = \frac{\partial^2 f}{\partial s^2} \frac{\partial s}{\partial z_1} \frac{\partial s}{\partial z_2} + \frac{\partial f}{\partial s} \frac{\partial^2 s}{\partial z_1 \partial z_2} \tag{8.120a}$$

$$\frac{\partial^2 f}{\partial z_1^2} = \frac{\partial^2 f}{\partial s^2} \left(\frac{\partial s}{\partial z_1} \right)^2 + \frac{\partial f}{\partial s} \frac{\partial^2 s}{\partial z_1^2} \tag{8.120b}$$

THEOREM 8.3—(TSIOTRAS AND CANUDAS-DE-WIT [48])
On a singular subarc, necessarily $\partial f/\partial s = 0$. □

It should be pointed out that the mere appearance of the singular control in the composite optimal trajectory is not ensured *a priori*, even in the case the local optimality of the control law is guaranteed by the satisfaction of Kelley's condition (8.117). For most automotive applications, however, the maximum braking torque is large enough, such that a singular subarc is always part of the optimal trajectory.

Notice that the expression of the singular control can also be written in the form

$$u_{\text{sing}} = \frac{z_1}{z_2} f(z_1, z_2) \tag{8.121}$$

and the control is in a purely *feedback* form (independent of co-states). This is the expression of the singular control during braking. The only assumption used in the derivation of (8.121) is that the friction force is a function of the slip coefficient s.

Maximum Friction Control

Assuming that the friction force is given as an algebraic map of the slip coefficient s, the control that keeps the friction force to its maximum value of the f vs. s curve can be calculated by imposing $\dot{s} = 0$, along with the conditions

$$\frac{\partial f}{\partial s} = 0, \qquad \frac{\partial^2 f}{\partial s^2} < 0 \tag{8.122}$$

Using the definition for the slip coefficient (assume $z_2 \neq 0$), one obtains

$$\dot{s} = \frac{z_2 \dot{z}_1 - z_1 \dot{z}_2}{z_2^2} \tag{8.123}$$

Thus, $\dot{s} = 0$ if and only if

$$z_2 \dot{z}_1 - z_1 \dot{z}_2 = z_2 u - z_1 f(z_1, z_2) = 0 \tag{8.124}$$

or that

$$u_{f_{\max}} = \frac{z_1}{z_2} f(z_1, z_2) \tag{8.125}$$

which is the same as the expression for the singular control in Equation (8.121), under the additional assumption that $\partial f/\partial s = 0$. Therefore, the singular control u_{sing} achieves $\dot{s} = 0$ and $\partial f/\partial s = 0$, i.e., it forces operation at the maximum of the friction vs. slip curve.

From (8.120b), the second condition in Equation (8.122) implies that

$$\frac{\partial^2 f}{\partial z_1^2} < 0$$

which is exactly Kelley's condition for optimality of the singular control.

Remark. It is clear that the singular control is constant. It can also be re-written as

$$u_{\text{sing}} = s^* f_{\max} + (1+\rho) f_{\max} \tag{8.126}$$

The last expression has the advantage that it does not require the calculation of the partial derivatives in Equation (8.120). It needs only the maximum value of the friction force $f_{\max}$, and the corresponding value of the slip coefficient s^*.

Dynamic Friction Model

The previous derivation was based on the assumption that the friction force between the tire and the ground can be accurately described as a static relation in terms of the slip coefficient. In fact, it is well-known that transient effects can be important. This drawback of static friction models can be addressed by using dynamic friction models like the ones discussed here. Next, we use the average lumped LuGre model from (8.43)-(8.44) in the optimal braking problem. The average lumped model is preferable over the distributed LuGre model since the latter one consists of an infinite number of states that make it very difficult to use for control design.

The system dynamics are given by

$$\dot{z}_1 = \dot{v}_r + (1+\rho)\dot{v} = u \tag{8.127}$$
$$\dot{z}_2 = \dot{v} = f(v_r, z_3) \tag{8.128}$$
$$\dot{z}_3 = \dot{\bar{z}} = v_r - \frac{\sigma_0 |v_r|}{g(v_r)} z_3 - \frac{\kappa_0}{L} |v_r + v| z_3 \tag{8.129}$$

where z_1 and z_2 as in (8.104) and where $z_3 = \bar{z}$, the internal friction state that characterizes the mean displacement of the bristles at the contact patch. The total friction force is given by

$$f(v_r, z_3) = \frac{F_n}{m}(\sigma_0 \bar{z} + \sigma_1 \dot{\bar{z}} + \sigma_2 v_r) \tag{8.130}$$

In (8.129), $g(v_r)$ is assumed to be given by

$$g(v_r) = \mu_c + (\mu_s - \mu_c) e^{-(v_r/v_s)^2} \tag{8.131}$$

The parameters of the friction model are the normalized longitudinal stiffness coefficient σ_0 [1/m], the normalized longitudinal damping coefficient σ_1 [s/m], the normalized viscous damping factor σ_2 [s/m], the normalized Coulomb and static friction μ_c and μ_s, and the Stribeck characteristic velocity v_s [m/s]. When uniform load distribution is assumed, κ_0 can be taken as a dimensionless constant in the interval $1 \leq \kappa_0 \leq 2$; see the discussion in Section 8.2.

The performance index to be minimized is given by (8.106). The final time t_f is free and the boundary conditions are

$$z_1(0) = z_{10}, \quad z_2(0) = z_{20}, \quad z_1(t_f) = z_2(t_f) = 0, \quad z_3(0), z_3(t_f) = \text{free} \tag{8.132}$$

In order to solve this minimization problem, we use an "optimal backstepping" approach [49]. To this end, we consider the reduced system

$$\dot{z}_2 = f(v_r, z_3) \tag{8.133}$$

$$\dot{z}_3 = v_r - \frac{\sigma_0 |v_r|}{g(v_r)} z_3 - \frac{\kappa_0}{L}|v_r + z_2|z_3 \tag{8.134}$$

$$f(v_r, z_3) = \frac{F_n}{m}(\sigma_0 z_3 + \sigma_1 \dot{z}_3 + \sigma_2 v_r) \tag{8.135}$$

where v_r is now the control input. If the optimal "control" v_r^* for the system (8.133)-(8.135) is known, then the optimal braking torque can be computed from (8.127). The full-system optimization problem requires knowledge of the boundary conditions for z_1. Since the boundary conditions of z_1 of the reduced-system will be, in general, different than the ones for the complete system, bang-bang subarcs will be required to satisfy these boundary conditions at $t = 0$ and $t = f_f$. We avoid the calculation of such bang-bang subarcs by making the realistic assumption that for large enough braking torque (typically the case), z_1 can be changed almost instantaneously via (8.127). Therefore, without loss of generality one can assume that the actual value of v_r matches the optimal ones from the reduced optimization system at $t = 0$ and $t = t_f$. This approach has the benefit of avoiding the numerically ill-conditioned calculation of the boundary subarcs which, at any rate, have an extremely small duration.

In order to proceed with the solution of the optimal control problem (8.133)–(8.135) subject to (8.106), we need to consider the natural control constraint $-v \le v_r \le 0$. From the definition of the longitudinal slip rate for braking, $s = \frac{\omega r}{v} - 1$, this constraint implies $-1 \le s \le 0$ or $0 \le \omega r \le v$. The constraint $v_r \in U = [-v, 0]$ is imposed by the typical braking systems, that is, the brakes can only apply a torque opposite to the angular velocity of the wheel.

Singular Perturbation Approach

As mentioned in [1] the LuGre friction model together with the equations of motion can be regarded as a singular perturbed system [32], [16]. In the LuGre tire friction model, we have that the parameter σ_0 is typically very big compared to the other terms of the internal state $\bar{z}$ equation. Moreover, the patch length L is small. Now, define $\ell = \sigma_0 L$ and let

$$\varepsilon = \frac{1}{\sigma_0}, \qquad \hat{z} = \frac{1}{\varepsilon}\bar{z}, \qquad \hat{z}_3 = \hat{z} \tag{8.136}$$

We can rewrite the system equations (8.133)–(8.135) in a singular perturbed form as follows:

$$\dot{z}_2 = \hat{f}(v_r, \hat{z}_3, \varepsilon) \tag{8.137}$$

$$\varepsilon \dot{\hat{z}}_3 = v_r - \frac{|v_r|}{g(v_r)}\hat{z}_3 - \frac{\kappa_0}{\ell}|v_r + z_2|\hat{z}_3 \tag{8.138}$$

$$\hat{f}(v_r, \hat{z}_3, \varepsilon) = \frac{F_n}{m}(\hat{z}_3 + \varepsilon \sigma_1 \dot{\hat{z}}_3 + \sigma_2 v_r) \tag{8.139}$$

We observe that the fast state of the system is the internal state of the friction model $\bar{z}_3$ and the slow state is the velocity of the vehicle z_2. This means that the friction develops faster than the vehicle changes its speed. Setting $\varepsilon = 0$, we derive the reduced model

$$\dot{z}_2 = \hat{f}(v_r, \hat{z}_{3ss}) \tag{8.140}$$

$$\hat{z}_{3ss} = \frac{v_r}{\frac{|v_r|}{g(v_r)} + \frac{\kappa_0 |v_r + z_2|}{\ell}} \tag{8.141}$$

$$\hat{f}(v_r, \hat{z}_{3ss}) = \frac{F_n}{m}(\hat{z}_3 + \sigma_2 v_r) \tag{8.142}$$

Observe that $\hat{z}_{3ss} = \sigma_0 \bar{z}_{ss}$ where $\bar{z}_{ss}$ is the steady state of the mean deflection of the bristles at the LuGre model. That is, $\bar{z}$ will reach $\bar{z}_{ss}$ very quickly. This allow us to use the reduced model (8.140)–(8.142) for our analysis.

The Hamiltonian for the optimization problem using the reduced system (8.140)-(8.142) is given by

$$\hat{\mathcal{H}} = z_2 + \lambda \hat{f}(v_r, \hat{z}_{3ss}) \tag{8.143}$$

The optimal controller must satisfy

$$v_r = \mathrm{argmin}_{v_r \in U}(\hat{\mathcal{H}}) = \mathrm{argmin}_{v_r \in U}(\hat{f}(v_r, \hat{z}_{3ss}))$$
$$= \mathrm{argmin}_{v_r \in U}\left(\frac{F_n}{m}(\hat{z}_{3ss} + \sigma_2 v_r)\right) \tag{8.144}$$

Observe that this equation implies that, for optimality, we require minimum (maximum negative) friction, a result that was also seen in the previous section where the Pacejka model was used. Recall that $\hat{f}(v_r, \hat{z}_{3ss}) = F_n(\hat{z}_{3ss} + \sigma_2 v_r)/m$. Figure 8.14 shows how $\hat{z}_{3ss} + \sigma_2 v_r$ changes with $v_r \in U$ for different v (or z_2). We observe that the minimizing v_r is either the one that makes

$$\frac{\partial(\hat{z}_{3ss} + \sigma_2 v_r)}{\partial v_r} = 0$$

or, when the velocity is small enough, $v_r = -v$ (when $s = -1$).

Numerical Example

To illustrate the previous analysis, we consider a numerical example of minimum braking distance for a vehicle with mass 1000 kg ($m = 250$ kg). The wheel radius is $r = 0.25\,\mathrm{m}$ and the wheel and drivetrain moment of inertia is $J = 1\,\mathrm{kg\,m}^2$. These values correspond to a value of $\rho = 15.625$. We first consider the case using Pacejkas' static friction model.

$$F_r/F_n = D\sin(C\arctan(Bs)) \tag{8.145}$$

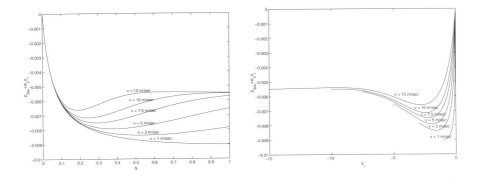

Figure 8.14 $\hat{z}_{3ss} + \sigma_2 v_r$ with longitudinal slip s and relative velocity v_r.

with values $D = 0.7, B = 7, C = 1.6$. The maximum normalized friction force can be computed directly from Equation (8.145) and is $F_r/F_n = 0.7$. The corresponding slip coefficient is $s^* = -0.2138$.

The initial conditions are given by $v(0) = r\omega(0) = 15$ m/s. To avoid the singularity of the friction model at $v = 0$ and $s = 0$, the final conditions are given by $v(t_f) = 0.1$ m/s and $r\omega(t_f) = 0.099$ m/s. The initial conditions correspond to a value of the slip coefficient $s(0) = 0$ and the final conditions correspond to a value of slip $s(t_f) = -0.01$. The maximum value of the braking torque is given by $1,500$ Nm which corresponds to $u_{\min} = -375$.

To numerically solve the associated two-boundary value optimization problem, a special FORTRAN code was written based on a root-solving Newton method using the subroutine hybrd of the MINPACK library [39]. All calculations were performed using double precision arithmetic to avoid numerical ill-conditioning. The results are shown in Figures 8.15–8.16. The optimal solution has three subarcs in this case. As expected from our analysis, during most of the trajectory a singular control law is used to achieve maximum friction force. The initial and final bang subarcs take care of the required boundary conditions (slip not at the maximum friction force). The distance traveled before complete stop is $\mathcal{J} = 16.382$ m. The total time to stop the vehicle under this applied braking torque profile (shown at the bottom of Figure 8.16(a)) is $t_f = 2.17$ s. The initial and final bang subarcs take place in a very short time interval. The duration of the initial and final subarcs are approximately 0.0123 and 3×10^{-4} s, respectively. That is, for all practical purposes, the initial and final bang subarcs can be replaced by two impulses at the initial and final parts of the trajectory. This justifies the reduced system approach followed for the dynamic model case.

Figure 8.15 shows the time histories of the states and the co-states. The angular velocity of the wheel and the corresponding co-state λ_1 exhibit a very fast transient during the initial part of the trajectory, as expected from the optimal torque profile. The optimality of the trajectory is verified by the time history of the switching function λ_1 in Figure 8.15. A more detailed

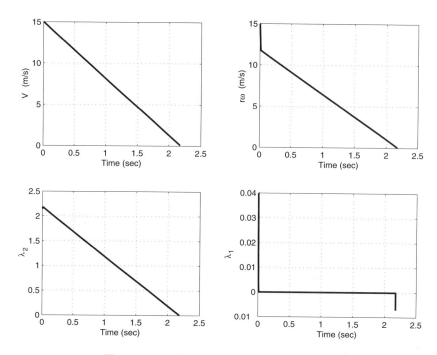

Figure 8.15 State and co-state time histories

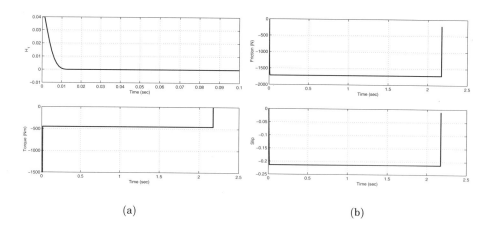

(a) (b)

Figure 8.16 (a) Switching function and braking torque history; (b) friction force and slip coefficient time histories

Friction Tire/Road Modeling,...and Optimal Braking Control

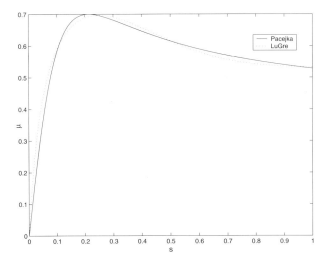

Figure 8.17 Fitting of Pacejka and LuGre models.

Table 8.1 Identified Parameters

Parameter	Value
σ_0	96.19 m^{-1}
σ_2	0 s/m
μ_c	0.52
μ_s	0.89
v_s	3.46 m/s

depiction of the switching function just before the entry to the singular subarc as shown at the top plot of Figure 8.16 (a). Finally, Figure 8.16 (b) shows the friction force and the slip coefficient history.

Next we compare the optimal controller designed using Pacejka's model (8.145) with the one designed using the LuGre model. The first step is to find the parameters of the LuGre model (σ_0, μ_s, μ_c etc.) such that the two models are comparable. To do so, we fit to Pacejka's $s - \mu$ plot, an $s - \mu$ plot constructed using (8.141)–(8.142) together with the definition of longitudinal slip, at an intermediate velocity of the simulation (7.5 m/s). We keep the same values for the Pacejka model coefficients. The fitting of the two curves is shown in figure 8.17. The parameters identified are shown in Table 8.1. We have used patch length $L = 0.2$ m, constant parameter $\kappa = 1.2$ and $\alpha = 2$. The optimal controller for the case of the LuGre friction model is given by (8.144). The simulations that follow are made using the

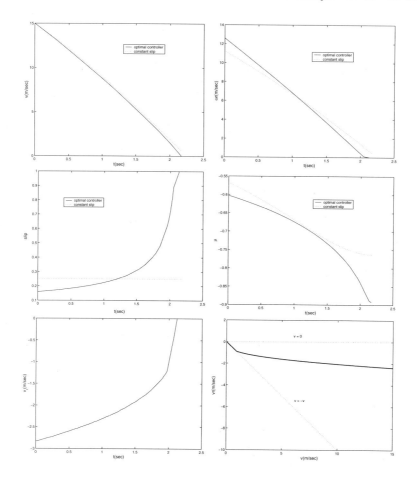

Figure 8.18 Simulation results of braking maneuver using the LuGre friction model: optimal and constant slip controllers

same parameters as for the static case: $m = 250\,\text{kg}$, $J = 1\,\text{kg m}^2$, $\rho = 15.625$ and initial velocity $v(0) = 15\,\text{m/s}$. Observe the way the controllers work. The optimal controller based on the static model, keeps the slip constant to its maximizing value, while in the LuGre model-based optimal controller, the slip $s = -v_r/v$ is changing to provide the maximum friction at all times. We observe that in the LuGre model case, v_r at some point hits the boundary of the constraint $-v \leq v_r \leq 0$ and we have $v_r = -v$. Then, v_r stays on the boundary of the constraint until the end. This is expected since, as seen previously, for small velocities the maximum friction appears at $s = 1$, i.e., $v_r = -v$.

Recall that the optimal braking controller maximizes friction at all times. Together with the optimal controller for the LuGre friction model case we present a simulation using the optimal controller designed for the Pacejka model (constant slip) but this time applied to the equations using the LuGre model. The constant slip selected is the one maximizing friction at the velocity that the steady state of the LuGre model was matched with the Pacejka model ($v = 7.5$ m/s). This slip is $s \approx 0.25$. As expected, it takes longer to stop the vehicle compared to using the optimal controller. The stopping distance using the optimal controller was 17.34 m while using the constant slip controller it was 17.81 m. These values are, of course, only indicative and depend strongly on the problem data.

8.7 Observed-based Emergency Braking Control

In this section, a control scheme for emergency braking of vehicles is designed. The scheme is based on the LuGre dynamic friction model and estimation of the tire/road friction. The control system output is the pressure in the braking system, and is calculated using only the wheel angular speed information. The controller utilizes estimated state feedback control to achieve a near-maximum deceleration. The state observer gain is calculated by using linear matrix inequality (LMI) techniques. This system has two advantages when compared with a typical anti-lock braking system (ABS): it generates less chattering during braking and produces a source of *a priori* information regarding safe spacing.

System Dynamics

In this section only the longitudinal dynamics of the vehicle is considered. We assume that the four wheels of the vehicle apply the same braking force. For simplicity, we also assume that the road has no slope and that the weight of the vehicle is distributed evenly among the four wheels. A quarter-vehicle model is used and we consider a modified lumped LuGre friction model as follows[7]:

$$\begin{cases} \dot{z} = -v_r - \theta \frac{\sigma_0 |v_r|}{h(v_r)} z \\ J\dot{\omega} = -rF_x - u_\tau \\ m\dot{v} = 4F_x - F_r \end{cases} \quad (8.146)$$

where z is the friction internal state, $v_r = v - r\omega$ is the relative velocity, $h(v_r) = \mu_c + (\mu_s - \mu_c)e^{-|\frac{v_r}{v_s}|^{1/2}}$, u_τ is the traction/braking torque, F_x the traction/braking force given by the tire/road contacting, F_r the rolling resistance, m the vehicle mass, J the tire rotational inertia, and the parameter θ is used to model the effect of different tire/road Coulomb friction coefficients. The braking force F_x is given by

$$F_x = F_n(\sigma_0 z + \sigma_1 \dot{z} - \sigma_2 v_r)$$

---

[7] We do not include the variation term κ in the lumped LuGre model in this section for simplicity. The same approach can be used if κ is included.

where F_x is a negative number in the case of braking and $F_n = mg/4$. By [51], the rolling resistance can be modeled as

$$F_r = \sigma_v mgv$$

where σ_v is the rolling resistance coefficient and g is the gravity constant.

Observer-based Braking Controller Design

Defining the state variables as

$$x_1 := \sigma_0 z, \quad x_2 := v, \quad , x_3 := v_r = v - r\omega$$

the system dynamics Equation (8.146) are rewritten as:

$$\begin{cases} \dot{\mathbf{x}} = A\mathbf{x} + B_1\theta\psi(\mathbf{x}) + B_2 u \\ y = C\mathbf{x} \end{cases} \tag{8.147}$$

where $\mathbf{x} = [x_1 \ x_2 \ x_3]^T$, where

$$A = \begin{bmatrix} 0 & 0 & -\sigma_0 \\ g & -g\sigma_v & -g(\sigma_2 + \sigma_1) \\ \alpha & -g\sigma_v & -\alpha(\sigma_2 + \sigma_1) \end{bmatrix},$$

$$B_1 = \begin{bmatrix} -\sigma_0 \\ -g\sigma_1 \\ -\alpha\sigma_1 \end{bmatrix}, \quad B_2 = \begin{bmatrix} 0 \\ 0 \\ \frac{r}{J}K_b \end{bmatrix}, \quad C = [\,0 \ \ \frac{1}{r} \ \ -\frac{1}{r}\,]$$

$$\psi(\mathbf{x}) = x_1 f(x_3), \quad u = P_b$$

where $f(x_3) = \frac{x_3}{h(x_3)}$, $\alpha = g(1 + \frac{mr^2}{4J})$, and $u_\tau = K_b P_b$, where K_b is the braking system gain and P_b the brake pressure, which is the controlled variable.

For most vehicles, we can measure the angular velocity of each wheel.

$$y = \omega = \frac{1}{r}(x_2 - x_3) \tag{8.148}$$

Since the internal state z, given by LuGre model, and the vehicle longitudinal velocity are unavailable, we must design an observer to estimate these states. Construct the following model-based nonlinear observer

$$\dot{\hat{\mathbf{x}}} = A\hat{\mathbf{x}} + B_1 \hat{\theta}\psi(\hat{\mathbf{x}}) + B_2 u + L(y - C\hat{\mathbf{x}}) + B_1 \mathcal{G} \tag{8.149}$$

where $\mathcal{G}$ is a tuning function to be determined subsequently.
The following assumptions are made for the system (8.147) and observer (8.149):

i. (A, B_1) is controllable and (A, C) is observable;
ii. $f(x_3)$ is positive and bounded and $f'(x_3)$ is bounded, i.e.

$$0 \leq f(x_3) \leq f_{max} \leq \rho_2 < \infty, \quad |f'(x_3)| \leq \rho_3 < \infty, \quad \forall \mathbf{x} \in \mathcal{D}_1 \subset \mathcal{R} \quad (8.150)$$

iii. The unknown parameter θ is bounded, i.e.

$$0 < \theta \leq \theta_{max} \quad (8.151)$$

iv. The map $w \mapsto \xi$ of the system

$$\begin{cases} \dot{\zeta} = (A - LC)\zeta + B_1 w \\ \xi = C\zeta \end{cases} \quad (8.152)$$

with $(A - LC)$ Hurwitz, is strictly passive; moreover, $\exists \rho_1 > 0$, a constant, and $\exists P = P^T > 0$ such that

$$(A - LC)^T P + P(A - LC) + (\rho_1^2 + \rho_4)I < 0 \quad (8.153)$$

as well as

$$PB_1 = C^T \quad (8.154)$$

where $\rho_4 = \frac{2\theta_{max}\rho_2}{r} > 0$.

THEOREM 8.4
Under Assumptions i.–iv., there exists an adaptive emergency braking controller that achieves

$$\hat{\lambda} \to \hat{\lambda}_{max}$$

asymptotically for the system (8.147) using measured angular velocity ω, where the estimated slip $\hat{\lambda} := \frac{\hat{x}_3}{\hat{x}_2} = \frac{\hat{v}-r\hat{\omega}}{\hat{v}}$ and $\hat{\lambda}_{max} := \hat{\lambda}_{max}(\hat{v}_r, \hat{v})$ is the longitudinal slip corresponding to the estimated maximum friction coefficient $\hat{mu}_{max}$ in the pseudo-static relationship between μ and λ.

Proof. Define $\tilde{\mathbf{x}} := \mathbf{x} - \hat{\mathbf{x}}$, $\tilde{y} := y - \hat{y} = C\tilde{\mathbf{x}}$ and $\tilde{\theta} := \theta - \hat{\theta}$, then the error dynamics for the system are

$$\dot{\tilde{\mathbf{x}}} = (A - LC)\tilde{\mathbf{x}} + B_1 \left[\theta\psi(\mathbf{x}) - \hat{\theta}\psi(\mathbf{x})\right] - B_1 \mathcal{G} \quad (8.155)$$

define the dynamic surface $\tilde{s}$ as $\tilde{s} := \hat{x}_3 - \hat{\lambda}_{max}\hat{x}_2$ and differentiate $\tilde{s}$

$$\begin{aligned}
\dot{\tilde{s}} &= \dot{\hat{x}}_3 - \dot{\hat{x}}_2\hat{\lambda}_{max} - \dot{\hat{\lambda}}_{max}\hat{x}_2 \\
&= \frac{r}{J}K_b P_b - \sigma_1(\alpha - g\hat{\lambda}_{max})f(\hat{x}_3)\hat{\theta} + \Big\{(\alpha - g\hat{\lambda}_{max})[\hat{x}_1 - (\sigma_2 + \sigma_1)\hat{x}_3 \\
&\quad - (1 - \hat{\lambda}_{max})g\sigma_v\hat{x}_2] + (l_3 - l_2\hat{\lambda}_{max})\tilde{y}\Big\} - \sigma_1(\alpha - g\hat{\lambda}_{max})\mathcal{G} - \dot{\hat{\lambda}}_{max}\hat{x}_2 \\
&= dK_b P_b + \beta_1(\hat{\mathbf{x}})\hat{\theta} + \beta_2(\hat{\mathbf{x}}) + \beta_3(\hat{\mathbf{x}})\mathcal{G} \quad (8.156)
\end{aligned}$$

where
$$d = \frac{r}{J}, \quad \beta_1(\hat{\mathbf{x}}) = -\sigma_1(\alpha - g\hat{\lambda}_{max})f(\hat{x}_3)\hat{x}_1,$$
$$\beta_2(\hat{\mathbf{x}}) = (\alpha - g\hat{\lambda}_{max})[\hat{x}_1 - (\sigma_2 + \sigma_1)\hat{x}_3]$$
$$+ (l_3 - l_2\hat{\lambda}_{max})\tilde{y} - \dot{\hat{\lambda}}_{max}\hat{x}_2 - (1 - \hat{\lambda}_{max})g\sigma_v\hat{x}_2,$$
$$\beta_3(\hat{\mathbf{x}}) = -\sigma_1(\alpha - g\hat{\lambda}_{max}),$$

and l_2, l_3 are the second and third elements of the gain vector $L \in \mathcal{R}^3$.

Consider the following Lyapunov function candidate

$$V = \frac{1}{2}\tilde{s}^2 + \frac{1}{2\gamma}\tilde{\theta}^2 + \tilde{\mathbf{x}}^T P \tilde{\mathbf{x}}$$

where $\gamma > 0$. Then

$$\dot{V} = \dot{\tilde{\mathbf{x}}}^T P \tilde{\mathbf{x}} + \tilde{\mathbf{x}}^T P \dot{\tilde{\mathbf{x}}} + \tilde{s}\dot{\tilde{s}} + \frac{1}{\gamma}\tilde{\theta}\dot{\tilde{\theta}}$$
$$= \tilde{\mathbf{x}}^T \left[(A - LC)^T P + P(A - LC)\right] \tilde{\mathbf{x}} + 2\tilde{\mathbf{x}}^T P B_1 \left[\theta\psi(\mathbf{x}) - \hat{\theta}\psi(\hat{\mathbf{x}})\right]$$
$$+ \frac{1}{\gamma}\tilde{\theta}\dot{\tilde{\theta}} + \tilde{s}\dot{\tilde{s}} - 2\tilde{\mathbf{x}}^T P B_1 \mathcal{G}$$

Notice that

$$\theta\psi(\mathbf{x}) - \hat{\theta}\psi(\hat{\mathbf{x}}) = \tilde{\theta}\psi(\hat{\mathbf{x}}) + \theta\left[\psi(\mathbf{x}) - \psi(\hat{\mathbf{x}})\right]$$

and use (8.154) to obtain

$$\dot{V} = \tilde{\mathbf{x}}^T \left[(A - LC)^T P + P(A - LC)\right] \tilde{\mathbf{x}} \qquad (8.157)$$
$$+ 2\tilde{y}\tilde{\theta}\psi(\hat{\mathbf{x}}) + 2\tilde{\mathbf{x}}^T P B_1 \theta \left[\psi(\mathbf{x}) - \psi(\hat{\mathbf{x}})\right] \qquad (8.158)$$
$$+ \frac{1}{\gamma}\tilde{\theta}\dot{\tilde{\theta}} + \tilde{s}\left[dK_b P_b + \beta_1(\hat{\mathbf{x}})\hat{\theta} + \beta_2(\hat{\mathbf{x}}) + \beta_3(\hat{\mathbf{x}})\mathcal{G}\right] - 2\tilde{\mathbf{x}}^T P B_1 \mathcal{G}$$

Letting the control input be

$$u = P_b = \frac{1}{dK_b}\left[-\beta_1(\hat{\mathbf{x}})\hat{\theta} - \beta_2(\hat{\mathbf{x}}) - \beta_3(\hat{\mathbf{x}})\mathcal{G} - \eta\tilde{s}\right]$$

Equation (8.156) becomes

$$\dot{\tilde{s}} = -\eta\tilde{s} \qquad (8.159)$$

Using (8.154) and (8.159) and letting

$$\dot{\hat{\theta}} = 2\gamma\tilde{y}\psi(\hat{\mathbf{x}}) \qquad (8.160)$$

we obtain from Equation (8.159)

$$\dot{V} = \tilde{\mathbf{x}}^T \left[(A - LC)^T P + P(A - LC)\right] \tilde{\mathbf{x}} + 2\tilde{y}\tilde{\theta}\psi(\hat{\mathbf{x}})$$
$$+ 2\tilde{\mathbf{x}}^T P B_1 \theta \left[\psi(\mathbf{x}) - \psi(\hat{\mathbf{x}})\right] + \frac{1}{\gamma}\tilde{\theta}\dot{\tilde{\theta}} - \eta\tilde{s}^2 - 2\tilde{y}\mathcal{G}.$$

Note that

$$\psi(\mathbf{x}) - \psi(\hat{\mathbf{x}}) = x_1 f(x_3) - \hat{x}_1 f(\hat{x}_3) = f(x_3)\tilde{x}_1 + \hat{x}_1 f'(x_3^*)(x_3 - \hat{x}_3)$$

where x_3^* is a value between x_3 and $\hat{x}_3$ derived by using the Mean Value Theorem for the smooth function $f(x) = x/h(x)$. Moreover, by (8.150), (8.151), and (8.160)

$$\begin{aligned}
\dot{V} &\leq \tilde{\mathbf{x}}^T \left[(A - LC)^T P + P(A - LC)\right] \tilde{\mathbf{x}} + 2\tilde{\mathbf{x}}^T C^T \theta_{max} \rho_2 \tilde{x}_1 \\
&\quad + 2\rho_3 \theta_{max} |\tilde{y}||\hat{x}_1||\tilde{x}_3| - \eta \tilde{s}^2 - 2\tilde{y}\mathcal{G} \\
&= \tilde{\mathbf{x}}^T \left[(A - LC)^T P + P(A - LC)\right] \tilde{\mathbf{x}} + \frac{1}{r}\theta_{max} \rho_2 (2\tilde{x}_1 \tilde{x}_2 + 2\tilde{x}_1 \tilde{x}_3) \\
&\quad + 2\rho_3 \theta_{max} |\tilde{y}||\hat{x}_1||\tilde{x}_3| - \eta \tilde{s}^2 - 2\tilde{y}\mathcal{G} \\
&\leq \tilde{\mathbf{x}}^T \left[(A - LC)^T P + P(A - LC)\right] \tilde{\mathbf{x}} \\
&\quad + \rho_4 \left(\tilde{x}_1^2 + \frac{1}{2}\tilde{x}_2^2 + \frac{1}{2}\tilde{x}_3^2\right) - \frac{\rho_4}{2}\left(|\tilde{x}_3| - \frac{\rho_3 r}{\rho_2}|\tilde{y}\hat{x}_1|\right)^2 \\
&\quad + \frac{\rho_4}{2}\tilde{x}_3^2 + \frac{\rho_4}{2}\left(\frac{\rho_3 r}{\rho_2}\right)^2 \hat{x}_1^2 \tilde{y}^2 - \eta \tilde{s}^2 - 2\tilde{y}\mathcal{G} \\
&\leq \tilde{\mathbf{x}}^T \left[(A - LC)^T P + P(A - LC) + \rho_4 I\right] \tilde{\mathbf{x}} - \eta \tilde{s}^2 \\
&\quad - \frac{\rho_4}{2}\left(|\tilde{x}_3| - \frac{\rho_3 r}{\rho_2}|\tilde{y}\hat{x}_1|\right)^2 + \tilde{y}\left[\frac{\rho_4}{2}\left(\frac{\rho_3 r}{\rho_2}\right)^2 \hat{x}_1^2 \tilde{y} - 2\mathcal{G}\right] \\
&\leq -\rho_1^2 \|\tilde{\mathbf{x}}\|^2 - \eta \tilde{s}^2 - \frac{\rho_4}{2}\left(|\tilde{x}_3| - \frac{\rho_3 r}{\rho_2}|\tilde{y}\hat{x}_1|\right)^2 + \tilde{y}\left[\frac{\rho_4}{2}\left(\frac{\rho_3 r}{\rho_2}\right)^2 \hat{x}_1^2 \tilde{y} - 2\mathcal{G}\right]
\end{aligned}$$

If we choose $\mathcal{G}$ such that

$$\mathcal{G} = \frac{\rho_4}{4}\left(\frac{\rho_3 r}{\rho_2}\right)^2 \hat{x}_1^2 \tilde{y} \qquad (8.161)$$

then

$$\dot{V} \leq -\rho_1^2 \|\tilde{\mathbf{x}}\|^2 - \eta \tilde{s}^2 - \frac{\rho_4}{2}\left(|\tilde{x}_3| - \frac{\rho_3 r}{\rho_2}|\tilde{y}\hat{x}_1|\right)^2 \leq 0$$

using Barbalat's Lemma, we can conclude that $\tilde{s} \to 0$, $\tilde{\mathbf{x}} \to 0$, as $t \to \infty$. Thus, by definition of $\tilde{s}$ and λ we have

$$\hat{\lambda} \to \hat{\lambda}_{max} \qquad \text{as } t \to \infty$$

$\square$

Remark. The adaptive nonlinear observer structure presented in this paper is similar to the scheme presented in [17]. [17] presented results that require an additional Lipschitz assumption on the function $\psi(x)$, and condition (8.154) has been replaced by $B_1^T P C^\perp = 0$ where $C^\perp$ is the projection on to null(C).

The tuning function $\mathcal{G}$ given by (8.161) is a linear function of $\tilde{y}$ and appears both in the observer and the control input. Compared with the tuning function in [11], Equation (8.161) does not require switching in the control input, and therefore produces a smoother control.

In addition, Assumptions i–iv must be satisfied in order for the system dynamics described by Equation (8.147) and, thus for the theorem to hold:

1. Regarding Assumption i, we can calculate the observability matrix

$$O = \begin{bmatrix} 0 & \dfrac{1}{r} & -\dfrac{1}{r} \\ \dfrac{g-\alpha}{r} & 0 & \dfrac{a}{r} \\ \dfrac{\alpha a}{r} & -\dfrac{g\sigma_v a}{r} & \dfrac{(g-\alpha)\left[\alpha(\sigma_2+\sigma_1)^2+\sigma_0\right]}{r} \end{bmatrix}$$

where $a = (\sigma_2+\sigma_1)(\alpha-g)$, thus rank$(O) = 3$, and (A, C) is an observable pair. Hence, Assumption i always holds;

2. To see that Assumption ii is always satisfied, we have

$$0 \leq f(x_3) = \frac{x_3}{h(x_3)} \leq \frac{x_3}{\mu_c} \leq \frac{\lambda_{max} v_{max}}{\mu_c} = \rho_2$$

and

$$|f'(x_3)| \leq \frac{1}{\mu_c}\left\{1+\left(\frac{\mu_s}{\mu_c}-1\right)\left[1+\frac{1}{2}\left(\frac{v_{max}}{v_s}\right)^{1/2}\right]\right\} = \rho_3$$

3. As for Assumption iv, we have to pick an observer gain L and a positive symmetric matrix P such that following optimization problem is feasible

$$\begin{cases} \max & \rho_1 \\ \text{s.t.:} & (A-LC)^T P + P(A-LC) + \rho_1^2 I + \rho_4 I < 0 \\ & PB_1 = C^T, \quad P = P^T > 0 \text{ and } \rho_1 > 0 \end{cases}$$

This can be calculated by linear matrix inequality (LMI) algorithms, such as those presented in [23].

Simulation Results and Discussions

In the following simulation example, we use the parameters from the LeSabre cars used by the California PATH program. These parameters are: $M = 1701.0\,\text{Kg}$, $\sigma_v = 0.005\,\text{N·s}^2/\text{m}^2$, $J = 2.603\,\text{Kg·m}^2$, $r = 0.323\,\text{m}$. We also take the road LuGre friction parameter in Equation (8.146) to be $\theta = 1$ and the braking gain $K_b = 0.9$. The nominal values of the parameters in the dynamic LuGre friction model are the same as those in [3].

We simulate an emergency braking maneuver with a vehicle initial velocity of $v = 30\,\text{m/s}$ and the designed observer-based controller. The initial condition for the observer dynamics is $\hat{\mathbf{x}}(0) = [0\ 29.5\ 0]^T$ and the true state is $\mathbf{x}(0) = [0\ 30\ 0.5]^T$; namely we use the measurement $r\omega$ ($= 29.5\,\text{m/s}$) as the initial condition for $\hat{v}$. Figure 8.19 shows the time responses of the real state vector $\mathbf{x}$ and estimated state vector $\hat{\mathbf{x}}$, while Figure 8.20(c) shows the time response of the estimated friction parameter $\hat{\theta}$. Figure 8.20(a) shows the time response of the controlled pressure P while Figure 8.20(b) shows the controlled sliding surface $\tilde{s}$. Figure 8.20(d) illustrates the difference, $\tilde{y}$, between the measurement y and output of observer $\hat{y}$. From these figures we can see that the estimated state $\hat{z}$ and parameter $\hat{\theta}$ converge to their respective true values quickly, and that the controlled input (pressure P) remains within its feasible domain, enabling the vehicle to come to a quick halt (decelerating at around $10\,\text{m/s}^2$). This example verifies the results of the previous section. However, the simulation results also reveal that the estimated states $\hat{v}$ and $\hat{v}_r$ do not converge to their true states during the braking process, remaining within a constant offset, even though the vehicle achieved its maximum estimated deceleration level, which is based on estimated states, as shown by Figure 8.20(b).

In what follows, we present a formal explanation of the above simulation results. From the state error dynamics (8.155), we find that

$$\dot{\tilde{y}} = -\frac{1}{r}[l_2 - l_3 + \sigma_1(g - \alpha)\mathbf{g}]\tilde{y} + f_1(\tilde{\mathbf{x}}), \qquad (8.162)$$

$$f_1(\tilde{\mathbf{x}}) = (g - \alpha)[(1 - \sigma_1\theta f(x_3)]\tilde{x}_1 - \sigma_v \tilde{x}_2$$
$$\qquad - (\sigma_1 + \sigma_2)\tilde{x}_3 - \sigma_1[\theta f(x_3) - \hat{\theta} f(\hat{x}_3)]\hat{x}_1$$

$$\mathbf{g} = \frac{\rho_4}{4}\left(\frac{\rho_3 r}{\rho_2}\right)^2 \hat{x}_1^2$$

In our example, we chose a relatively large value for the gain L with $l_2 > l_3$ ($L = [-400\ -60\ -500]^T$). As a consequence, $\tilde{y} \to 0$ quickly. Similarly, we can also assume that $\tilde{\theta} \to 0$ quickly, due to our choice of a high adaptation gain ($\gamma = 200$) and the presence of persistence of excitation, which we observed in the numerical example. The analysis for absence of persistence of excitation is in progress.

Using the approximation $\tilde{y} \approx 0$ and $\tilde{\theta} \approx 0$, we now analyze the dynamics of the state errors (8.155) and obtain

$$\dot{\tilde{\mathbf{x}}} = \bar{A}(x_3)\tilde{\mathbf{x}} \qquad (8.163)$$

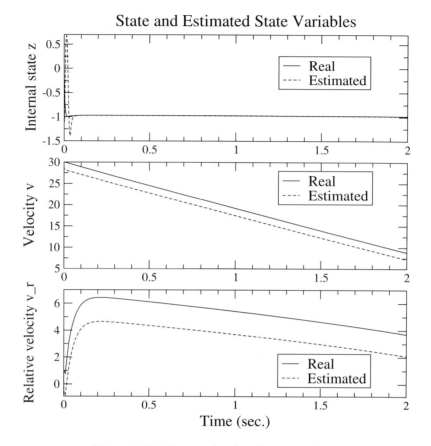

Figure 8.19 Estimated and real state variables

where

$$\bar{A}(x_3) = \begin{bmatrix} -\sigma_0 \theta f(x_3) & 0 & 0 \\ g[1 - \sigma_1 \theta f(x_3)] & -g\sigma_v & -g\sigma_2 \\ \alpha[1 - \sigma_1 \theta f(x_3)] & -g\sigma_v & -\alpha\sigma_2 \end{bmatrix}$$

Notice that $\sigma_0 = 280$, $\hat{\theta} \approx 1$ and $\frac{v}{\mu_c} \geq f(x_3) > \frac{v_r}{\mu_s}$. We can therefore conclude that $\tilde{x}_1 \to 0$ quickly with a decay rate of around $\sigma_0 \theta f(x_3)$ during the beginning of the braking process, due to the fact that v_r is large. This explains why the estimated state $\hat{x}_1$ converges quickly to the real state x_1. In the case of the state estimates $\hat{x}_2$ and $\hat{x}_3$, from Equation (8.163), we find that the eigenvalues of matrix $\bar{A}(x_3)$ associated with these two states are

$$s_{2,3} = \frac{-(g\sigma_v + \alpha\sigma_2) \pm \sqrt{(g\sigma_v)^2 + (\alpha\sigma_2)^2 + 4g^2\sigma_v\sigma_2}}{2}$$

Friction Tire/Road Modeling,...and Optimal Braking Control

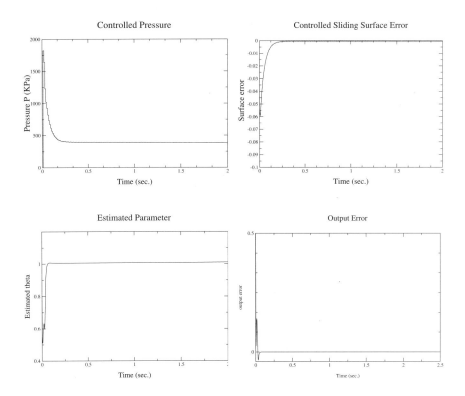

Figure 8.20 (a) Controlled braking pressure P; (b) sliding surface $\tilde{s}$; (c) Estimated friction parameter $\hat{\theta}$; (d) measurement error $\tilde{y}$ (rad/s)

Since σ_v and σ_2 are very small,

$$-1 \ll s_{2,3} < 0, \qquad \forall t \geq 0$$

The rate of decay for $\tilde{x}_2$ and $\tilde{x}_3$ is small and the eigenvector associated with s_2 is around $w_2 \approx [0 \ \ 1 \ \ 1]^T$. Figure 8.21 shows a sketch of the trajectory portrait of the approximate nonlinear system (8.163). For any initial condition $P_0 = (\tilde{x}_1(0), \tilde{x}_2(0), \tilde{x}_3(0)) \in \mathcal{R}^3$, the flow trajectory will quickly approach the $\tilde{x}_2 \times \tilde{x}_3$ plane because of the rapid convergence of $\tilde{x}_1$ (s_1 is large). Moreover, the trajectory will converge to w_2 on the $\tilde{x}_2 \times \tilde{x}_3$ plane if $\tilde{x}_2(0) > 0$ and $\tilde{x}_3(0) > 0$, as shown in Figure 8.21. Thus, if we pick

$$\tilde{x}_2(0) \geq 0, \qquad \tilde{x}_3(0) \geq 0, \tag{8.164}$$

then

$$\max\{\tilde{x}_2(0), \tilde{x}_3(0)\} \geq \tilde{x}_2(t) \approx \tilde{x}_3(t) \geq 0, \quad \forall \, t \geq t_0$$

where t_0 is fairly small and depends on the convergence rate and initial conditions of $\tilde{x}_1(t)$.

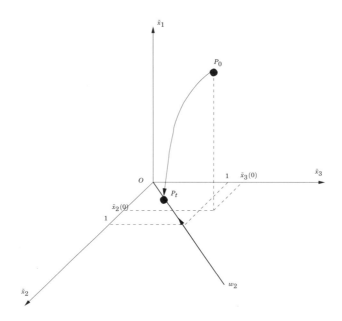

Figure 8.21 A schematic trajectory plot for nonlinear system $\dot{\tilde{\mathbf{x}}} = \bar{A}(x_3)\tilde{\mathbf{x}}$

When the angular velocity is the only measurement available to the control system and a state observer is designed, the angular velocity estimation error can be expressed as

$$\tilde{\omega} = \omega - \hat{\omega} = \frac{1}{r}(x_2\overset{.}{-}x_3) - \frac{1}{r}(\hat{x}_2 - \hat{x}_3) = \frac{1}{r}(\tilde{x}_2 - \tilde{x}_3) \tag{8.165}$$

It should be noticed that $\tilde{\omega} = 0$ does not imply that $\tilde{x}_2 = \tilde{x}_3 = 0$ but that $\tilde{x}_2 = \tilde{x}_3$. Vector w_2 in Figure 8.21 belongs to the surface where $\tilde{\omega} = 0$. It is important to remark that this limitation obeys the relation between x_2 and x_3 and is not dependent on the dynamic friction model or observer structure that was used in this paper.

Note that, by Theorem 8.1, we obtained $\hat{\lambda} \to \hat{\lambda}_{max}$ and $\lambda \to \lambda_{max}$, due to the fact that $\hat{\mathbf{x}} \to \mathbf{x}$. However, since the states $\hat{x}_3$ and $\hat{x}_2$ converge slowly, there will be some error between λ and $\hat{\lambda}$. This error can be estimated as follows.

$$\begin{aligned} 0 < \lambda(t) - \hat{\lambda}(t) &= \frac{x_3}{x_2} - \frac{\hat{x}_3}{\hat{x}_2} = \frac{x_2\tilde{x}_3 - x_3\tilde{x}_2}{x_2\hat{x}_2} \\ &\approx \frac{(x_2 - x_3)\tilde{x}_2}{x_2\hat{x}_2} = \left(1 - \frac{x_3}{x_2}\right)\frac{\tilde{x}_2}{\hat{x}_2} \end{aligned} \tag{8.166}$$

$$\leq (1 - \lambda(t))\frac{\lambda(0)x_2(0)}{\hat{x}_2(t)}, \quad \forall t \geq t_0 \tag{8.167}$$

Note that, in general, $\lambda(0) \leq 3\%$ during normal driving conditions before braking. As a consequence, the slip estimate error is small. Similarly, it can be shown that $\lambda_{max} - \hat{\lambda}_{max}$ will also be small after the state $\tilde{x}_1 \to 0$. Therefore, the proposed control system will achieve a near-maximum deceleration level, in spite of the fact that state estimation errors $\tilde{x}_2$ and $\tilde{x}_3$ converge slowly.

Remark. It should be noticed that by the controller design, we can always guarantee the underestimation of longitudinal slip $\lambda(t)$ as shown in Equation (8.167). This is a very important property, which helps to preserve the vehicle safety and stability on the highway, since the safe inter-vehicle space depends on the estimation of maximum acceleration and deceleration.

8.8 Conclusions

In this paper, we have revisited the problem of characterizing the friction at the tire/surface interface for wheeled vehicles. We have reviewed the major models used in the literature, namely, static, dynamic, lumped and distributed models. We have shown that static friction models are inadequate for describing the transient nature of friction build-up. Dynamic friction models are necessary to capture such transients during abrupt braking and acceleration phases. We propose a new dynamic friction model that accurately captures friction transients, as well as velocity-dependent characteristics and tire/road properties. The model is developed by extending the well-known LuGre point friction model to the case of a contact patch at the tire/surface interface. We have also shown how the friction force for lateral/cornering or combined longitudinal/lateral motion can also be modelled.

We have presented a method to estimate on-line the changes in road condition using only wheel rotational velocity information. To achieve this goal, we have used a lumped friction model with internal dynamics that, on one hand provide a more accurate description of the contact friction, and on the other hand, allows us to characterize road condition variations via a single parameter. We have presented a model-based observer that ensures asymptotic tracking of this parameter, hence tracking of changing road conditions. The observer works under mild conditions, *e.g.*, a non-vanishing evolution of the slip rate[8]. This condition is quite natural (it implies that the vehicle should operate away from the ideal pure rolling condition). The observer presented here has been derived in a general framework allowing to also study the case when the vehicle velocity is not measurable. In particular, Assumption A2-ii in Section 8.5 allows for this possibility.

We have shown that the optimal braking strategy for a wheeled vehicle is to operate at the maximum point of the friction/slip curve. This strategy has been used extensively in the past in the design of ABS systems. One of the interesting results of our analysis is that the optimal "max-friction" control is singular. Maximum torque control is used to match the

---

[8]When the slip is zero, the system is not observable. Assumption i in Section 8.7 precludes this case.

boundary conditions. The total optimal trajectory is therefore composed of a sequence of bang-bang and singular arcs. For typical automotive application, where the applied torque is high, the bang-bang subarcs can be safely approximated by impulses to achieve the specified boundary conditions of the slip.

We have discussed emergency braking control under unknown tire/road conditions and system states, based on a dynamical friction model and the assumption that the only available measurable signal to the controller is the wheel angular velocity. The braking pressure controller is determined based on the estimation of the system state variables and the unknown friction parameters. The simulation results show that the vehicle can be stopped quickly with near-maximum deceleration by applying this controller. The asymptotic convergence of the estimated states and parameters estimates has been proven. Moreover, it was also shown that the friction properties can be estimated and near maximum deceleration achieved, in spite of the slow convergence rate of the vehicle velocity and wheel relative velocity error estimates.

Acknowledgements

The first two authors would like to acknowledge support from CNRS and NSF (award No. INT-9726621/INT-9996096), for allowing frequent visits between the School of Aerospace Engineering at the Georgia Institute of Technology and the Laboratory of Automatic Control at Grenoble, France. These visits led to the development of the dynamic friction model presented in this paper. The first author would like to thank M. Sorine and P. A. Bliman for interesting discussions on distributed friction models, and to Michel Basset and Gerard Gissinger, for their collaboration. The first and the third authors would also like to acknowledge the support of Renault Research Division, who partially support some of the results shown in the paper. The second author gratefully acknowledges partial support from the US. Army Research Office under contract No. DAAD19-00-1-0473. The fourth and fifth authors acknowledge the support of the California PATH program under the grant MOU 373.

8.9 References

[1] F. Altpeter, F. Ghorbel, and R. Longchamp. Relationship between two friction models: A singular perturbation approach, in *Proc. 37th IEEE Conf. Decision and Control*, Tampa, FL, 1998.

[2] L. Alvarez, Yi, J., and R. Horowitz. Adaptive Emergency Brake Control in AHS with Underestimation of Friction Coefficient. In *Proc. American Control Conference*, 574–579, Chicago, IL, 2000.

[3] Yi, J., L. Alvarez, R. Horowitz, and C. Canudas de Wit, Adaptive Emergency Braking Control in Automated Highway System Using Dynamic Tire/Road Friction Model. In *Proc. 39th IEEE Conf. Decision and Control*, Sydney, Australia, 2000.

[4] N. Barabanov and R. Ortega. Necessary and Sufficient Conditions for Passivity of the LuGre Friction Model, *IEEE Trans. Automatic Control*, 45(4), 830–832, 2000.

[5] J. Bernard and C. L. Clover. Tire Modeling for Low-Speed and High-Speed Calculations, *Society of Automotive Engineers*, Paper 950311, 1995.

[6] E. Bakker, L. Nyborg and H. Pacejka. Tyre Modelling for Use in Vehicle Dynamic Studies, *Society of Automotive Engineers*, Paper 870421, 1987.

[7] P. A. Bliman, T. Bonald, T. and M. Sorine, Hysteresis Operators and Tire Friction Models: Application to Vehicle Dynamic Simulations, *Proc. of ICIAM'95*, Hamburg, Germany, 3–7 July, 1995.

[8] A. Browne and D. Whicker. Design of Tire Tread Elements for Optimum Thin Film Wet Traction, *SAE paper*, Nb. 770278, Detroit, MI, USA, 1977.

[9] M. Burckhardt. ABS und ASR, Sicherheitsrelevantes, Radschlupf-Regel System, *Lecture Scriptum*, University of Braunschweig, Germany, 1987.

[10] M. Burckhardt. *Fahrwerktechnik: Radschlupfregelsysteme*, Vogel-Verlag, Germany, 1993.

[11] C. Canudas de Wit and R. Horowitz. Observers for Tire/Road Contact Friction using only wheel angular velocity information. In *Proc. 38th IEEE Conf. Decision and Control*, Phoenix, AZ, 1999.

[12] C. Canudas de Wit, R. Horowitz and P. Tsiotras. Model-Based Observers for Tire/Road Contact Friction Prediction, In *New Directions in Nonlinear Observer Design*, H. Nijmeijer and T.I Fossen (Eds), Springer-Verlag, Lectures Notes in Control and Information Science, May 1999.

[13] Canudas de Wit, C., Olsson, H., Åström, K.J., and P. Lischinsky. A New Model for Control of Systems with Friction, *IEEE Trans. Automatic Control*, 40(3), 419–425, 1995.

[14] C. Canudas de Wit and P. Tsiotras. Dynamic Tire Friction Models for Vehicle Traction Control, *Proc. IEEE Conference on Decision and Control*, 3746–3751, Phoenix, AZ, 1999.

[15] C. Canudas de Wit, P. Tsiotras, E. Velenis, M. Basset, and G. Gissinger, Dynamic Friction Models for Road/Tire Longitudinal Interaction, *Submitted to Vehicle System Dynamics*, May 2001.

[16] H. Chow and P. Kokotovic. Near-optimal feedback stabilization of a class of nonlinear singularly perturbed systems, *SIAM Journal on Control and Optimization*, 16(5), September 1978.

[17] Y. M. Cho and R. Rajamani. A Systematic Approach to Adaptive Observer Synthesis for Nonlinear Systems. *IEEE Trans. Automatic Control*, 42(4), 534–537, 1997.

[18] X. Claeys and J. Yi and L. Alvarez and R. Horowitz and C. Canudas de Wit and L. Richard. A Simple 3D Parametric Dynamic Tire/Road Friction Model for Vehicle Simulation and Control, *Submitted to the 2001 ASME-IMECE*, New York, USA, November 2001.

[19] X. Claeys, J. Yi, L. Alvarez, R. Horowitz, C. Canudas de Wit and L. Richard. Tire Friction Modeling under Wet Road Conditions, *American Control Conference*, Washington DC, USA, June 2001.

[20] C. L. Clover and J. E. Bernard. Longitudinal Tire Dynamics, *Vehicle System Dynamics*, 29, 231–259, 1998.

[21] P. R. Dahl. Solid Friction Damping of Mechanical Vibrations, *AIAA Journal*, 14(12), 1675–1682, 1976.

[22] J. Deur. Modeling and Analysis of Longutudinal Tire Dynamics Based on the LuGre Friction Model, In *Proc. IFAC Conference on Advances in Automotive Control*, 101–106, Karlsruhe, Germany, 2001.

[23] L. El Ghaoui, R. Nikoukhah, and F. Delebecque. LMITOOL: a Package for LMI Optimization. In *Proceedings of 34th IEEE Conference of Decision and Control*, New Orleans, LA, 1995.

[24] L. O. Faria, J. T. Oden, B. T. Yavari, W. W. Tworzydlo, J. M. Bass, and E. B. Becker. Tire Modeling by Finite Elements, *Tire Science and Technology*, 20(1), 33–56, 1992.

[25] G. Gim and P. E. Nikravesh. A Unified Semi-Empirical Tire Model with Higher Accuracy and Less Parameters, *SAE International Congress and Exposition*, Detroit, MI, 1999.

[26] G. Gim and P. E. Nikravesh. An Analytical Model of Pneumatic Tyres for Vehicle Dynamic Simulations. Part I: Pure Slips, *Int. J. Vehicle Design*, 11, 589-618, 1990.

[27] J. Harned, L. Johnston, L. and G. Scharpf. Measurement of Tire Brake Force Characteristics as Related to Wheel Slip (Antilock) Control System Design, *SAE Transactions*, 78, Paper 690214, 909–925, 1969.

[28] W. Horne and F. Buhlmann. A Method for Rating the Skid Resistance of Micro/Macrotexture Characteristics of Wet Pavements, Frictional Interaction of Tire and Pavement, *American Society for Testing and Materials*, STP 793, 191-218, 1983.

[29] H. Kelley, R. Kopp, and H. Moyer. Singular extremals, in *Topics in Optimization*, Academic Press, New York, 1967.

[30] U. Kiencke and A. Daiß. Estimation of Tyre Friction for Enhaced ABS-Systems, In *Proceedings of the AVEG'94*, 1994.

[31] U. Kiencke. Realtime Estimation of Adhesion Characteristic Between Tyres and Road. In *Proc. IFAC World Congress*, 1, 1993.

[32] P. Kokotovic, R. O'Malley, and P. Sannuti. Singular perturbations and order reduction in control theory—An overview, *Automatica*, 12, 123–132, 1976.

[33] Y. Liu and J. Sun. Target Slip Tracking Using Gain-Scheduling for Antilock Braking Systems, In *Proc. American Control Conference*, 1178–1182, Seattle, WA, 1995.

[34] J. P. Maurice, M. Berzeri, and H. B. Pacejka, Pragmatic Tyre Model for Short Wavelength Side Slip Variations, In *Vehicle System Dynamics*, 31, 65–94, 1999.

[35] W. Meyer and J. Walter. Frictional Interaction of Tire and Pavement, *ASTM Publication*, ASTM Publication Code Number 04-793000-37, 1983.

[36] C.F. Moore. *The Friction of Pneumatic Tyres*, Elsevier Scientific Publishing Co., New York, 1975.

[37] D. Moore. Drainage Criteria for Runway Surface Roughness, *Royal Aeronautical Society*, 1964.

[38] D. Moore. The Friction and Lubrification of Elastomers, *Pergamon Press*, 1972.

[39] J. J. Moré, B. Garbow, and K. E. Hillstrom. User Guide for MINPACK-1, Technical Report ANL-80-74, Argonne National Laboratory, Applied Mathematics Division, Argonne, IL, 1980.

[40] W. R. Pasterkamp and H. B. Pacejka. The Tire as a Sensor to Estimate Friction, *Vehicle Systems Dynamics*, 29, 409–422, 1997.

[41] H. B. Pacejka and R. R. Sharp. Shear Force Developments by Pneumatic Tires in Steady-State Conditions: A Review of Modeling Aspects, *Vehicle Systems Dynamics*, 20, 121–176, 1991.

[42] M. Pottinger and T. Yager. The Tire Pavement Interface, *ASTM Publication*, ASTM Publication Code Number 04-929000-27, 1985.

[43] W. Ramberg and W. R. Osgood. Description of Stress-Strain Curves by Three Parameters, *Technical Note 902*, National Advisory Committee for Aeronautics, Washington, DC, 1943.

[44] S. Rohde. On the Combined Effects of Tread Element Flexibility and Pavement Microtexture on Thin Wet Traction, *SAE paper*, Nb. 770277, Detroit, MI, USA, 1977.

[45] S. Rohde. On the Effect of Pavement Microtexture on Thin Film Traction", *International Journal of Mechanical Science*, 18, 95-101, 1975.

[46] M. Sargin. Stress-Strain Relationship for Concrete and the Analysis of Structural Concrete Sections, *SM Study 4*, Solid Mechanics Division, University of Waterloo, Canada, 1971.

[47] M. Sorine and J. Szymanski. A new all-vehicle-speed dynamic tire model, *IFAC Symp. Control in Transportation Systems 2000*, Braunschweig, Germany, June 2000.

[48] P. Tsiotras and C. Canudas-de-Wit. On the Optimal Braking of Wheeled Vehicles, *Proc. American Control Conference*, Chicago, IL, June 28–30 2000.

[49] P. Tsiotras and H. J. Kelley. Drag-Law Effects in the Goddard Problem, *Automatica*, 27, 481–490, 1991.

[50] A. van Zanten, W. D. Ruf, and A. Lutz. Measurement and Simulation of Transient Tire Forces, *International Congress and Exposition*, Detroit, MI, *SAE Technical Paper Series 890640*, 1989.

[51] J. Y. Wong. *Theory of Ground Vehicles*, John Wiley, New York, 1993.

9

Nonlinear Observer Control of Internal Combustion Engines with EGR

E. Hendricks

Abstract

The increased requirements of engine control systems with respect to accuracy, functionality and emission levels have led to a new generation of control strategies. In contrast to earlier systems, these control systems are based on dynamic physical engine models (mean value engine models, MVEMs) and nonlinear estimation. In fact, the new second generation engine control units (ECUs) are just going into production and represent the first mass market application of nonlinear observers.

The purpose of this paper is to review critically the design principles behind some of the newest ECUs. An attempt will also be made to indicate the general direction of development of the newest systems and possible new applications for this methodology.

9.1 Introduction

In the past five years, significant strides have been made in constructing models for internal combustion engines, in particular mean value engine models (MVEMs). These accomplishments, along with the rapid fall in the price of microprocessors and the increasing strictness of emissions legislation have lead to the use of MVEMs directly in Engine Control Units (ECUs). These newer second generation (based on modern control theory) control strategies are now in production and display a new level of functionality and performance. This is both with respect to control accuracy and ease of calibration. It is the purpose of this paper to review critically some of these production nonlinear observers and to suggest possible improvements of them.

Engine observers are generally realized in two different ways in current ECUs: as open loop (or feedforward) observers or closed-loop observers. Examples of both types will be given in what follows.

It should be stated here that the controllers that will be reviewed critically were, at the time of their introduction, based on what was known at the time when they were finalized for production. Since these times, new innovations have been made that have significantly changed the background for making ECUs. These innovations have included more complete physical knowledge of engines/observers as well as improvements in microprocessors and data storage capacity. Production applications are also strongly limited by cost and time concerns, and this has to be balanced against what can be accomplished in the laboratory at a research level.

The controllers to be reviewed are based on papers in which the author and co-workers have contributed as will become apparent. The author and co-workers have, however, not been involved in the production development of these controllers. These have been produced independently, using only ideas from papers published in the open literature.

9.2 Torque Control Feedforward Observer

The Bosch ME7 torque-based controller was recently presented as a new concept in engine control, which is now in production [6], [4], [14]. It is based on an engine model published in the literature some years before [29], [2], as confirmed by [6]. Since the publication of the Bosch paper, two others from different manufacturers have appeared that are based on the same concept: IAV Automotive Engineering, Inc. [15] and Ricardo, Inc. [25].

The torque control concept used in the controllers above emerges from simple manipulations of the engine model in [29] and [2]. The crank-shaft speed state equation from these papers can be written as

$$\dot{n} = -\frac{1}{In} \left[-(P_f + P_p + P_b) + H_u \eta_i m_f(t - \Delta \tau_d) \right]$$
$$= \frac{1}{In} \left[-\frac{2\pi n}{60}(Q_f + Q_p + Q_b) + H_u \eta_i m_f(t - \Delta \tau_d) \right] \quad (9.1)$$

where the Qs are torques, $P = \omega \cdot Q = 2\pi n/60$ and where $\Delta\tau_d$ is the injection-torque time delay. This equation can be solved for the static fuel flow required for a given requested load torque, $Q_{b,req}$, if the crank shaft speed derivative is set equal to zero:

$$m_f(t - \Delta\tau_d) = \frac{2\pi n}{60\, H_u\, \eta_i} \left(Q_{b,req} + Q_f + Q_p \right) \tag{9.2}$$

The airflow required is then

$$m_{ap} = \lambda_{des}\, L_{th}\, m_f(t - \Delta\tau_D) \tag{9.3}$$

where λ_{des} is the desired value of the normalized air/fuel ratio and $\lambda_{des} = 1$ for stoichiometric operation with three way catalysts (TWCs). If the engine operator's pedal command is interpreted as a torque command via $Q_{b,req}$ and a drive-by-wire throttle is used, equations 9.2 and 9.3 can be used to determine how far the throttle should be opened to provide the requested torque. This assumes, of course, some detailed knowledge of the engine being operated.

In order to use Equations (9.2) and (9.3), it is necessary to know the frictional and pumping torques as well as the indicated efficiency as a function of the input and state variables of the engine. As shown in [29] and [2] the frictional and pumping powers (or torques) can be expressed as functions of the manifold pressure and crank shaft speed. Moreover the indicated efficiency can be expressed as the product of four independent terms:

$$\eta_i = \eta_{in}(n) \cdot \eta_{im}(m_i) \cdot \eta_{i\theta}(\theta) \cdot \eta_{i\lambda}(\lambda) \tag{9.4}$$

where m_i is the air charge per stroke and where some changes in the original expressions have been introduced to correspond to those in [6]. In [29], semi-physical expressions (derived from a simplified theory) are given for these terms and these expressions are more or less used in the Bosch controller though they are commonly used as mapped, not equation, functions.

A number of advantages accrue from using the torque-based strategy. One of the most important of these is that this approach makes it possible to structure the entire controller using one overall goal. It is immediately possible to use the same strategy for both diesel and SI engines, with or without EGR, with or without turbocharger, and also with direct gasoline injection. The torque control goal is also a great advantage with respect to system calibration as there is only one single overall calibration goal instead of a collection of them. In addition, this common goal gives the operators what they are most interested in—i.e., responsiveness. It is possible to adjust the "feel" of the system by adjusting the pedal to torque mapping, so many different driving styles can be accommodated and, if required, tailored to individual drivers [15].

Since the strategy is only dependent on using knowledge of the mapped characteristics of the engine and does not use feedback internally, it is a feedforward solution and hence the system depends on an open-loop observer. Moreover, a stationary state or static solution is used to the construct the observer—thus, the observer is also non-dynamic. Unfortunately, this means that the torque control cannot operate correctly in a transient mode. This is in part because of the unavoidable injection/torque time delay but also because of the absence of a dynamic description.

The time delay can be made small with respect to the operator's command using a drive-by-wire throttle, so the main problem is that the torque control system is not capable of reproducing exactly the actual pedal commands. This may be seen in Figure 3 in [6] and, in the throttle/torque plots shown in [25, Figures 12, 13, 14 and 15]. It is also stated specifically that this is the case in [6], [15], and [25]. On rapid tip-in, there is initially an overshoot in the actual engine torque with respect to the command, then an undershoot and finally, on fast tip-out, the requested torque follows somewhat after the pedal command.

Dynamic, Open-loop Air Flow Estimation

A dynamic problem that has to be dealt with in a ECU is the estimation and/or measurement of the air charge or the port air mass flow. Knowledge of this quantity is necessary in order to determine how much fuel is to be injected. While it is not stated explicitly in [6] how this determination is made, it is implied that this is done as in [5]. What is suggested in the latter paper is that the cylinder air charge be found by using an open loop dynamic observer. This is accomplished as follows.

The starting point for the observer is the manifold pressure state equation, which is

$$\dot{p}_i = \frac{R\,T_i}{V_i}\,(-m_{ap} + m_{at} + m_{EGR}) = f_{Ip}(\alpha, p_i, n, T_i, m_{EGR}) \qquad (9.5)$$

where the subscript "I" indicates that this model is for the isothermal manifold filling MVEM (IMVEM) [19], [16]. Exhaust gas recirculation (EGR) has been added to the original state equation in the form in which it is thought that it was originally used in [6]. An assumption implicit in writing down Equation (9.5) is that the ambient, intake manifold and EGR temperatures are the same [19].

The expressions used for the port and throttle air mass flows are

$$m_{at}(\alpha, p_i, p_a) = A\,(1 - \cos(\alpha))\,\sqrt{p_a\,\rho_a}\,\,g(p_i, p_a) \qquad (9.6)$$

where A is the cross-sectional area of the throttle body, ρ_a is the ambient air density,

$$g(p_i, p_a) = \begin{cases} \sqrt{\dfrac{2\kappa}{\kappa-1}\left[\left(\dfrac{p_i}{p_a}\right)^{\frac{2}{\kappa}} - \left(\dfrac{p_i}{p_a}\right)^{\frac{\kappa+1}{\kappa}}\right]} & \dfrac{p_i}{p_a} \geq \left(\dfrac{2}{\kappa+1}\right)^{\frac{\kappa}{\kappa-1}} \\ \left(\dfrac{\kappa}{\kappa+1}\right)^{\frac{1}{2}} \left(\dfrac{2}{\kappa+1}\right)^{\frac{1}{\kappa-1}} & \text{otherwise} \end{cases} \quad (9.7)$$

and

$$m_{ap}(n, p_i) = \dfrac{V_d}{120\, R\, T_i}\, e_v \cdot p_i \cdot n \qquad (9.8)$$

where the volumetric efficiency, e_v, is not used explicitly in the expression in [5]. The expression for the throttle air mass flow above does not agree with that suggested by the author and co-authors [2] but this is a less important point here.

The quantity $e_v \cdot p_i$, proportional to the air charge per stroke, can be expressed as the equation

$$e_v \cdot p_i = s_i\, p_i + y_i \qquad (9.9)$$

where s_i (positive) and y_i (negative) are constants (or are weakly dependent on the crank shaft speed). The physical explanation for this is derived in [2] and used in [5] and [6].

In order to use the expression in Equation (9.5) in the ECU, a transition is made into the crank angle domain and the resulting equation discretized to obtain a difference equation with a sampling interval, which is in crank angle increments, $4\pi/z$, where z is the number of cylinders. This difference equation is used to predict the manifold pressure for calculation of the port air mass flow using the speed-density relation, Equation (9.8) above.

The overall result is that Equation (9.5) is used as a slowly (under-) sampled, open-loop observer for the manifold pressure. It is difficult to see how this will work for a real engine over its operating life without some trimming. as the volumetric efficiency of an engine is temperature, pressure and wear-dependent. It should also be pointed out that the varying sampling frequency is harmonically related to that of the main noise source which is the pumping fluctuations. This is, in particular, a problem with for four cylinder engines.

In [5], it is suggested that the manifold pressure state equation can be used with measurements of a hot wire MAF (mass air flow) meter if the throttle air mass flow m_{at} is replaced with MAF sensor measurements, m_{atm}, in Equation (9.5). In general, this will lead to difficulties because of the injection/combustion time delay and also the MAF sensor response time. This is at best 200 to 50 ms (low to high flow operating points). This problem is also implied in statements in [6], [25] and [5]. Moreover, common hot wire

MAF sensors are not accurate for high pumping fluctuation levels. These problems and the resulting measurement inaccuracies have been discussed in some detail and illustrated with measurements in [21].

Air Charge Estimation with EGR

An important facility which is included in all modern engines is the provision for exhaust gas recirculation to minimize NOx emissions. This is built into the observer as the term m_{EGR} in Equation (9.5). From the description in [6] it is apparent that this is accomplished as follows. The actual gas quantity mass inducted per stroke is

$$m_i = \frac{30}{n} m_{ap} \tag{9.10}$$

including the EGR flow. This means that the fresh air charge per stroke is

$$\begin{align}
m_i &= \frac{30}{n}(m_{ap} - m_{EGR}) \tag{9.11}\\
&= \frac{30}{n}\left(\frac{V_d}{120\,R\,T_i} e_v \cdot p_i - \frac{m_{EGR}}{n}\right) n \tag{9.12}\\
&= \frac{V_d}{4\,R\,T_i} s_i \left(p_i + \frac{y_i}{s_i} - \alpha_{off}\, m_{EGR}\right) \tag{9.13}
\end{align}$$

using Equation 9.9, where

$$\alpha_{off} = \frac{120\,R\,T_i}{s_i\,V_d} \tag{9.14}$$

Obviously, Equation (9.13) can be interpreted in terms of partial pressures for the fresh charge and the EGR. The effect of the EGR is to introduce a negative offset in the linear air charge/stroke expression, Equation (9.9): see Figure 10 in [6].

Equation (9.13) is only approximately correct because Equation (9.5) is not correct in the presence of EGR. The large heat energy contained in the EGR flow has been ignored. Also, transient adiabatic compression and expansion effects have been ignored. This has been shown in [19] and [16]. The air charge prediction error involved can be significant, 6–12%. This problem is further exacerbated by the response times of typical thermistor manifold air temperature sensors which are on the order of 3 to 6 s.

Equation (9.5) is for an isothermal manifold filling MVEM. A more correct adiabatic MVEM model for the intake manifold filling dynamics with EGR is expressed as the coupled nonlinear differential equations:

$$\begin{align}
\dot{p}_i &= \frac{\kappa\,R}{V_i}(-m_{ap}T_i + m_{at}T_a + m_{EGR}T_{EGR})\\
&= f_{Ap}(\alpha, p_i, n, T_a, T_i, m_{EGR}, T_{EGR})
\end{align} \tag{9.15}$$

and
$$\begin{aligned}\dot{T}_i &= \frac{RT_i}{p_i V_i}\left[-m_{ap}(\kappa-1)T_i + m_{at}(\kappa T_a - T_i) + m_{EGR}(\kappa T_{EGR} - T_i)\right] \\ &= f_{AT}(\alpha, p_i, n, T_a, T_i, m_{EGR}, T_{EGR})\end{aligned} \quad (9.16)$$

where the subscript A indicates that this model is for the adiabatic MVEM (AMVEM) and where the heat transfer between the intake manifold surface and the intake air has been neglected [19], [16]. In general, this is approximately true for typical engines because of the small heat exchange coefficient between the intake manifold and the intake air. The validity of Equations (9.15) and (9.16) has been confirmed by direct measurement [19].

The IMVEM and the AMVEM have been published and compared in a paper, which is available on the World Wide Web as well as a downloadable Matlab^{TM}/Simulink package, including both models [17].

Thus, the use of the approximations inherent in the static method of handling EGR in the torque-based system of [6] introduces large errors in the air charge calculation, and this implies long calibration times and complicated software to compensate for them. It also suggests significant emission penalties [27].

Transient Fuel Film Compensation

If the torque control system is to work on a port injected engine, it is necessary to use a transient fuel film compensation algorithm. If this is done, then it will be difficult to fuel the engine properly during fast transients. This is because the compensation algorithm is difficult to adjust for proper low-emission engine operation at all operating points and temperatures. This is probably at least part of the explanation for the overshoot on fast tip-ins in the responses mentioned above in [25]: on large tip-ins, the engine is overfuelled, as then it will operate correctly and this does not impact too strongly on the performance on the mild US or European driving cycles. It does, however, impact negatively emissions performance at highway cruising speeds and passing accelerations.

9.3 Closed-loop Observer

Apart from the question of torque control, closed-loop observers may be profitably used in ECUs for the estimation of the air charge, to compensate for the deficiencies of sensors, to reduce the effects of measurement noise, and to give first order compensation for changes in the engine parameters due to wear. Such observers may even be used to estimate the instantaneous torque delivered by the engine. To the author's knowledge, this was first shown in the papers [20] and [30]. Here, an extended Kalman filter was used for engine state and engine torque estimation.

The first production application of this technology was made by Delphi Automotive Systems, Inc. in 1998 [9]. This work is based on the paper [30], as confirmed by [3], and was first demonstrated publicly in a vehicle in 1998 [10]. The overall strategy is called a pneumatic state estimator (PSE) and thermal state estimator (TSE) in the original paperm which will here be shorted to PTSE for the purposes of discussion. The original production ECU EGR treatment strategy has recently been improved by Muller [11], part as the basis of his master thesis work [24].

Engine State Estimation

The PTSE is intended for air/fuel ratio control so that it concentrates on the problem of finding the correct cylinder air charge based on (noisy) manifold pressure and ambient temperature measurements and the speed-density equation. The fundamental underlying concept is a continuous integration of Equation (9.5) which is provided with an innovation to make a constant gain extended Kalman filter. That is

$$\dot{\hat{p}}_i = f_{Ip}(\alpha, \hat{p}_i, n, T_a, \cdot) + K_p(p_{im} - \hat{p}_i) \tag{9.17}$$

where the variable with the hat is an estimate, the m subscripts indicate a measurement and the K is the Kalman gain. In addition to Equation (9.17), there are five other pressures that are estimated dynamically in the same way in the PTSE. The Kalman gains in these estimators are adjusted using empirical methods, implying that the PTSE is in reality a pole placement, deterministic observer. This is in contrast to the work in [30] where these were found using engine noise models.

In the production ECU, the equations mentioned above are integrated using Euler integration with a constant sample time. The Kalman filter equations above are provided with integral terms for the innovations to insure that the state estimates are centered around the mean values of the states in the steady state.

Different temperatures in the engine are estimated using algebraic expressions based on simple physics. In the case of the intake manifold temperature, this implies it has to be determined from an equation like (9.16). This is a dynamic equation so this is not immediately possible. Assuming, however, that the required quantities can be measured or estimated, one can use a steady state solution to Equation (9.16).

$$0 = \frac{RT_i}{p_i V_i}\left[-m_{ap}\left(\kappa - 1\right)T_i + m_{at}\left(\kappa T_a - T_i\right) + m_{EGR}\left(\kappa T_{EGR} - T_i\right)\right] \tag{9.18}$$

This implies that

$$T_i = \kappa \frac{m_{at}T_a + m_{EGR}T_{EGR}}{m_{ap}(\kappa - 1) + m_{at} + m_{EGR}} \tag{9.19}$$

is the estimated intake manifold temperature. It is not known that the intake manifold temperature is actually calculated in this way in [9], but this is a reasonable guess. This solution, of course, ignores the actual effective time constant in (9.16).

Because an approximate physical instead of an empirical model is the basis for the PTSE, it is easy to adapt to new engines and several original equipment manufacturers (OEMs) have adopted it for this reason. The physical foundation of the PTSE makes it a physically modular system where subsystems can be understood intuitively, isolated and tested individually. Empirical calibration, which is very time consuming and expensive, is thus eliminated. In addition, a sensitivity analysis is possible for the overall system.

It is simpler to calibrate the PTSE system for scheduling of atmospheric pressure changes, purge, throttle, and idle speed valves and EGR [12], [11] than conventional systems. Further, diagnostics for valve flow, sensor measurement, and undesired pressure drops are relatively simple. In addition to these functions, temperature estimates at a number of locations in the engine are made available for other ECU functions. An automated calibration procedure has been developed for the PTSE.

The main obvious disadvantage of the PTSE is that it requires modeling of a number of subsystems with relatively great detail and accuracy. This detail means, however, that once the modelling is done, the control system will be valid over the entire temperature range in which the engine control system is to operate. The automatic calibration procedure is, however, a great help in minimizing the calibration/identification work necessary to find all of the parameters required for the model.

Running closed-loop estimators of six or more variables means that it is relatively heavy computationally, and that efficient programming is necessary to fit it into the ECU memory and CPU. It should also be mentioned that what is on offer is a pure state estimator, not a state and parameter estimator. This means that there can be some drift of the estimator over the life of an engine. It is not obvious that the pressure state estimator alone as described is capable of managing this problem.

Air Charge Estimation/Measurement

It is clear that the Delphi system will share some of the main transient estimation problems as the Bosch system above with respect to the error introduced by not using Equation (9.16). This error will be somewhat reduced by the introduction of the innovation term in the closed-loop observer, but the slow response of typical temperature sensors is not compensated for. Some compensation is also obtained from the EGR estimator suggested in [11], but a complete discussion of this point is outside the scope of this paper.

As mentioned above, a common sensor in use for the throttle air mass flow is a hot wire sensor. The PTSE as currently described in the literature is not compatible with the use of such a sensor as originally presented. It is supposed, however, that some facility has since been provided for the use of this kind of sensor, though this author is not aware of this possibility at the present time.

As has been pointed out recently, the use of a pressure sensor makes it difficult to predict correctly the cylinder air charge when the engine is used with a turbocharger. This is because when the waste gate opens, the exhaust back pressure changes significantly, and this changes the effective volumetric efficiency of the engine [13]. A solution to this problem, suggested in this paper, is to estimate the volumetric efficiency on-line and turn off the adaption during large pressure transients. Another less sophisticated solution might be to map the term $e_v \cdot p_i$ (which is a simple, compact function [2]) with and without the wastegate being open. The ECU, detecting rapid pressure spikes at high manifold pressures, could then change from one air charge per stroke table to another as required.

Transient Fuel Film Compensation

If the PTSE is used on a port injected engine then there will also be a problem with the transient fuel film compensation algorithm as with the Bosch system. As with the other system, ignorance of the instantaneous intake manifold temperature will lead to calibration difficulties with the compensator especially if this has to be done over a large temperature range. It is thought that this problem has been addressed in later work but it is not clear how this has been done.

9.4 Possible Improvements

From the critical discussion above, it is clear that there are several problems which should be considered with respect to the production use of nonlinear estimators:

1. Open vs. closed loop estimators;
2. Manifold pressure and/or port air mass flow estimation;
3. Manifold temperature estimation;
4. Hot wire (or hot film) air meter deficiencies;
5. Improved fuel film compensation;
6. Dynamic torque estimation.

The purpose of this section is to make suggestions about possible solutions of these questions. The solutions to be suggested are not the most advanced or the most accurate: they are only meant to be compatible with the capacity of current ECUs and current OEM attitudes and desires.

Air Charge Estimation with EGR

While there are some questions about the accuracy of the Delphi PTSE, it is obvious that the use of an extended Kalman filter is a great benefit to enhancing the accuracy of it. Generally speaking, closed loop observers will always perform better that feedforward systems. Thus the answer to Problem 1 above is clear: a closed-loop observer is to be preferred.

From the papers [19] and [16] and the observations above, it is obvious that the adiabatic model is to be preferred to the isothermal model. This suggests that an APTSE (adiabatic PTSE) should be used as the basis of the extended Kalman filter for an SI engine with EGR. This means that the state estimator should be the coupled set of equations:

$$\dot{\hat{p}}_i = f_{Ap}(\alpha, \hat{p}_i, n, T_a, \hat{T}_i, \cdot) + K_p(p_{im} - \hat{p}_i) + K_{pT}(T_{im} - \hat{T}_i) \quad (9.20)$$

and

$$\dot{\hat{T}}_i = f_{AT}(\alpha, \hat{p}_i, n, T_a, \hat{T}_i, \cdot) + K_{Tp}(p_{im} - \hat{p}_i) + K_T(T_{im} - \hat{T}_i) \quad (9.21)$$

were the Ks are constant Kalman gains. The use of both innovations is necessary because the intake manifold pressure and temperature are strongly coupled. The innovations should also be used as integral terms in the observer to reduce the steady state errors to zero. Integration steps for the two coupled differential equations above can be the same: the effective time constant of the second equation is of the order of twice that of the first equation [16]. The APTSE above is the suggested solution to Problems 2 and 3 above.

In fact, a system like that in Equations (9.20) and (9.21) has already been tested and the results published, though in a more advanced form [18]. This controller is constructed around a predictive observer based mainly on the two equations above, and it requires the on-line evaluation of the algebraic Riccati equation. Thus, it is at present too complex and expensive for production applications. The air/fuel ratio control exhibited by this laboratory controller is quite accurate even for very large and fast throttle angle, manifold pressure and crankshaft speed transients. The suggested controller, using the APTSE, will be approximately this accurate if the following provisions are made.

Measurements with Semiconductor Pressure and Thermistor Temperature Sensors

Presently available semiconductor intake manifold pressure sensors have rise times somewhat smaller than the minimum throttle opening time which is approximately 50 ms [21]. Current thermistor temperature sensors are much slower, as indicated above. In order to avoid the finite rise times of each sensor, an ETC must be used. As the manifold pressure sensor is measuring the most important state and as it is the fastest sensor, the ETC must not open the throttle before a time has past that is sufficient for the pressure sensor to respond correctly, approximately 50 ms, and the delay time for opening the fuel injectors. After this, it is to follow the pedal's commands explicitly.

The pedal commands should drive the observer directly so that this signal is available before the command reaches the fuel injectors. Then, the observer will have the correct estimate of the instantaneous air mass flow available at the time when this air mass flow passes the fuel injector, at least approximately. The Kalman gain K_p should therefore be large in order that the observer follow the pressure change. As the temperature measurement is slow, it cannot be trusted until the engine is in the steady state. This means that the Kalman gain K_T should be small or zero at the start of the fast transient and should only be changed to a finite value after a time corresponding to the settling time of the temperature sensor, about five times its time constant. The time constant for a rising input is faster than that for a falling input: the Kalman gain must be adjusted for this difference. The Kalman gains corresponding to the correlation of the pressure and temperature signals, K_{pT} and K_{Tp}, must also be adjusted accordingly. In this way, the Kalman filter will trust the pressure measurement as it becomes available but will ignore the temperature sensor measurement except in the steady state or when it is slowly changing.

With these provisions, the APTSE should work approximately as well as the predictive observer of [18].

Measurements with Hot Wire MAF Sensors

One of the difficulties with hot wire or hot film MAF sensors is that they to not measure directly one of the most important states of the engine. This means that it is difficult to include their measurement directly in an observer based on Equations (9.15) and (9.16) above. The solution to this problem is to model the throttle air mass flow as a state.

This is possible using a technique originally presented in [23] and requires the derivation of a differential equation for the throttle air mass flow. Assuming that IMVEM is a reasonable approximation to the AMVEM for the throttle air mass flow and without going into detail, the result of the derivation is

$$\dot{m}_{at}(\alpha, n, m_{at}) = m_{pTa} f_\alpha(\alpha, m_{at}) \dot{\alpha} \qquad (9.22)$$
$$+ \; m_{pTa} f_\beta(\alpha, m_{at}) \left(-m_{ap}(m_{at}, n) + m_{at} m_{EGR} \right)$$

or

$$\dot{m}_{at}(\alpha, n, m_{at}) = f_{Im}(\dot{\alpha}, \alpha, m_{at}, T_i, n) \qquad (9.23)$$

The extra functions involved, $f_\alpha(\cdot)$, $f_\beta(\cdot)$ and $m_{ap}(\cdot)$, can be derived analytically directly from those of the IMVEM. For the details see [23]. The prediction of Equation (9.23) is fairly accurate (in spite of being for the IMVEM) because the internal physics of the intake manifold do not impact too strongly on the throttle air mass flow. The equation immediately above has to be combined with one for the manifold temperature to be complete,

Equation (9.16). The overall observer for the intake manifold is then

$$\dot{\hat{m}}_{at} = f_{Im}(\dot{\alpha}, \alpha, \hat{m}_{at}, \hat{T}_i, n) + K_m(m_{atm} - \hat{m}_{at}) + K_{mT}(T_{im} - \hat{T}_i) \quad (9.24)$$
$$\dot{\hat{T}}_i = f_{AT}(\alpha, \hat{p}_i, n, T_a, \hat{T}_i) + K_{Tm}(m_{atm} - \hat{m}_{at}) + K_T(T_{im} - \hat{T}_i) \quad (9.25)$$

where the Ks are again Kalman gains. Integral terms for the innovations may be added in order to reduce the steady state error in the mean to zero. If the Kalman gain, K_m, is made to be time varying it will be possible to compensate somewhat for the response time of the hot wire sensor.

On fast throttle angle transients or at high pulsation levels, the Kalman gain, K_m, should be reduced to zero. This will insure that the poor response of the sensor does not influence the model predictions. When the flow does not change too rapidly, the gain should be adjusted so that the filter uses the measurements to a greater degree. The Kalman gains, K_{mT} and K_{mT}, have to be adjusted accordingly. The gain K_T should be adjusted as indicated above in Section 4.2. This is a suggestion for the simple solution to Problem 4 above.

It is clear that this method of compensating for the deficiencies of the sensor is only approximate, and will require a reasonable amount of empirical adjustment to work properly. In the end, the current hot wire and hot film sensors should be replaced with some which are faster, more accurate, and stable. This will be the only way of meeting future emissions restrictions.

Transient Fuel Film Compensation

Fuel film compensation in port injected engines has been a problem for a number of years, and has not yet been solved in a satisfactory and convenient manner, especially over wide temperature ranges. In general, calibration of the (usually) feedforward nonlinear compensators that are available is difficult because the fuel film dynamic system is only experimentally accessible in transients. The level of compensation must also change with the throttle input amplitude and speed, and with the ambient and intake manifold air temperatures as well as the engine coolant temperature. Turbocharging and EGR insertion multiply these difficulties.

There are available two nonlinear compensators for the two most common nonlinear models, [1], [22]. Neither of these is entirely satisfactory because the underlying models are probably incorrect as they both ignore the liquid fuel that enters the combustion chamber. That this is the case is shown clearly in [31] and [8]. Nevertheless, the available compensators can be used with some degree of success if the parameters of these models are adjusted properly as a function of operating point. At least, they can significantly improve the air-fuel ratio control accuracy of observer-based engine control systems [30]. This is a suggestion for the interim solution of Problem 5 above.

Recently, work has been carried out to make the conventional compensators adaptive and this technique, even though it is in preliminary state,

[26], [7], would seem to be the correct method of attack in the long run. Port injected engines will be very common for a number of years to come because of their relatively low cost. They will, therefore, be a must for developing markets, for example, China and India.

Dynamic Engine Torque Estimation

The IMVEM observer, which is described in [30], is a full-order observer which means that it estimates simultaneously two states: the intake manifold pressure and the crankshaft speed. The structure of the observer for the crankshaft speed is such that the innovations for the manifold pressure and crank shaft speed enter the Kalman filter model in parallel with the pumping and frictional torques (or, if desired, powers). The relevant observer equation is

$$\dot{\hat{n}} = \frac{1}{I\hat{n}}\{-\frac{2\pi\hat{n}}{60}[Q_f(\hat{p}_i,\hat{n}) + Q_p(\hat{p}_i,n) + Q_b(\hat{n})] \\ + I\hat{n}[K_{np}(p_{im}-\hat{p}_i) + K_n(n_m-\hat{n})] + H_u\,\eta_i(\cdot)m_f(t-\Delta\tau_d)\} \quad (9.26)$$

as is easily seen from Equation (9.1), where the m subscripts indicate measurements and the hats again indicate estimates. The innovations are given in the second line in the equation.

This means that the innovations are effectively the load torque or power and if they are simply read out of the model, they will effectively be a fast dynamic torque estimate. In [30], it is stated that estimates with an accuracy of $5 - 10\%$ can be obtained. With the improvements which have been made in MVEM modelling, it is thought that this accuracy can be improved to say $2 - 5\%$ which is sufficient for a dynamic torque control strategy. Experiments are currently being conducted to check this conclusion. Such a torque estimator would be more compact, more easily calibrated, and much easier to construct than the stochastic and frequency analysis estimators which can be found in the literature, see, for example, [28].

This is a first suggestion for the dynamic torque estimation question put forward above.

9.5 Conclusions

A critical review has been given for two types of model/observer-based engine control systems. These two types of systems represent the current production cutting edge in engine control systems. One of the applications, the one from Delphi, is one of the first mass market application of nonlinear observers. It has been made clear that both of these applications have clear advantages and drawbacks with respect to ease of calibration and performance. However, in spite of what criticism can be made of the details of the implementations, they both represent a clear departure from earlier practice and a clear advance with respect to non-dynamic, table-based, cut and try strategies.

Again, it must be emphasized that production applications are not the same as experimental laboratory concept models. Elegance and functionality must be balanced against engineering and component costs. And it is difficult to imagine an industry that is more sensitive to cost concerns than the automotive industry.

9.6 Nomenclature

The following symbols are used in this paper:

t	time (s)
α	throttle plate angle (degrees)
n	engine speed (rpm/1000 or krpm)
p_a	ambient pressure (bar)
T_a	ambient temperature (degrees Kelvin)
p_i	absolute manifold pressure (bar)
T_i	intake manifold temperature (degrees Kelvin)
T_{EGR}	EGR temperature (degrees Kelvin)
P_f	engine friction losses (kW)
P_p	engine pumping losses (kW)
P_b	engine load power (kW)
H_u	fuel heating value (kJ/kg)
η_i	indicated efficiency
I	engine moment of inertia ($= I_{ac} \cdot (\pi/30)^2 \cdot 1000\,\text{rpm}$)
I_{ac}	actual engine moment of inertia (kg·m^2)
m_{at}	air mass flow past throttle plate (kg/s)
m_{ap}	air mass flow into intake port (kg/s)
m_{EGR}	EGR mass flow (kg/s)
e_v	volumetric efficiency based on manifold conditions
V_d	engine displacement (liters)
V_i	manifold + port passage volume (m^3)
R	gas constant (here $287 \cdot 10^{-5}$)
κ	ratio of the specific heats = 1.4 for air
λ	normalized air/fuel ratio
L_{th}	stoichiometric air/fuel ratio (14.67)

9.7 References

[1] C. F. Aquino. Transient A/F control characteristics of the 5 liter central fuel injection engine. *SAE Technical Paper 910494*, 1981.

[2] E. Hendricks, A. Chevalier, M. Jensen, S. C. Sorenson, D. Trumpy, and J.

Asik. Modelling of the intake manifold filling dynamics. *SAE Technical Paper 960037*, 1996.

[3] P. Maloney, and P. Olin, Delphi). Private communication, 1998.

[4] J. Gerhardt, H. Honninger and H. Bischof. New approach of functional and software structure. *SAE Technical Paper*, (980801), 1998. Also in SAE Special Publications, SP-1357.

[5] N. F. Benninger and G. Plapp (Bosch). Requirements and performance of engine management systems under transient conditions. *SAE Technical Paper 910083*, 1991.

[6] J. Gerhardt, N. Benninger, and W. Hess (Bosch). Torque-based system structure of an electronic engine management system as a new base for drive train systems (Bosch ME7 system). Technical report, 6. Aachener Kolloqium, Fahrzeug- und Motorentechnik, Aachen, Germany, October 1997.

[7] J. E. Pakkala and J. B. Burl. Fuel evaporation parameter identification during si cold start. *SAE Technical Paper 2001-01-0552*, 2001. Also in SAE Speciel Publications, Sp-1585.

[8] E. W. Curtis, C. F. Aquino, D. Trumpy, and G. C. Davis. A new port and cylinder wall wetting model to predict transient Air/Fuel excursions in a port fuel injected engine. *SAE Technical Paper 961186*, 1996.

[9] P. Maloney P. Olin (Delphi). Pneumatic and thermal state estimators for production engine control and diagnostics. *SAE Technical Paper 980517*, 1998.

[10] P. Maloney, P. Olin, and J. Water (Delphi). IFAC Workshop: Advances in automotive control, Mohigan State Park, Londonville OH. Technical report, IFAC, February 1998.

[11] M. Muller, P. M. Olin, and B. Schreurs (Delphi). Dynamic EGR estimation. *SAE Technical Paper 2001-01-0553*, 2001. and SAE Special Publications, SP-1585.

[12] P. Olin, and P. Maloney, P. (Delphi). Barometric pressure estimator for production engine control and diagnostics. *SAE Technical Paper 1999-01-0206*, 1999. Also in SAE Special Publications, SP-1418.

[13] P. Andersson and L. Eriksson. Air-to-cylinder observer on a turbocharged si-engine with wastegate. *SAE Technical Paper 2001-01-0262*, 2001. Also in SAE Special Publications, SP-1585, 2001.

[14] B. Mencher, H. Jessen, L. Kaiser, and J. Gerhardt. Preparing for CARTRONIC. *SAE Technical Paper 2001-01-0268*, 2001.

[15] A. Greff and T. Gunther. A new approach to a multi-fuel, (IAV Eng.), torque based ECU concept using automatic code generation. *SAE Technical Paper 2001-01-0267*, 2001.

[16] E. Hendricks. Isothermal vs. Adiabatic Mean Value SI Engine Models. *IFAC Workshop on Advances in Automotive Control, Karlsruhe Germany*, 28-30 March 2001.

[17] E. Hendricks. A Generic Mean Value Engine Model for Spark Ignition Engines. Technical report, SIMS 2000, DTU, Sept. 18 - 19, Lyngby, Denmark, 2000. This paper and the Matlab/Simulink models can be found at the web address www.iau.dtu.dk/~eh/.

[18] A. Chevalier, C.W. Vigild, and E. Hendricks. Predicting the port air mass flow of SI engines in Air/Fuel ratio control applications. *SAE Technical Paper 2000-01-0260*, 2000. Also in SAE Special Publications, SP-1500.

[19] M. Fons, M. Muller, A. Chevalier, C. Vigild, S.C. Sorenson, and E. Hendricks. Mean value engine modelling of an SI engine with EGR. *SAE Technical Paper 1999-01-0909*, 1999. lso in SAE Special Publications, SP-1419.

[20] T. Vesterholm and E. Hendricks. Continuous SI engine observers. Technical Report TP14-15:30, *American Control Conference (ACC)*, Boston, MA., June 1991.

[21] E. Hendricks, A. Chevalier, and M Jensen. Event based engine control: Practical problems and solutions. *SAE Technical Paper 950008*, 1995. Also in Special Publications, SP-1082.

[22] E. Hendricks, T. Vesterholm, P. Kaidantzis, P. Rasmussen, and M. Jensen. Nonlinear Transient Fuel Film Compensation (NTFC). *SAE Technical Paper 930767*, 1993.

[23] E. Hendricks, J. Poulsen, M.B. Olsen, P.B. Jensen, M. Fons, and C. Jepsen. Alternative observers for SI engine Air/Fuel ratio control. Technical report, *Proc. 35th IEEE Conf. on Decision and Control*, 2806-2811, Kobe, Japan, 1996.

[24] M. Muller. Mean value modelling of turbocharged spark ignition engines. Master's thesis, Inst. for Automation, Lab. for Energetics, Technical University of Denmark, 1998.

[25] N. Heintz, M. Mews, G. Stier, A.J. Beaumont, and A.D. Noble An approach to Torque-Based engine management systems. *SAE Technical Paper 2001-01-0269*, 2001.

[26] N.P. Fekete, U. Nester, I. Gruden, and J.D. Powell. Model-based Air-Fuel ratio control of a lean multicylinder engine. *SAE Technical Paper*, (950846), 1995. Also in SAE Special Publications, SP-1086.

[27] E. Achleitner, W. Hosp, A. Koch, and W. Schurz (Siemens). Electronic engine control system for gasoline engines for lev and ulev standards. *SAE Technical Paper*, (950479), 1995.

[28] B. Lee, G. Rizzoni, Y. Guezennec, and A. Soliman. Engine control using torque estimation. *SAE Technical Paper 2001-01-0995*, 2001. Also in SAE Special Publications, SP-1585.

[29] E. Hendricks and S.C. Sorenson. Mean value modelling of spark ignition engines. *SAE Technical Paper 900616*, 1990.

[30] E. Hendricks, T. Vesterholm, and S.C. Sorenson. Nonlinear, closed loop, SI engine control observers. *SAE Technical Paper*, (920237), 1992. Also in SAE Special Publications, SP-898.

[31] K. Saito, K. Sekiguchi, N. Imatake, K. Takeda, and T. Yaegashi. A new method to analyze fuel behavior in a spark ignition engine. *SAE Technical Paper 950044*, 1995.

10

Idle Speed Control Synthesis using an Assume-guarantee Approach

A. Balluchi   L. Benvenuti   M. D. Di Benedetto
A. L. Sangiovanni-Vincentelli

Abstract

The goal of an idle control for automotive engines is to maintain the engine speed within a given range, robustly with respect to load torque disturbances acting on the crankshaft. Mean value models have been used in the past to design idle control algorithms. However, the behavior of the torque generation process and the dynamics of the power train are not captured with enough accuracy to guarantee that the idle control specifications as given by car makers are met. We use a cycle-accurate hybrid model to overcome these obstacles. To tackle the complexity of the controller design, the system is decomposed in three parts. For each part in isolation, a control law is derived for a simplified model, assuming that the other parts can be controlled to yield appropriate inputs. The overall control strategy is then applied to the system. Hence, the correct interaction of the feedback loops is formally verified using an assume-guarantee approach, to ensure that the behavior of the controlled system meets the given specifications.

10.1 Introduction

The synthesis of an idle control strategy for an internal combustion engine is among the most challenging problems in engine control. The objective is maintaining the engine speed as close as possible to the value that minimizes fuel consumption, while preventing the engine from stalling. The difficulty lies in the unpredictable load variations coming from the intermittent use of devices powered by the engine, such as the air conditioning system and the steering wheel servo-mechanism. A survey on different engine models and control design methodologies for the idle control is given in [7]. Both time-domain (*e.g.*, [4]) and crank-angle domain (*e.g.*, [11]) mean-value models have been proposed in the literature. More recently, cycle-accurate models capturing periodic engine speed variations due to torque fluctuations, were investigated in [10]. Multivariable control [9]), ℓ_1 control [4], H_∞ control [5], μ-synthesis [6], sliding mode control [8], and LQ-based optimization [1] have been applied to idle control on a variety of models. However, a fully satisfactory solution has still to emerge. In this paper, we use a hybrid model to describe the cyclic behavior of the engine, thus capturing the effect of each spark ignition on the generated torque and the interaction between the discrete torque generation and the continuous power train and air dynamics. We consider a traditional spark ignition engine without gas direct injection (GDI). The torque generated by each cylinder and applied to the engine crankshaft can be assumed to be a function of the spark ignition time, and of the air-fuel mixture mass loaded in the cylinder during the intake phase. Since the air-fuel ratio is assumed to be constant (at the stoichiometric value), then the mixture mass is controlled by the throttle plate position and is subject to the dynamics of the cylinder filling. Hence, the available controls for the idle problem are: the spark ignition time and the position of the throttle valve, which regulates the air inflow[1]. A hybrid model of the plant is obtained from the general model of an internal combustion engine presented in reference [3], by using nonlinear expressions and model parameters identified on a commercial car in collaboration with our industrial partner, Magneti Marelli Powertrain S.p.A.. In [3] and [2], the problem of maintaining the crankshaft speed within a given range was formalized as a "safety" specification for the hybrid closed–loop system[2]. For a simplified engine hybrid model, where some nonlinear expressions were linearized and no actuation delay was considered, the idle control problem was solved by computing analytically the set of all hybrid states for which there exists a hybrid control strategy meeting the specification. The class of all "safe" controllers obtained by the procedure was called the *maximal controller*. Despite the simplification of the model, the expression

---

[1]The effect of a spark command on the torque generation is "stronger" than the one of a throttle plate command, since air inflow is subject to both manifold dynamics and delay due to mix compression. Hence, sudden loads can be much better compensated with spark ignition than with air inflow, while air inflow can be used to control the engine in steady state.

[2]A safety specification requires the system to stay within a set of specified safe states.

of the maximal controller obtained in [3], [2] is quite complex. Consequently, the implementation of a particular controller extracted from the maximal controller is quite expensive in terms of computing time and memory.

To solve the problem in ways that are industrially feasible, we have to take into account nonlinearities, actuation delays, and yet we have to keep a close eye on implementation costs. The approach we propose in this paper is to select semi-heuristically an easy-to-implement controller and then verify that it satisfies the specifications.

Verifying the system by simulation and prototyping is certainly possible but we cannot guarantee that the system will satisfy the specifications in all operating conditions. In fact, it is often the case that idle control needs extensive empirical adjustments in the car.

In this paper, we present a first cut for a methodology based on a divide and conquer approach to obtain the controller and formal techniques to guarantee that the controller is contained in the maximal controller and, hence, that it satisfies the specifications.

In our case, the closed-loop system is viewed as being composed of three subsystems (intake manifold, cylinders and power train subsystems). The control law is derived by simplifying substantially each of the subsystems so that the hybrid nature of the model is ignored and the subsystems are linearized and discretized with a fixed sampling time. The control law is synthesized so that the constraints are satisfied in this simplified domain. Then, this control law is applied to the fully-fledged system. Of course, at this point, there is no *guarantee* that the control law will yield a closed-loop system that satisfies the constraints even though the decomposition and the corresponding simplifications have been made so that we do not wander far from the original model. The basic idea of our methodology is to verify each subsystem in isolation, assuming that the behavior of the other subsystems satisfies an appropriate set of conditions. Then, the consistency of the assumptions on the behaviors is verified, knowing that each of the subsystems works correctly in isolation. The subsystems are the same as the ones identified for the derivation of the control law. However, the behavior of the closed-loop power train subsystem is still too complex to verify formally, since it contains the hybrid power train model. Hence, the hybrid power train model is itself decomposed in the continuous part and the discrete part and, then, verified following the same divide and conquer approach. Note that this approach is a particular case of the assume-guarantee paradigm widely used in formal verification.

To summarize, our methodology consists of dividing the overall system in a set of interconnected subsystems, simplifying the model for each, derive a control law for the decomposed, simplified system, and finally formally verify that the control law (possibly modified to take into account some of the simplifications made) satisfies the constraints of the fully fledged model. To the best of our knowledge, this approach is novel in hybrid control.

The paper is organized as follows: in Section 10.2, a description of the hybrid model of the engine in the idle region of operation is recalled. In

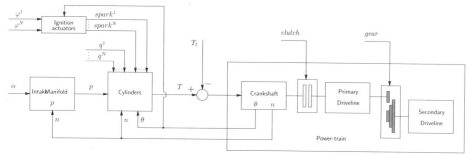

Figure 10.1 Model of the engine

Section 10.3, an idle speed controller is proposed and in Section 10.4, the verification results for the behavior of the closed-loop hybrid system are reported.

10.2 Plant Hybrid Model

In this section, we briefly describe a hybrid model of a *four*-stroke internal combustion engine. This model has been obtained from the hybrid model presented in reference [3] by specifying nonlinear expressions and model parameters on the basis of the experimental data obtained from a commercial car at idle speed provided us by Magneti Marelli Powertrain (Italy).

The model is composed of four interacting blocks, namely, the intake manifold, the cylinders, the power train, and the actuators, as shown in Figure 10.1.

The *intake manifold* pressure dynamics are a continuous-time process controlled by the throttle-valve position α. Denoting by p the pressure, manifold dynamics are modelled as

$$\dot{p}(t) = a_p(p(t))\, p(t) + b_p(p(t))\, s(\alpha(t)) \tag{10.1}$$

where $s(\alpha)$ is the *equivalent throttle area*, given in terms of throttle angle α. Parameters a_p and b_p depend in a strongly nonlinear fashion on the geometric characteristics of the manifold, on the physical characteristics of the gas and atmosphere, and on the current value of the pressure p. The dynamics of actuation of the throttle valve is modelled by a linear first-order dynamical system:

$$\dot{\alpha}(t) = a_\alpha \alpha(t) + b_\alpha V(t) \tag{10.2}$$

where V is the motor input voltage that is assumed to be a discrete time signal produced with a sampling period τ_A.

The *cylinders*-block models the torque generation. The torque T generated by each piston at each cycle depends on the thermodynamics of the air-fuel mixture combustion process. The profile of T depends on the phase of the cylinder, the piston position, the mass m of air and q of fuel loaded during the intake phase, and on the spark ignition timing. For idle speed values, the quantity m of air loaded into each cylinder at the end of the intake run can be assumed to depend only on the value of the intake manifold pressure at the intake end time t_{int} as

$$m = k_m(p(t_{int}))\, p(t_{int}) \tag{10.3}$$

Assuming that fuel injection is regulated by an inner control loop that maintains the air-fuel ratio to stoichiometry, the profile of the torque T generated by each cylinder can be described by a piecewise constant function that is assumed to be zero everywhere except in the expansion phase, in which

$$T(t) = G\, m\, \eta(\varphi) \tag{10.4}$$

where the gain G is a constant parameter, φ is the spark advance[3], and $\eta(\varphi)$ is the ignition efficiency.

When the gear is neutral and the clutch is released, the secondary driveline is disconnected from the engine so that the *power train* is described only in terms of the crankshaft speed n and position θ by the continuous-time system

$$\dot{n}(t) = a_n\, n(t) + b_n\, (T(t) - T_l(t)) \tag{10.5}$$
$$\dot{\theta}(t) = k_n\, n(t) \tag{10.6}$$

where a_n, b_n are constant parameters, T is the torque produced by the engine, and T_l represents a load torque acting on the crankshaft.

The ignition *actuators* must produce the spark ignition at every cycle, according to the current decision of the control algorithm, and synchronously with the crankshaft position θ. Since ignition control takes time to actuate, then it has to be decided sufficiently in advance to make sure that it is properly delivered to the plant. The spark is, in general, ignited with a different spark advance at every cycle. The value of the spark advance must then be computed at the end of the intake stroke, so that the ignition subsystem can be programmed to ignite the spark at the proper time.

In conclusion, the behavior of a four-cylinder in-line engine and power train can be represented using only one discrete state, as shown in figure 10.2. In fact, the engine kinematics are such that, at any time, each cylinder is in a different stroke of the engine cycle and only one cylinder is generating the torque T. Every half of a crankshaft run, *i.e.*, when $\theta = 180$, the control law $f_\varphi(\cdot)$ is computed to obtain the spark ignition to be applied to the cylinder that is entering the compression stroke at the current dead center. The throttle valve control $f_a(\cdot)$ is computed at fixed frequency $1/\tau_A$ and will affect the amount of air loaded by the cylinder that is performing the intake stroke at the current sampling time. The torque $T - T_l$ is the total torque acting on the crankshaft.

---

[3]The spark ignition time is commonly defined in terms of the spark advance that denotes the difference between the angle of the crankshaft when the cylinder is at the end of the compression stroke and the one at the time of ignition.

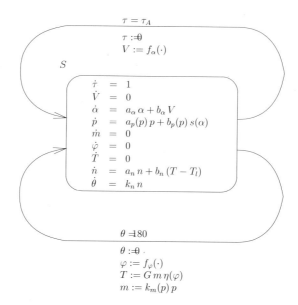

Figure 10.2 Engine hybrid model at idle speed

10.3 Idle Speed Control Design

The specification for the idle speed control is to maintain the crankshaft revolution speed around a nominal value n_0, so that it never exits a range $n_0 \pm \Delta n$, with $\Delta n > 0$. This specification has to be achieved for any value of the load torque disturbance T_l within a lower bound zero and an upper bound $T_l^M > 0$.

The proposed control is a state feedback control. While the spark advance feedback $f_\varphi(\cdot)$ is computed at dead-center times, the throttle feedback $f_\alpha(\cdot)$ is set at fixed-frequency sampling times.

Let $\{t_i\}$ denote the sequence of dead-center times and let t_i denote the current dead-center time.

According to the engine hybrid model described in the previous section and shown in Figure 10.2, the spark advance control $\varphi(t_i)$, chosen at time t_i, will affect the value of the torque $T(t_{i+1})$ that drives the crankshaft from time t_{i+1} to time t_{i+2}. The amount of driving torque $T(t_{i+1})$ depends also on the value of the load disturbance $T_l(t)$, acting on the same time interval $[t_{i+1}, t_{i+2})$. The result of the action of this torque will be a given value of crankshaft speed $n(t_{i+2})$ at time t_{i+2}. Let $\{t_j^A\}$ denote the sequence of sampling times t_j^A of the throttle valve control feedback $f_\alpha(\cdot)$. The control actions $\alpha(t_j^A)$ for $t_j^A \in [t_i, t_{i+1})$ drive the manifold dynamics during an intake stroke starting from the dead-center time t_i to the dead-center time t_{i+1}. The amount of mass loaded by the cylinder at time t_{i+1} depends on these control actions. Note that the corresponding torque will be produced only from time

Idle Speed Control Synthesis using an Assume-guarantee Approach

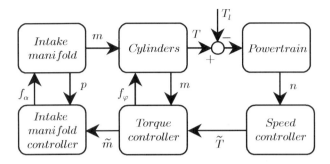

Figure 10.3 Control scheme.

t_{i+2} to time t_{i+3}, due to the delay introduced by the compression stroke.

Controller structure

Due to the complexity of the system to control, our strategy is to decompose the system into three nested subsystems and to devise an appropriate control for each of them in isolation, assuming that the variables that connect the subsystem to the others satisfy an appropriate assumption. The interconnect variables are subject to control and the assumption is chosen so that the controller can indeed make them true when applied to the appropriate subsystem. Of course, there is no *a priori* guarantee that the control law derived with these heuristics satisfies the constraints when applied to the hybrid model. The goal of the verification step is to prove that the strategy pays off: the constraints are indeed satisfied.

The controller is composed of three nested loops:

- an engine speed control in the outer loop, which is responsible for the generation of torque values the cylinders are requested to produce;
- a torque control in the middle loop, which regulates the torque produced by the engine to the desired value and is implemented by the spark advance feedback $f_\varphi(\cdot)$;
- a manifold control in the inner loop, which is responsible for regulation of the mass loaded by the cylinders and implemented by the throttle valve feedback $f_\alpha(\cdot)$.

The overall control scheme is shown in Figure 10.3.

Speed and Torque Control Loops

In this section, we detail the simplifications and show how the control law is derived. The hybrid nature of the torque generation process is approximated by means of a fixed frequency $n_0/30 = 1/\tau_0$ discrete-time process and the driveline dynamics is discretized with sampling period τ_0 so that it reduces to:

$$n(k\tau_0 + \tau_0) = a_n^d n(k\tau_0) + b_n^d \left(T(k\tau_0) - T_l(k\tau_0) \right)$$

with $a_n^d = e^{a_n \tau_0}$ and $b_n^d = (e^{a_n \tau_0}) - 1)(b_n/a_n)$. The robustness of the closed-loop system behavior with respect to this approximation has to be checked in the verification phase. Moreover, assuming that the middle loop will be able to produce the requested torque with a one-step delay, we have

$$T(k\tau_0) = \widetilde{T}(k\tau_0 - \tau_0) \tag{10.7}$$

where $T(k\tau_0)$ is the driving torque during the expansion phase from time $k\tau_0$ to $k\tau_0 + \tau_0$ and $\widetilde{T}(k\tau_0 - \tau_0)$ is the reference torque computed at the end of the intake phase by the speed controller. This torque depends on the mass of air

$$m(k\tau_0 - \tau_0) = k_m(p(k\tau_0 - \tau_0)) \, p(k\tau_0 - \tau_0)$$

loaded in the cylinder entering the compression phase and on the spark advance $\varphi(k\tau_0 - \tau_0)$ that will be applied to the same cylinder.

We design an engine speed control feedback such that the torque $\widetilde{T}(k\tau_0 - \tau_0)$ emulates that of a PI controller, represented as follows:

$$\widetilde{T}(z) = K_p \left(1 + \frac{\tau_0 z}{\tau_I (z-1)}\right) \left(n(z) - n_0 \frac{z}{z-1}\right) \tag{10.8}$$

This controller is able to asymptotically compensate constant torque disturbances. The gain K_p and the time constant τ_I are selected as to stabilize the system dynamics

$$n(k\tau_0 + \tau_0) = a_n^d n(k\tau_0) + b_n^d \widetilde{T}(k\tau_0 - \tau_0) \tag{10.9}$$

around the nominal engine speed n_0, when the disturbance torque T_l is constant.

The spark advance is then computed by simply inverting th equation

$$\widetilde{T}(k\tau_0 - \tau_0) = G \, m(k\tau_0 - \tau_0)) \, \eta \, (\varphi(k\tau_0 - \tau_0))$$

taking into account the saturation $\varphi \in [\varphi_{min}, \varphi_{max}]$. If the computed spark advance is not saturated, then it will be able to produce the desired torque, as in (10.7), on the basis of the mass of air loaded at time $k\tau_0 - \tau_0$.

Furthermore, the middle-loop control has to generate the sequence of reference values $\widetilde{m}(k\tau_0 - \tau_0)$ that is given as input to the manifold control inner loop. This reference value, when produced, will affect the torque generated in the time interval $[k\tau_0 + \tau_0, k\tau_0 + 2\tau_0]$ and is computed by inverting equation

$$\widetilde{T}(k\tau_0 - \tau_0) = G \, \widetilde{m}(k\tau_0 - \tau_0) \, \eta \, (\varphi_0) \tag{10.10}$$

where the parameter φ_0, with $\varphi_{min} < \varphi_0 < \varphi_{max}$, is the nominal spark advance to be used during idle control. The tuning of the control parameter

φ_0 to value smaller than φ_{max} allows some degree of reacting to fast positive torque requests using the spark advance input. Similarly, setting φ_0 to a value greater than φ_{min}, allows to reaction to fast negative torque requests.

Note that the mass of air loaded into the cylinder at time $k\tau_0 - \tau_0$, i.e., at the end of the intake phase, depends on the throttle control applied on the time interval $[k\tau_0 - 2\tau_0, k\tau_0 - \tau_0]$.

Intake Manifold Control Loop

The objective of the intake manifold control is that of making the cylinders load the amount of mass $\widetilde{m}$ specified by the torque control.

For the design of the intake manifold control, we assume that the reference signal $\widetilde{m}$ is produced at a fixed frequency $1/\tau_0$, and τ_A is such that $\tau_0 = N_A \tau_A$ for an appropriate integer N_A. The robustness of the closed-loop system with respect to this assumption has to be checked in the verification phase.

The intake manifold and cylinder filling dynamics are discretized with a sampling period equal to the throttle valve control sampling time τ_A. By inverting Equation (10.3), a sequence $\widetilde{p}$ of manifold pressure setpoints is obtained from the sequence of desired mass $\widetilde{m}$ given by (10.10). Hence, a PID-controller that produces convergence of the manifold pressure p to the target signal $\widetilde{p}$ is used for the throttle piecewise-constant control feedback $f_\alpha(\cdot)$:

$$V(z) = \hat{K}_p \left(1 + \frac{\tau_A z}{\hat{\tau}_I(z-1)} + \frac{\hat{\tau}_D(z-1)}{\tau_A z} \right) (p(z) - \widetilde{p}(z)) \tag{10.11}$$

Simulations

Simulation results are reported in Figure 10.4. The crankshaft revolution speed has to be maintained around the nominal value $n_0 = 800$ rpm with maximum excursion of ± 50 rpm so that $\tau_0 = 30/n_0 = 37.5$ ms. The value of the power train parameters are $a_n = -1.531$ and $b_n = 95.49$; those of the intake manifold are: $\tau_A = \tau_0/4$, $a_p = -20.94$, $b_p = 1821$, $a_\alpha = -10.3$, $b_\alpha = 9.8$ and $\varphi_0 = 12$. Finally, the PID controllers are defined by the following time constants and gains: $K_p = 0.0781$, $\tau_I = 234.5$ ms, $\hat{K}_p = 0.0613$, $\hat{\tau}_I = 135.7$ ms, $\hat{\tau}_D = 29.5$ ms. Figure 10.4 shows relevant signals in a simulation of the closed-loop system, obtained with a load torque T_l that is a step of amplitude 5 Nm applied at time $t = 3.2$ s.

10.4 Closed-loop System Behavior Verification

In this section, we need to show that the hybrid feedbacks described in Sections 10.3, 10.3, when applied to the full hybrid system model described in Section 10.2, achieve the task of maintaining the engine speed within the specified range for any action of the load disturbance, provided that the initial condition of the hybrid system belongs to a specified set. In other

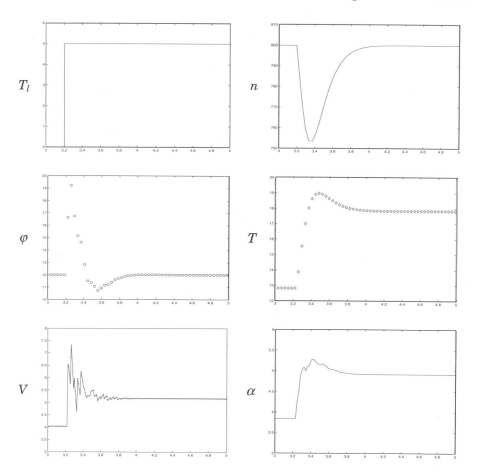

Figure 10.4 Simulation results.

words, we need to show that a given set of initial conditions is a robust invariant set for the closed-loop system obtained by applying the proposed idle speed control to the hybrid engine model. This result guarantees that, if the hybrid system state is steered inside this set (that may not be the maximal one), then the proposed idle feedback control can be activated and the idle speed specification will be met under any load disturbance.

In [2], the maximal robust invariant set for an engine at idle speed was computed analytically for a simplified engine hybrid model, where the nonlinear expressions were linearized and no actuation delays were considered. The main objective there was to establish the best performances achievable by a given engine without adding any detail and/or constraints on the idle speed controller. The approximated linear expressions were used to make the analytical computation feasible, and the largest set of initial conditions for which at least an idle speed controller exists was derived.

In this paper, we added details and constraints on the controller, specifying actuator delays and dynamics in the hybrid model of the engine. Hence, the verification task is more complex and cannot be carried out analytically. We need to show that, from a given set of initial conditions, the controller synthesized in the previous section achieves the idle speed specification. Due to the added details on actuations characteristics and due to the choice of a particular controller, this set of initial conditions will be necessarily contained in the maximal robust invariant set for idle speed control computed for the nonlinear model.

Performing formal verification of the behavior of the entire hybrid closed-loop system with respect to the given specification is prohibitively complex. Hence, we decided to apply an assume-guarantee paradigm to verify the correctness of the proposed controller. The verification is based on the same decomposition of the system used in the design of the feedback controls. Then, we consider first the power train closed-loop system, then the cylinders closed-loop system, and finally the intake manifold closed-loop system.

Engine Speed Feedback Formal Verification

Applying the assume-guarantee paradigm, the correctness of the behavior of the speed control outer loop is verified *assuming that the torque control middle loop will produce the requested values of torque with a single phase delay*.

First, we analyze the evolution of the continuous dynamics (10.5)–(10.6). To this end, we compute in the configuration subspace (n, T) a set $\mathcal{P}$ such that if, at each dead center, the configuration is inside the set $\mathcal{P}$, then the continuous evolution of $n(t)$ will satisfy the system specification $n(t) \in [n_0 - \Delta, n_0 + \Delta]$ also for any t between dead center times under any action of the torque disturbance T_l. The set $\mathcal{P}$ is obtained by backwards integration of Equations (10.5)-(10.6) over a stroke for $T_l \in [0, 5]$ Nm. Figure 10.5 shows the set $\mathcal{P}$.

Hence, the speed controller produces a correct behavior if, for any evolution of the closed-loop system, at each dead center the engine torque T and the engine speed n belong to the set $\mathcal{P}$. Let z_k^I denote the state of the PI engine speed controller. Under the assumption that the torque control middle loop will produce the requested values of torque with a single phase delay, the evolution of variables (n_k, T_k, z_k^I) for the closed-loop system[4] is

---

[4]The index k refers to the k-th dead center.

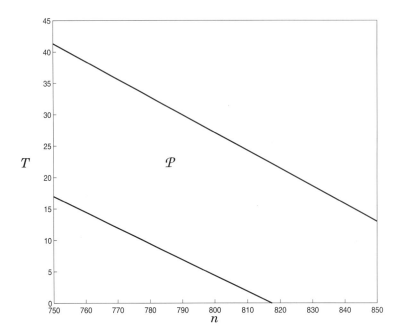

Figure 10.5 The set $\mathcal{P}$.

given by

$$n_{k+1} = e^{a_n t^\star} n_k + \frac{b_n}{a_n}\left(e^{a_n t^\star} - 1\right)(T_k - T_l) \tag{10.12}$$

$$T_{k+1} = \widetilde{T}_k = z_k^I - K_P \frac{\tau_I + \tau_0}{\tau_I}(n_k - n_0) \tag{10.13}$$

$$z_{k+1}^I = z_k^I - K_P \frac{\tau_0}{\tau_I}(n_k - n_0) \tag{10.14}$$

In (10.12), $t^\star$ denotes the dead-center time that is modeled as a bounded unknown disturbance in the range $[30/(n_0+\Delta), 30/(n_0-\Delta)]$. This formalization allows us to verify the robustness of the closed-loop system with respect to the fact that the engine speed controller is designed for the discrete-time approximation of the process (10.5) using $t^\star = \tau_0$. At each dead center the engine torque T_k and the engine speed n_k belong to the set $\mathcal{P}$, if there exists a not empty robust invariant set $\mathcal{M}_\mathcal{P}$ contained in $\mathcal{P} \times \mathbb{R}$, and the PI-controller initial configuration z_0^I is chosen such that $(n_0, T_0, z_0^I) \in \mathcal{M}_\mathcal{P}$.

Then, to complete the verification of the speed controller we have to compute a robust invariant set $\mathcal{M}_\mathcal{P}$ for the dynamics of Equations (10.12), (10.13) and (10.14) contained in $\mathcal{P} \times \mathbb{R}$. Such an invariant set is required to be robust with respect to both the unknown load torque T_l and the unknown dead center times $t^\star$.

Figure 10.6 shows the maximal invariant set $\mathcal{M}_\mathcal{P}$ for the dynamics of Equations (10.12), (10.13) and (10.14) contained in $\mathcal{P} \times \mathbb{R}$, robust with

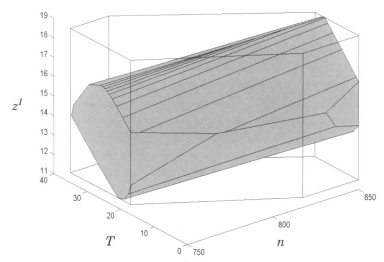

Figure 10.6 Maximal invariant set $\mathcal{M}_{\mathcal{P}}$ contained in $\mathcal{P} \times \mathbb{R}$, robust with respect to the load torque disturbance T_l

respect to the load torque disturbance T_l only. This set has been computed by applying backwards reachability analysis starting from the set $\mathcal{P} \times \mathbb{R}$ under any action of the disturbance T_l.

Further elaboration of this set to ensure robustness with respect to the disturbance t^* is currently under investigation.

Notice that, since the assumed behavior for the torque closed-loop subsystem is that of producing the desired torque $\tilde{T}$ with a single phase delay, the specification to be verified for the middle subsystem does not depend on the evolution of the outer control loop. This allows us to formally verify that if the intake manifold control inner loop will exhibit an input-output behavior that is contained in the stream depicted in Figure 10.7, in terms of requested mass $\tilde{m}(k-1)$ and loaded mass during the intake strokes $m(k)$, then the torque control middle loop is able to produce the requested behavior between the desired torque $\tilde{T}$ and the engine torque T.

Finally, in the last step of the assume-guarantee approach, we will consider the verification of the intake manifold control inner loop to show that the behavior of the throttle valve control is correct in the sense that it provides the assumed behavior specified in Figure 10.7. An objective of this verification is to show that the discrete representation of the pressure dynamics, used in the design of the throttle valve control, models sufficiently well the continuous evolution of the manifold pressure. A second issue is the verification that the interaction between the two loops with different

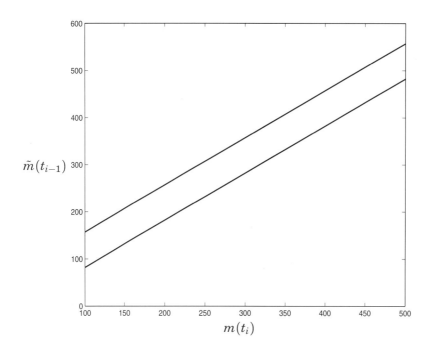

Figure 10.7 Assumed behavior for the intake manifold closed-loop subsystem

triggers (the throttle control running at fixed frequency $1/\tau^A$ and torque control synchronized with the engine dead centers) produces a correct behavior.

10.5 Conclusions

The idle speed control problem has been formalized as a safety specification for a hybrid model of the engine and the power train. The hybrid model describes the intake manifold and cylinder-filling dynamics, the torque generation process and the power train, as well as throttle and spark ignition actuators. An idle speed controller has been designed which exploits the decomposition of the system into three components: the engine speed controller, the torque controller and the intake manifold controller. For the design of each controller, a simplified model of the plant wass used. Each controller was synthesized assuming that the other subsystems can be controlled to give the appropriate input-output behavior.

The control law so obtained was then applied to the hybrid system modeled in its entirety, and simulation results were obtained. Since there was no guarantee that the behavior of the closed-loop system will satisfy the specifications, we apply formal verification techniques to demonstrate that the control law does indeed meet the constraints. Since formal verification is a very complex task, we applied an assume-guarantee approach that views the system as composed of the three subsystems for which the control law was derived. The assume-guarantee principle allows us to obtain a set of formal verification problems that are solvable within the domain of present formal verification tools.

10.6 References

[1] M. Abate and V. Di Nunzio. Idle speed control using optimal regulation. Technical Report 905008, SAE, , 1990.

[2] A. Balluchi, A., L. Benvenuti, M. D. Di Benedetto, G. M. Miconi, U. Pozzi, T. Villa, H. Wong-Toi, and A. L. Sangiovanni-Vincentelli. Maximal safe set computation for idle speed control of an automotive engine. In N. Lynch and B. H. Krogh, editors, *Hybrid Systems: Computation and Control*, volume 1790 of *Lecture Notes in Computer Science*, 32–44. Springer-Verlag, New York, 2000.

[3] A. Balluchi, L. Benvenuti, M. D. Di Benedetto, C. Pinello, and A. L. Sangiovanni-Vincentelli. Automotive engine control and hybrid systems: Challenges and opportunities. *Proceedings of the IEEE*, 88, "Special Issue on Hybrid Systems" (invited paper)(7):888–912, 2000.

[4] K. R. Butts, N. Sivashankar, and J. Sun. Application of ℓ_1 optimal control to the engine idle speed control problem. *IEEE Trans. on Control Systems Technology*, 7(2):258–270, 1999.

[5] C. Carnevale and A. Moschetti. Idle speed control with H_∞ technique. Technical Report 930770, SAE, 1993.

[6] D. Hrovat and B. Bodenheimer. Robust automotive idle speed control design based on μ-synthesis. In *Proc. IEEE American Control Conference*, 1778–1783, San Francisco, CA, 1993.

[7] D. Hrovat and J. Sun. Models and control methodologies for IC engine idle speed control design. *Control Engineering Practice*, 5(8),, 1997.

[8] L. Kjergaard, S. Nielsen, T. Vesterholm, and E. Hendricks. Advanced nonlinear engine idle speed control systems. Technical Report 940974, SAE, 1994.

[9] C. H. Onder and H. P. Geering. Model-based multivariable speed and air-to-fuel ratio control of a SI engine. Technical Report 930859, SAE, 1993.

[10] D. Shim, J. Park, P. P. Khargonekar and W. B. Ribbens. Reducing automotive engine speed fluctuation at idle. *IEEE Trans. on Control Systems Technology*, 4(4):404–410, 1996.

[11] S. Yurkovich and M. Simpson Crank-angle domain modeling and control for idle speed. *SAE Journal of Engines*, 106(970027):34–41, 1997.

11

Fault Diagnosis of Switched Nonlinear Dynamical Systems with Application to a Diesel Injection System

D. Förstner   J. Lunze

Abstract

For the on-line monitoring of safety- and emission-relevant parts of automotive systems diagnostic methods are needed that take into account the parameter uncertainties and nonlinearities of automotive systems and the real-time restrictions of on-board diagnosis. This paper concerns the diagnosis of switched nonlinear dynamical systems, which is applied to a Diesel injection system. A model-based diagnostic method is presented that processes the events generated by the output signals when passing predefined thresholds. For nominal and faulty behaviour, a model in the form of a non-deterministic automaton is set up. The diagnostic algorithm uses this model to compare the measured event sequences with the nominal or faulty behaviour in order to determine faults from inconsistencies. The trade-off between the complexity of the algorithm and the accuracy of the diagnostic result can be found by varying the model depth, which is the number of recent events stored in the model state. The successful application of this method is demonstrated for the power stage of a common-rail Diesel injection system.

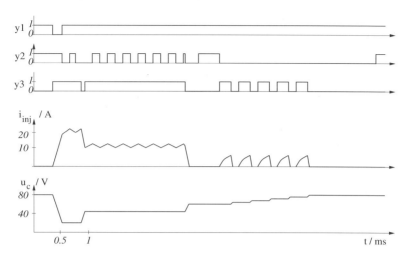

Figure 11.1 Input and output signals of the power stage

11.1 Introduction

Diagnosis of Automotive Systems

The increasing demands in monitoring safety-relevant functions and in observing the legal emission limits of automotives during operation has led to the introduction of powerful diagnostic algorithms into the automotive control system. This paper focuses on applications where the automotive components can be dealt with as *switched nonlinear systems*. Such systems are typical for automotive applications because, as Figure 11.1 shows, the behaviour of such systems is described by a mixture of discrete-valued and continuous-valued signals. In the figure, three binary control signals depicted in the upper part of the graph are applied to an electronic component whose two continuous-valued state signals are given in the lower part. Changes of the input signals switch the system dynamics.

Such systems cannot be handled by the diagnostic approaches that have been elaborated in the recent years, because these *model-based diagnostic systems* assume that the system to be diagnosed is described by a unique differential equation. These methods use well-known principles of state observation [6], parameter estimation [10], or parity relation checks [7] to find possible faults. Hence, these approaches are limited to linear or "slightly" nonlinear systems for which the differential equation includes analytical functions that can be presented in a closed form.

Diagnosis Based on Qualitative Modelling

In this paper, a *qualitative representation* of the system is used to overcome these drawbacks. It combines qualitative model-based diagnostic methods developed in artificial intelligence [8] with the input-output view of systems theory. A discrete-event model in form of a nondeterministic automaton is used for describing the event sequences that are generated by a continuous-valued dynamical system. Similar approaches have been reported, for instance, in [1], [15], [16], and in a discrete-time setting in [12], [17]. The main contribution of this paper is to use non-deterministic automata for switched nonlinear systems and to apply the resulting diagnostic method to an automotive system.

Diagnostic methods based on qualitative models allow us to deal with nonlinear, hybrid, or uncertain systems. Furthermore, the complexity of the diagnostic system is reduced by using qualitative instead of quantitative information. If faults change the dynamical behaviour qualitatively, the use of qualitative instead of detailed quantitative information is sufficient to solve the diagnostic task.

Figures 11.2 and 11.3 summarise the basic idea of the qualitative model-based diagnostic method described in the following. The continuous-variable input and output signals of the system under consideration are transformed into a sequence of quantised values (Figure 11.2), each of which represents an interval. If the border between two intervals is passed, an event occurs, which is symbolised by the bars in the lower part of the figure. The diagnostic method uses only this sequence of events.

Figure 11.3 shows the formalisation of this way of diagnosis. As the signals of the given dynamical system are not measured directly but only through quantisers, a *quantised system* is considered. A diagnostic algorithm checks whether the measured event sequences are consistent with the nominal behaviour of the quantised system. In order to perform this task, the diagnostic algorithm needs a model of the quantised system. In this paper, the quantised system is modelled by a non-deterministic automaton [11]. If the measured event sequences are consistent with the model, no fault occurs. If, however, inconsistencies are detected, the system is known to be faulty. A diagnostic algorithm that has access to models of the faulty quantised system can also identify which fault is present.

Note that the quantised system is a *hybrid system*, where the internal signals like the input u and state x are continuous-valued whereas the quantised signals are discrete-valued. The consideration of the given systems in the quantised system framework has the aim to uniformly describe all appearing signals as discrete-valued signals. Therefore, a uniformly discrete-event model can be used, which has a much lower complexity compared to the description of the hybrid system. The abstraction step, where the non-deterministic automaton is set up for a given set of differential equations describing the switched nonlinear continuous-variable system together with the quantisers, is used to transform the given information about the system into a model that is well-suited for diagnosis. The abstraction step is the main focus of this paper.

This paper extends this diagnostic method to quantised systems where the continuous-variable system is a switched nonlinear system. The main result concerns the modelling of quantised switched systems by means of

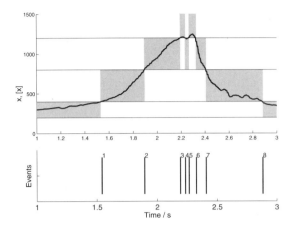

Figure 11.2 Signal quantisation

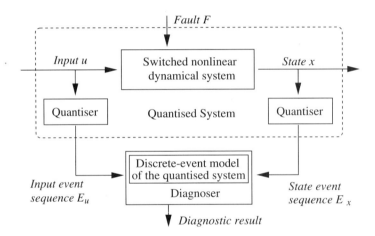

Figure 11.3 Quantised system within a model-based diagnostic scheme

the non-deterministic automaton. It will be shown how such a model can be abstracted from the differential equation together with the switching rules and the description of the quantisers. A parameter, called *model depth*, fixes the number of events that are used to define the state of the non-deterministic automaton. This parameter gives some freedom in choosing the model accuracy.

Structure of the paper

The qualitative, discrete-event modelling of switched nonlinear systems is introduced in Section 11.2. This model has to possess the properties *completeness* and *accuracy* (Section 11.3) which determine the correctness and accuracy of the diagnostic result obtained by means of this model if the principle of consistency-based diagnostic (Section 11.4) is applied.

Fault Diagnosis...Diesel Injection System

The *non-deterministic automaton* used to represent the qualitative behaviour of the switched nonlinear system is introduced in Section 11.5. The model state represents a short sequence of previously passed events. By changing the length of this sequence (*model depth*), the accuracy of the model can be varied. Methods to set up a qualitative model for a given application are briefly reviewed.

The *diagnostic algorithm* that adopts the consistency-based principle to the non-deterministic automaton is presented in Section 11.6.

The results are illustrated in Section 11.7 by an *automotive application*. It shows how the qualitative modelling and the diagnostic results are obtained by the methods presented in this paper for the power stage of a common-rail Diesel injection system.

11.2 Discrete-event Behaviour of Switched Nonlinear Systems

Switched Nonlinear System

As shown in Figure 11.3, the core of the system under consideration is a switched nonlinear dynamical system. It is assumed that the continuous-time behaviour is given by the differential equation

$$\dot{x}(t) = \begin{cases} f_1(x(t), F) & \text{if } u(t) = 1 \\ f_2(x(t), F) & \text{if } u(t) = 2 \\ \vdots & \vdots \\ f_U(x(t), F) & \text{if } u(t) = U \end{cases} \tag{11.1}$$

$$x(0) = x_0$$

with state vector $x \in \mathbb{R}^n$ and input $u \in \{1, 2, ..., U\}$. The dynamics of the system are switched according to the discrete input value. It is assumed that the state trajectory is continuous and that the input does not change its value infinitely often within a finite time interval. In the time interval $[0, T_h]$, the system follows the state trajectory denoted by $x_{[0,T_h]}$.

The system behaviour depends on the fault $F \in \mathcal{F}$, where $\mathcal{F}$ is the set of faults considered and $F_0 \in \mathcal{F}$ symbolises the faultless system.

Signal Quantisation

The right quantiser in Figure 11.3 introduces a partition of the state space $\mathbb{R}^n$ into the sets $Q_x(z)$, $z \in \mathcal{N}_x = \{1, \ldots, n_z\}$. The quantised value of the state, which is synonymously called the qualitative value of the state $x(t)$ at time t, is given by the index z of the set $Q_x(z)$ to which the state belongs:

$$[x(t)] = z \iff x(t) \in Q_x(z) \tag{11.2}$$

The change of the qualitative value $[x(t)]$ from i to j is called an event e. If the event occurs at time t_k, the new qualitative state value is given by $z_k = [x(t_k)] = j$. Thus, an event e_k represents the same information as the pair of qualitative states (z_{k-1}, z_k) between the quantised state value jumps:

$$e_k = (z_{k-1}, z_k) \tag{11.3}$$

In the following, event sequences are described by the sequence of discrete values that the corresponding signal assumes.

The quantised system is considered in the time interval $[0, T_h]$, where the continuous-variable system follows the trajectory $x_{[0,T_h]}$ and the quantiser generates the state event sequence

$$E_x(1 \ldots H_x) = \mathrm{Quant}(x_{[0,T_h]}) = (z_0, z_1, \ldots, z_{H_x}) \tag{11.4}$$

H_x is the number of state events that the system generates within a given time horizon.

The quantisation principle applies also to the input $u(t)$. As the input space $\{1, 2, \ldots, U\}$ is already discrete, each qualitative input $v \in \mathcal{N}_u = \{1, 2, \ldots, n_v\}$ refers unambiguously to a discrete quantitative input u. The input $u(t)$ is a piecewise constant function. It corresponds to the sequence of input events

$$E_u(1 \ldots H_u) = \mathrm{Quant}(u_{[0,T_h]}) = (v_0, v_1, \ldots, v_{H_u}) \tag{11.5}$$

As the sequences of state and input events are to be investigated together, the joined sequence of qualitative input and state events is built, which equals a quantisation of the vector $s(t) = \left(u(t)^\top, x(t)^\top\right)^\top$. The qualitative value $[s(t)]$ of this vector is symbolised by the 2-vector $q(t)$, which represents the qualitative input $v(t) = u(t)$ and the qualitative state $z(t) = [x(t)]$. The joined event sequence

$$\begin{aligned} E(1 \ldots H) &= \mathrm{Quant}\begin{pmatrix} u_{[0,T_h]} \\ x_{[0,T_h]} \end{pmatrix} \\ &= \left(\begin{pmatrix} v_0 \\ z_0 \end{pmatrix} \begin{pmatrix} v_1 \\ z_1 \end{pmatrix} \begin{pmatrix} v_2 \\ z_2 \end{pmatrix} \cdots \begin{pmatrix} v_H \\ z_H \end{pmatrix} \right) \\ &= (q_0, q_1, \ldots, q_H) \end{aligned} \tag{11.6}$$

describes the behaviour of the quantised system. Hence, the quantised system has a discrete-event behaviour.

Each change of the qualitative value q represents either an input event, a state event, or a synchronous state and input event. In supervisory control, it can often be assumed that changes in the input occur synchronously to state events, which facilitates the modelling of such systems. In most diagnostic applications to automotive systems, however, input and state events occur asynchronously [3]. In the application example presented in Section 11.7, the binary input values of the power stage change synchronously due to passed thresholds of the output signals but also asynchronously due to internal signals of the central control unit that determines the mode of the power stage.

Fault Diagnosis...Diesel Injection System 251

Non-Determinism of the Quantised System

As only the qualitative values are measured, the event sequence E generated by the quantised system cannot be uniquely predicted [14]. The reason for this is given by the fact that for a given initial qualitative state z_0, the initial state x_0 of the system (11.1) is not exactly known but merely restricted to the set $Q_x(z_0)$, cf. Equation (11.2). Depending on $x(0) \in Q_x(z_0)$ and the fault F, the system may produce one sequence of the set

$$\mathcal{B}_S(z_0, F, H) = \{E(1\ldots H) \mid \exists u_{[0,T_h]}, x_{[0,T_h]} : \quad (11.7)$$

$$E = \text{Quant}\begin{pmatrix} u_{[0,T_h]} \\ x_{[0,T_h]} \end{pmatrix}, x_{[0,T_h]} \text{ satisfies (11.1) for } u_{[0,T_h]} \text{ and } F\}.$$

Implications of the Switched System Assumption

The assumed system structure (11.1) has the following consequences for the qualitative modelling:

- as the input assumes a value of a finite set, the question of how the system behaves qualitatively for a given qualitative input requiresus only to investigate one of the discrete values at a time. Hence, the modelling step decomposes to U steps where a specific input is considered.

- during a time interval where the input u is constant, the system is described by an autonomous nonlinear differential equation.

Both facts facilitate the analysis of the quantised system behaviour. In the general case where a qualitative input represents an infinite (but bounded) set of quantitative values, more complicated methods would have to be applied, because in this case the qualitatively known input causes further non-determinism.

11.3 Requirements on Models Used for Diagnosis

The quantitative description (11.1) and the quantisers (11.2) are by definition a precise representation of the quantised system. However, it is a very complex description that is not suitable for on-board diagnostis. Instead, a representation of the quantised system is needed which is compact and easy to evaluate during the diagnostic process. The nondeterministic automaton presented in Section 11.5 is one form of such a model, which satisfies the requirement to have a model of low complexity.

The model used in diagnosis has to own two basic properties: completeness and accuracy. Both are necessary to obtain correct and precise diagnostic results. These properties and the corresponding requirement on the model to be set up in Section 11.5 are formulated now by means of the notion of the system behaviour $\mathcal{B}_S(z_0, F, H)$, which is the set of all discrete-event trajectories that the quantised system can generate. Likewise, $\mathcal{B}_M(z_0, F, H)$ denotes the model behaviour:

Complete Model

A complete model generates all possible event sequences that the quantised system may produce. Hence, for a *complete* model, the relation

$$\mathcal{B}_M(z_0, F, H) \supseteq \mathcal{B}_S(z_0, F, H) \tag{11.8}$$

holds for each initial qualitative state z_0, fault F, and time horizon H.

Model Accuracy

In most applications, the discrete-event model will also generate event sequences that cannot occur in the quantised system. The existence of such *spurious solutions*

$$\mathcal{B}_{spurious}(z_0, F, H) := \mathcal{B}_S(z_0, F, H) \setminus \mathcal{B}_M(z_0, F, H) \tag{11.9}$$

is a well-known phenomenon of qualitative modelling [11]. If two different models with behaviour $\mathcal{B}_M^A$ and $\mathcal{B}_M^B$ are given, they may generate different sets of spurious solutions. If the hierarchy relation

$$\mathcal{B}_M^A(z_0, F, H) \supset \mathcal{B}_M^B(z_0, F, H) \supseteq \mathcal{B}_S(z_0, F, H) \tag{11.10}$$

holds for all H, z_0 and F, then the model B is termed more *accurate* than model A. Consequently, model B generates less spurious solutions than model A. Note that the model accuracy is a property defined for two models, whereas the completeness relates each single model to the quantised system.

11.4 Consistency-based Diagnosis

This section explains the diagnostic task to be solved and the main idea of its solution. It is assumed that an unknown fault $F \in \mathcal{F}$ has occurred at time $t \leq 0$ and is present until the diagnostic algorithm is stopped. $E(1 \ldots H)$ denotes the sequence of input events and state events of the quantised system for a given time horizon H. The main idea of *consistency-based diagnosis* is to compare the observed sequence with the behaviour of the model. The diagnostic task is to answer the question whether the model of a specific fault F can generate the event sequence $E(1 \ldots H)$, i.e., whether the relation

$$E(1 \ldots H) \in \mathcal{B}_M(z_0, F, H) \tag{11.11}$$

holds. The diagnostic result is denoted by the set $\mathcal{F}_H$ that includes all faults F for which the measured event sequence belongs to the model behaviour:

$$\mathcal{F}_M(E(1 \ldots H)) := \{F \mid E(1 \ldots H) \in \mathcal{B}_M(z_0, F, H)\}. \tag{11.12}$$

Fault Diagnosis...Diesel Injection System

This set is an estimate of the fault set

$$\mathcal{F}_S(E(1\ldots H)) := \{F \mid E(1\ldots H) \in \mathcal{B}_S(z_0, F, H)\} \tag{11.13}$$

that should be found by the diagnostic algorithm. This set covers all faults for which the quantised system can indeed produce the event sequence $E(1\ldots H)$.

Correctness of the Diagnostic Result

As the diagnostic method only yields the set $\mathcal{F}_M$ instead of $\mathcal{F}_S$, the main question is: what can be concluded about possible faults if $\mathcal{F}_M$ is known? An immediate answer to this question is obtained if the discrete-event model is complete as required in the preceding section [4]. According to Relation (11.8), the following relation holds for all sequences $E(1\ldots H)$:

$$F \notin \mathcal{F}_M(E(1\ldots H)) \;\Rightarrow\; F \notin \mathcal{F}_S(E(1\ldots H)). \tag{11.14}$$

Hence, if a fault can be rejected based on a complete model, it can surely not be present in the quantised system. The relation (11.14) yields the basis for the detection and identification of faults:

- Fault detection: if a sequence $E(1\ldots H))$ is given and

$$F_0 \notin \mathcal{F}_M(E(1\ldots H)) \tag{11.15}$$

  holds where F_0 denotes the nominal case, then a fault is present within the quantised system.

- Fault identification: if for a given sequence $E(1\ldots H))$ and a specific fault $\bar{F}$ the relations

$$\begin{aligned}&\bar{F} \in \mathcal{F}_M(E(1\ldots H)) \\ &F \notin \mathcal{F}_M(E(1\ldots H)) \qquad \text{for all } F \in \mathcal{F}, F \neq \bar{F}\end{aligned} \tag{11.16}$$

  hold, then $\bar{F}$ is present in the quantised system provided that $\mathcal{F}$ describes all possible faults (closed-world assumption).

If the model is not complete, a false alarm may occur, *i.e.*, a fault may be detected for a faultless measurement, or a fault may be identified that is not present. *Correct* diagnostic results can, thus, only be obtained for complete models.

Accuracy of the Diagnostic Result

If the diagnosis is performed with two different models A and B with B being more accurate than A according to (11.10), the following relations are valid for the diagnostic results $\mathcal{F}_M^A$ and $\mathcal{F}_M^B$, *cf.* [4]:

$$\forall E(1\ldots H): \; F \notin \mathcal{F}_M^A(E(1\ldots H)) \;\Rightarrow\; F \notin \mathcal{F}_M^B(E(1\ldots H)) \tag{11.17}$$

$$\forall F \in \mathcal{F}: \exists E(1\ldots H) \not\in \mathcal{B}_S(z_0, H, F): \quad F \in \mathcal{F}_M^A(E(1\ldots H)) \tag{11.18}$$
$$F \not\in \mathcal{F}_M^B(E(1\ldots H))$$

The consequence of these two relations are:

- each fault excluded by using model A will also be excluded by using model B.

- there exists at least one event sequence for each fault F for which the diagnostic algorithm using model B can prove that F does not occur in the system, whereas the diagnosis using model A cannot exclude this fault.

As faults can be detected and identified faster (or at all) with model B, this shows that a model that represents the system behaviour more accurately than another yields a more *accurate* diagnostic result. Hence, the hierarchy of models relates to a hierarchy of diagnostic results. The best model would be a model that represents the system behaviour exactly, the worst model is the trivial model.

11.5 Representation of Quantised Systems by means of Automata

Non-deterministic Automaton

A compact discrete-event representation of the quantised system by a non-deterministic automaton

$$N(F) = (\Omega, \mathcal{N}_u, \mathcal{N}_x, R(F), \omega_0) \tag{11.19}$$

should be used during the diagnosis. The automaton has the state set Ω, the output set $\mathcal{N}_x$, the input set $\mathcal{N}_u$, and the initial model state ω_0. Whereas the state set has to be defined yet, the sets $\mathcal{N}_x$ and $\mathcal{N}_u$ are given by the signal quantisation. For each fault $F \in \mathcal{F}$, a non-deterministic automaton $N(F)$ has to be set up. The sets Ω, $\mathcal{N}_x$ and $\mathcal{N}_u$ are the same for all models. The dynamical behaviour of the automaton is described by the state transition relation, which depends on the fault F:

$$R(F) \subseteq \Omega \times \mathcal{N}_x \times \Omega \times \mathcal{N}_u \tag{11.20}$$

The automaton can step from the current state ω_k to the successor state ω_{k+1} provided that the input v_k is present if

$$(\omega_{k+1}, z_{k+1}, \omega_k, v_k) \in R \tag{11.21}$$

holds. When the automaton performs this transition, it assumes z_{k+1} as the output. By recursive evaluations, the automaton passes a sequence of model states and, hence, generates a sequence of qualitative states tht depends on the assumed sequence of qualitative inputs.

Model State

The model state ω_k can be interpreted in different ways. A common interpretation is $\omega_k = z_k$, where the automaton states correspond directly to the quantised state values of the system. However, the non-deterministic automaton represents the discrete-event behaviour more accurately if it captures additional information about the past qualitative behaviour in its model state. Therefore, the model state is extended to the following form:

$$\omega_k = \left(\begin{pmatrix} v_{k-d} \\ z_{k-d} \end{pmatrix}, \ldots, \begin{pmatrix} v_{k-1} \\ z_{k-1} \end{pmatrix}, \begin{pmatrix} * \\ z_k \end{pmatrix} \right), \quad k \geq d \quad (11.22)$$

ω_k captures a short sequence of previously passed state or input events. The parameter d is called *model depth* and denotes the number of events represented in the model state. The symbol $*$ means that no information is given about this element. For $d = 0$, the state-based definition $\omega_k = z_k$ is obtained.

Initially, only the qualitative state z_0 is assumed to be known. Therefore, the set of model states is supplemented by model states where qualitative inputs and qualitative states assume the "don't know" symbol $*$. Then

$$\omega_0 = \left(\begin{pmatrix} * \\ * \end{pmatrix}, \ldots, \begin{pmatrix} * \\ * \end{pmatrix}, \begin{pmatrix} * \\ z_0 \end{pmatrix} \right) \quad (11.23)$$

can always be assigned to a given z_0. The set of possible state event sequences can be generated recursively by multiple evaluations of the automaton's transition relation for a given initial qualitative state z_0, fault F, and time horizon H:

$$\mathcal{B}_M(z_0, F, H) \quad (11.24)$$
$$= \{ E(1 \ldots H) = \left(\begin{pmatrix} v_0 \\ z_0 \end{pmatrix} \ldots \begin{pmatrix} v_H \\ z_H \end{pmatrix} \right) \mid$$
$$(\omega_{k+1}, z_{k+1}, \omega_k, v_k) \in R(F), \; k = 0 \ldots H-1 \quad (11.25)$$
$$\omega_0 \text{ satisfies (11.23) for } z_0 \}$$

The number $|\Omega|$ of model states and the number $|R|$ of transitions grow exponentially with d. Therefore, a trade-off has to be found between the model size and the model accuracy. In [2], it is shown that the automaton size can be reduced by joining equivalent model states. By this reduction the exponential growth can be prevented, in general. Nevertheless, the model will grow monotonically with d.

Determination of the State Transition Relation of the Automaton

The crucial task when using the presented model-based diagnostic method is to set up the transition relation R of the automaton. Four methods are available to solve this task:

- Knowledge acquisition: the transition relation is set up manually by an expert. This method is only feasible for small systems with only few qualitative inputs and states (small numbers n_v, n_z);

- Abstraction: If a quantitative system description (11.1) is given, the transition relation can be determined by an abstraction procedure that analyses the quantitative system dynamics [3];

- Identification: a set of measurements that covers the possible behaviours for each concerned fault can be used to set up the transition relation of the automaton. A qualitative identification algorithm is presented in [5];

- Model composition: if automata for linked components are available they can be composed. The partitioning of the linked signals need to be equal in the considered component models.

As mentioned above, the assumed system structure simplifies the qualitative modelling. Due to the discrete input set only a limited set of inputs and their effect on the quantised system dynamics has to be analysed. In the general case with continuous inputs, intervals of input values have to be considered that require interval calculations or the approximation of the input set by a large number of discretised input values. According to equation (11.1), in this paper the system dynamics are assumed to be autonomous as long as no input event occcurs. Such a lack of an input dependency of the system dynamics is the reason why the analysis of quantised autonomous systems is easier than the analysis of quantised non-autonomous systems.

11.6 A Diagnostic Algorithm for Quantised Systems

To determine the diagnostic result $\mathcal{F}_M$ defined by eqn. (11.12) based on the non-deterministic automaton representation, it has to be tested whether changes in the input or output are consistent with the transition relations $R(F)$ of the automata. The diagnosis is performed in a recursive way. It starts with no information about the occurrence of a fault. Therefore, it is assumed that all faults $F \in \mathcal{F}$ may have occurred. Furthermore, it is assumed that the initial qualitative state z_0 is known.

The diagnostic algorithm given in Figure 11.4 determines $\mathcal{F}_M(E(1 \ldots H))$ recursively for given $\mathcal{F}_M(E(1 \ldots H-1))$ as follows:

$$\mathcal{F}_M(E(1 \ldots H)) = \{F \mid F \in \mathcal{F}_M(E(1 \ldots H-1)) \ \wedge \qquad (11.26)$$
$$(\omega_H, z_H, \omega_{H-1}, v_{H-1}) \in R(F) \}.$$

The first part of Equation (11.26) concerns the case that fault F could not be excluded by using the measurement data with time horizon $H-1$. Then it is tested whether the newly observed qualitative state z_H satisfies the

Fault Diagnosis...Diesel Injection System

Given: Non-Deterministic automaton $N(F)$ for each fault $F \in \mathcal{F}$
Initial qualitative state z_0
Initial qualitative input v_0
Time horizon $\bar{H}$

1. Initialisation: $H = 0$, $\mathcal{F}_M(()) = \mathcal{F}$, ω_0 according to (11.23) for z_0.
2. Wait for the next event, set $H := H + 1$,
   measure v_H and z_H.
3. Test $(\omega_H, z_H, \omega_{H-1}, v_{H-1}) \in R(F)$ for all $F \in \mathcal{F}_M(E(1\ldots H-1))$
   with ω_H being the new model state.
4. Determine $\mathcal{F}_M(E(1\ldots H))$ by Equation (11.26).
5. If $H < \bar{H}$ proceed with step 2.

Result: $\mathcal{F}_M(E(1\ldots H))$ for $H = 1, 2, \ldots, \bar{H}$.

Figure 11.4 Diagnostic algorithm based on non-deterministic automata

state transition relation $R(F)$ with the qualitative input v_{H-1} and model state ω_{H-1}. If such a transition $(\omega_H, z_H, \omega_{H-1}, v_{H-1})$ exists, the new model state ω_H is known, too. Note that the transition and, hence, the new model state is unique despite the non-determinism of the model. This is due to the fact that during the diagnostic process the qualitative state has not to be predicted (where non-determinism occurs) but is given by the measured event sequence. If a fault F has already been excluded due to measurements obtained until time step $H-1$, the fault F is not a solution of the diagnostic problem with time horizon H, either.

The value of $\mathcal{F}_M(E(1\ldots H))$ can be determined for one measured input-state pair after another. Thus, the diagnostic algorithm given in Figure 11.4 can be used on-line. This is the main advantage compared to diagnosing the quantised system with the precise representation that is composed of the differential equation and the definition of the quantisers. The automaton as concise representation makes an on-line application of the diagnosis possible. In [2], this diagnostic algorithm has been extended by a state observation. Then, no information about z_0 and no initialisation of ω_0 is needed.

11.7 Automotive Application: Fault Diagnosis of a Power Stage

The presented discrete-event modelling and diagnosis approach was applied to an automotive application example. The considered electronic power stage (Figure 11.5) is a hybrid system with discrete components (parts of the CPU and a logical control unit) and analog components (resistors, inductances, capacities, diodes). It supplies an electromagnetic valve of a Diesel engine system.

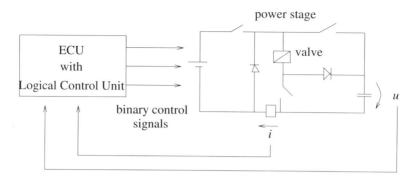

Figure 11.5 Power stage of a Diesel injection system

The power stage shows a typical hybrid behaviour, which is depicted in Figure 11.1. The system has three digital input signals (upper plots) and two continuous output signal (lower plots). The depicted behaviour is periodically repeated with a cycle time of about 10 to 100 ms. The binary control signals switch transistors within the power stage. Consequently, the structure and, hence, the dynamics of the system depends on the binary inputs. The linear power stage components lead to a linear system behaviour until the input switches and until the diodes of the circuit do not restrict the signal evolution. In summary, the power stage is a switched nonlinear dynamical system with switching transistors and nonlinear diodes.

The quantisations of the two continuous signals are shown by the dotted lines in Figure 11.6 where a nominal and a faulty behaviour are depicted. The quantisation intervals are selected in such a way that the qualitative behaviour captures the main characteristics of the nominal system. An event is generated if the discrete input value changes or if a borderline of the state signals is passed. The event times are marked by vertical bars in the bottom graphs of each lower subplot of Figure 11.6.

Discrete-event Modelling of the Power Stage

The non-deterministic automata $N(F)$ have been constructed for the normal case and five fault modes (short circuit, open circuit, diode fault, resistance deviation, capacitance deviation) by abstracting the state transition relation from a given differential equation for given fault and input. The values of the three digital signals are considered as binary code for the discrete values $\{0, 1, \ldots 7\}$. Hence, eight qualitative input values are used. The continuous values of the current signal and the voltage signal are divided into five intervals each leading to a set of 25 qualitative states. The automata have been set up for the model depths $d = 0$, $d = 1$, and $d = 2$. The size of the reduced model state sets are $|\Omega_0| = 25$, $|\Omega_1| = 2332$, $|\Omega_2| = 2996$, the size of the transition relations are $|R_0| = 2983$, $|R_1| = 17796$, $|R_2| = 18159$. This shows that due to the growing size of the models, it is advisable to use a small model depth d.

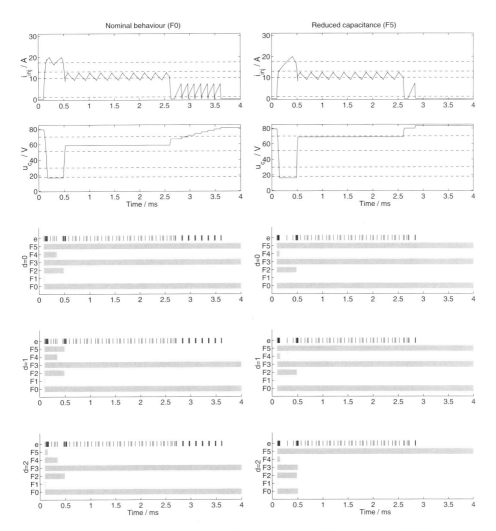

Figure 11.6 Fault identification at the power stage of the common-rail injection system

Diagnostic results

Figure 11.6 depicts two diagnostic runs. It shows the evolution in time of the given signal values and the diagnostic results. The grey boxes denote for each model (numbered from 0 for the normal case and from 1 to 5 for the faults) that the event sequence has been consistent with the model so far. Hence, a fault is a candidate of the really occurring fault as long as the bar is grey. For the time interval, in which the bar is white, the corresponding fault is known not to occur. If the bar for the fault F_0 is white, the system is known to be faulty. If for a time instant only one grey bar occurs, the fault is unambiguously identified.

In the left scenario, a normal behaviour was given. The right scenario concerns a power stage with reduced capacity. No false alarm occurs as the models used are complete. For growing model depth, faults are faster detected and can better be distinguished. This illustrates relations (11.17) and (11.18). Note however, that also for a large model depth $d = 2$, the nominal case F_0 cannot be identified uniquely in the left scenario, for instance. The example shows the feasibility of the presented diagnostic approach. Furthermore, it demonstrates that indeed a trade-off has to be found between model size and accuracy of the diagnostic result.

11.8 Conclusions

A diagnostic method for quantised switched systems has been presented and applied to a power stage of a Diesel injection system. The main idea for reducing the complexity of the diagnostic task and for making the diagnostic algorithm applicable under real-time constraints is the use of a non-deterministic automaton to concisely represent the quantised system. The requirements on this model and methods for obtaining such a model have been discussed. It has been shown that, independently of the specific model form, the model must be complete in order to facilitate correct diagnostic results. Compared to similar models that have been proposed in the literature, the model used here has the advantage that the model state is defined with respect to a parameter called the model depth that denotes the number of previously passed events stored in the automaton state. This parameter can be used to adapt the model accuracy and to choose the model complexity to the system under consideration. The practical application has shown how such a trade-off can be found for the Diesel injection system.

For applications to automotive systems of a large size, the method presented here has to be extended into two directions. First, temporal information about the event sequence has to be included into the discrete-event model. The non-deterministic automaton used here can only generate the event sequence that can appear at the output of the quantised system, but it does not say anything about the temporal distance of these events. Timed descriptions like semi-Markov processes provide the basis for this extension [13]. Second, large-scale systems cannot be described by a unique automaton but have to be modelled in a component-oriented way, which results in an automata network. Therefore, the diagnostic method has to be extended so that an automata network rather than a single automaton is concerned. Preliminary results on decomposing the diagnostic algorithm with qualitative models have been described in [9] and [18].

11.9 References

[1] P. Baroni, G. Lamperti, P. Pogliano, and M. Zanella. Diagnosis of large active systems. *Artificial Intelligence*, 110:135–183, 1999.

[2] D Förstner. *Qualitative Modellierung für die Prozeßdiagnose und deren*

Anwendung auf Dieseleinspritzsysteme. Ph.D. thesis, Technical University Hamburg–Harburg, 2001.

[3] D. Förstner and J. Lunze. Discrete-event abstraction of quantised systems with asynchronous input and state events. In *Proc. 4th International Conference on Automation of Mixed Processes: Hybrid Dynamic Systems (ADPM)*, 55–60, Dortmund, Germany, 2000.

[4] D. Förstner and J. Lunze. Discrete-event models of quantised systems for diagnosis. *Int. J. Control*, 74(7):690–700, 2001.

[5] D. Förstner and J. Lunze. Fault-detection of a diesel injection system by qualitative modelling. In *3rd IFAC Workshop Advances in Automotive Control*, pages 273–279, Karlsruhe, Germany, 2001.

[6] P. M. Frank. Analytical and qualitative model-based fault diagnosis—A survey and some new results. *Eur. J. Control*, 2:6–28, 1996.

[7] J. J. Gertler. *Fault Detection and Diagnosis in Engineering Systems*. Marcel Dekker, 1998.

[8] W. Hamscher, J. de Kleer, and L. Console, editors. *Readings in Model-Based Diagnosis*. Morgan-Kaufmann, San Mateo, 1992.

[9] B. Heiming. *Parallele Prozessdiagnose auf der Grundlage einer qualitativen Systembeschreibung*. Ph.D. Thesis. Technical Report Reihe 8, Nr. 817, VDI-Verlag, 2000.

[10] R. Isermann. Process fault detection based on modeling and estimation methods—A survey. *Automatica*, 20(4):387–404, 1984.

[11] J. Lunze. Qualitative modelling of linear dynamical systems with quantized state measurements. *Automatica*, 30:417–431, 1994.

[12] J. Lunze. Diagnosis of quantised systems. In *4th Symposium on Fault Detection, Supervision and Safety for Technical Processes (SAFEPROCESS)*, pages 28–39, Budapest, 2000.

[13] J. Lunze. Diagnosis of quantized systems based on a timed discrete-event model. *IEEE Trans. Systems, Man and Cybernetics*, SMC-30:322–335, 2000.

[14] J. Lunze, B. Nixdorf, and J. Schröder. On the nondeterminism of discrete-event representations of continuous-variable systems. *Automatica*, 35(3):395–406, 1999.

[15] K. B. Ramkumar, P. Philips, H. A. Preisig, W. K. Ho, and K. W. Lim. Structured fault-detection and diagnosis using finite-state automaton. In *Proc. 24th Annual Conf. of the IEEE Industrial Electronics Society*, pages 1667–1672, Aachen, Germany, 1998.

[16] M. Sampath, R. Sengupta, S. Lafortune, K. Sinnamohideen, and D. C. Teneketzis. Failure diagnosis using discrete-event models. *IEEE Trans. Control Systems Technology*, 4(2):105–124, 1996.

[17] F. Schiller and J. Schröder. Combining qualitative model-based diagnosis and observation within fault-tolerant systems. *AI Communications*, 12(1, 2):79–98, 1999.

[18] J. Schröder. *Modelling, State Observation and Diagnosis of Quantised Systems*. Ph.D. thesis, Technical University Hamburg-Harburg, 2001.

12

Modelling the Dynamic Behaviour of Three-way Catalytic Converters during the Warm-up Phase

G. Fiengo   L. Glielmo   S. Santini

Abstract

New regulations for emission control require the improvement of the system composed of a spark ignition internal combustion engine and three-way catalytic converter (TWC). In particular, an important problem is to minimize harmful emissions during the transient warm-up phase where the TWC is not working yet and, hence, a large amount of pollutants are emitted in the air.

Towardd this goal, here we present a dynamical thermochemical TWC model simple enough for the design and test of warm-up control strategies. The model is obtained through an asymptotic approximation of a more detailed model, *i.e.*, by letting the adsorption coefficient between gas and substrate tend to infinity. Further, we present a fast integration algorithm based partly on a 'method of lines' space-discretization, partly on the 'method of characteristics' for 'quasi-linear' hyperbolic partial differential equations, the separation being allowed by a two-time-scale analysis of the system.

The model has been identified, through a purposely designed genetic algorithm, and validated with experimental data.

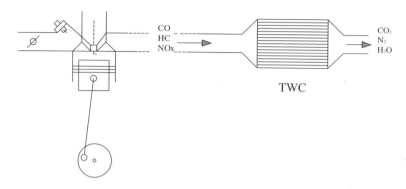

Figure 12.1 Emission treatment system and major pollutant components.

12.1 Motivations

Automotive regulations require the tight control of pollutant emissions. In order to determine whether or not a vehicle meets emissions requirements, full legislated tests have been specified in the USA and Europe, which totally obviates this issue (respectively, Federal Test Procedure, FTP, and Economic Commission for Europe, ECE, cycles)[1]. The vehicle under test is placed on a chassis dynamometer and driven through a specific dynamic drive cycle, which includes idling, acceleration and decelerations at various rates, and cruises. The tailpipe emissions are collected into bags during the test; the mass of each emission component is then measured and divided by the length of the test to obtain the grams per miles or km.

The three-way catalytic converter (TWC) is used to post-treat the spark ignition internal combustion engine (SI-ICE) exhaust in order to fulfil those tests. The feedgas, passing through the TWC, is converted into relatively harmless gases (see Figure 12.1). As is well known, the converter chemically enables the removal of carbon monoxide, oxides of nitrogen, and hydrocarbons. Notice that to let the catalysed reactions proceed simultaneously with satisfactory efficiency in a warmed-up converter (above 600 K), the mixture supplied to the catalyst has to be stoichiometric. At lower temperatures, the necessary chemical reaction cannot occur and the TWC is said to *light off* when total unburnt hydrocarbons conversion efficiency reaches 50 %. Notice that, government regulations also require alerting the driver if the TWC is not working properly (On Board Diagnosis, OBD II, European On Board Diagnosis, EOBD).

In current technology gasoline vehicle, an oxygen sensor (λ-sensor), placed in the exhaust pipe ahead of the catalyst, is used as a feedback control signal for the fuel injection system in order to ensure stoichiometry of the mixture when the TWC is hot; no specific emission control is performed at cold start and/or warm-up.

To meet new emission regulations that include a cold start and warm-up

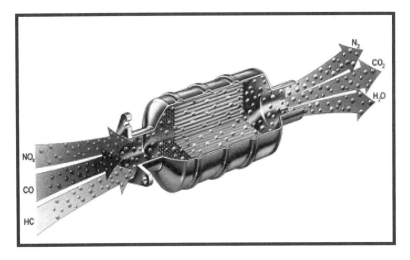

Figure 12.2 TWC.

phase in the emission tests and require onboard diagnosis, a new generation of sophisticated controllers will be developed. The design of these more sophisticated real-time applications requires simple, reduced order models of TWCs during the transient thermal phase.

To this aim, here we present some dynamical models of TWCs that are different for accuracy and are computational complexity suitable for the design and test of warm-up control strategies and for diagnosis. It has to be pointed out that TWC models will be also a key point in view of the future introduction of new pollutant sensors for real-time closed-loop emission regulations in commercial vehicle.

12.2 Basics of the TWC

In this section the basics of the TWC working will be described. By this way, it will be possible to focus the attention on the different mechanism participating to the emission reduction in order to asses their relevance in TWC modelling problems.

In an afterburner catalyst, reactants flow through the catalytic bed (matrix) and reaction products are carried away by the gas stream. Usually, modern converters are of the monolithic type (tubular reactor): exhaust gas flows through the reticular structure of a honeycomb ceramic block, maximizing the exposed surface area, and is adsorbed by the catalytic surface where it reacts (Figure 12.2). The ceramic block is usually covered by a thin coating of platinum, rhodium or palladium.

Notice that the transport phenomena between gas and solid phase are of importance in modelling the reactor and in deriving the overall reaction expression [15]. The phenomena involve a number of sequential steps including transport process between the gas stream and catalytic sites as well as the steps in the reactions at the catalyst sites. As the action of a catalyst has the objective of increasing the rate of chemical reactions, in many conditions the mass transport step is the rate-controlling stage of the process and, usually, cannot be neglected in a physically sound model of thr TWC.

Another key point in the TWC modelling problem is the definition of a chemical scheme able to represent the catalysed process inside the converter, especially during transients. The definition of a possible reaction scheme is always based on a compromise between minimization of the tunable parameters and model representability in real-world operating conditions. In fact, due to many factors such as the great number of components present in the converter feed and the complexity of the catalytic-radical mechanism, a complete and accurate chemical scheme could be very complicated to write and implement. Moreover, the analytical description of chemical phenomena occurring inside the TWC should be sufficiently flexible to match the behavior of a *large variety* of catalyst formulations and washcoats.

Most models employ simplified reaction schemes and empirical rate expressions, of Langmuir-Hinshelwood type [9], [34], where reaction rates depend on concentrations and temperature at the catalytic sites *along* the reactor. Unfortunately, the lack of precise kinetic measurements, especially during the transient warm-up phase, makes the choice of the suitable expressions and the *tuning* of their parameters a crucial aspect for model reliability.

The catalytic oxidation reactions are highly exothermic, producing considerable heat at the active sites. Thus, thermal conduction through the porous substrate layer and convective heat transfer from the external surface of the catalyst to the gas stream can be observed in the TWC. It has to be pointed out that the heat transfer between the solid and the gas phase and the internal temperature of the catalyst are important, because reactions are strongly affected by this internal temperature at sites within the device [5], [24], [25]. In this way, a correlation between thermal phenomena and reaction rates should be in terms of this internal temperature rather than the temperature of the gas flowing over the catalyst surface. During the cold start and the following warm-up temperature plays a key role in the determination of the light-off time. Note that before light-off temperature changes in the catalyst are mainly caused by thermal energy adsorption from the feedgas, while, after the light-off time, they are caused by a combination of thermal and chemical process.

Starting from a Physics-based TWC Model

The working of a TWC, deriving from the coupling between thermal and chemical phenomena, can be described by mass and energy balance differential equations. This work takes as a starting point a simple monodimensional partial differential equation (PDE) model, where the non-uniform

Three-way Catalytic Converters

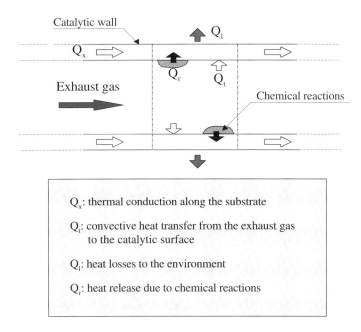

Figure 12.3 A representation of TWC substrate thermo-dynamics.

flow distribution at the monolith face is neglected (Figure 12.3) [2]:

$$0 = u_m(t,x)\rho_g(t,x)c_g\frac{\partial T_g}{\partial x} + hG_A(T_g - T_s) \quad (12.1\text{a})$$

$$0 = u_m(t,x)\frac{\partial X_g}{\partial x} + K_D G_A(X_g - X_s) \quad (12.1\text{b})$$

$$(1-\varepsilon)\rho_s c_s \frac{\partial T_s}{\partial t} = \underbrace{(1-\varepsilon)\lambda_s \frac{\partial^2 T_s}{\partial x^2}}_{Q_x} + \underbrace{hG_A(T_g - T_s)}_{Q_t}$$

$$- \underbrace{h_{\text{ext}} S_{\text{ext}}(T_s - T_{\text{ext}})}_{Q_l} - \underbrace{\Delta H^T R'(X_s, T_s)}_{Q_r} \quad (12.1\text{c})$$

$$(1-\varepsilon)\frac{\partial X_s}{\partial t} = K_D G_A(X_g - X_s) - R(X_s, T_s) \quad (12.1\text{d})$$

We consider p chemical components participating in q catalytic reactions and the above equations describe, respectively, the energy equilibrium in the gas phase, the mass equilibrium in the gas phase, the energy equilibrium in the solid phase and the mass equilibrium in the solid phase. Subscripts g and s stand for gas and substrate (*i.e.*, the reactive surface), respectively; T is temperature expressed in [K]; X is the p-vector of component concentrations [mol/m^3]; $K_D = \text{diag}(k_{D,1}, \ldots, k_{D,p})$ is the diagonal matrix of adsorption coefficients between the gas phase and the substrate for the various components; R' is the q-vector of specific reaction rates for the chemical reactions and R is the p-vector of specific reaction rates for the components, both depending on substrate temperature and concentrations [18]; ΔH^T is the q-vector of reaction enthalpies; the independent variables t and x are, respectively, the time and the axial position along the monolith; the various other coefficients are illustrated in Appendix 12.A.

In the chemical model, six chemical components (hence $p = 6$) are employed: carbon monoxide CO, nitrogen monoxide NO, hydrogen H_2, oxygen O_2, propylene C_3H_6 and methane CH_4. The last two are unburned hydrocarbons and summarise many different components difficult to discern from one to another with conventional analyzers. Following [20], [25], we divide *a priori* the total HC emitted by the engine into 86% *fast* reacting hydrocarbons, represented by C_3H_6, and 14% *slow* reacting hydrocarbons, represented by CH_4.

All the above components take part into five chemical reactions ($q = 5$) occurring on the catalytic surface, which determine the oxidation of CO, H_2, HC, and the reduction of NO [5]:

$$CO + 0.5O_2 \rightarrow CO_2 \tag{12.2a}$$
$$C_3H_6 + 4.5O_2 \rightarrow 3CO_2 + 3H_2O \tag{12.2b}$$
$$CH_4 + 2O_2 \rightarrow CO_2 + 2H_2O \tag{12.2c}$$
$$H_2 + 0.5O_2 \rightarrow H_2O \tag{12.2d}$$
$$CO + NO \rightarrow CO_2 + 0.5N_2 \tag{12.2e}$$

The specific reaction rates for the components R and for the chemical reactions R' are nonlinear functions of the temperature and the concentrations of the chemical components on the reactive surface. In this study we utilised the expression in [5] suitably simplified so as to reduce the computing load.

12.3 A Two-time-scale Infinite-adsorption Model of TWC

As already mentioned, the minimization of harmful emissions during the transient warm-up phase, where the TWC is not working yet and, hence, a large amount of dangerous emissions is emitted in the air, is still an open problem. Toward this goal, a TWC model suitable for the validation of warm-up control strategies is presented in this section.

Infinite-adsorption Model

The model (12.1), quite cumbersome for purposes of simulation and control design, has been modified as follows. Consider first that the TWC emission reduction mechanism is based mainly on the coupling of two phenomena: a mass transfer between the exhaust gas and the catalytic surface, and a catalysed chemical process on the substrate (Equations (12.1b), (12.1d)). A more detailed analysis of the device [15], confirmed by our numerical simulations with model (12.1), shows that at low temperatures the chemical process is far slower than the mass transfer phenomenon, while it is faster at high temperatures. This suggests that a possible asymptotic approach toward order reduction is to consider the mass transfer phenomenon to be infinitely fast by letting the adsorption parameters $k_{\text{D},i}$ tend to infinity (see the Appendix 12.A for more mathematical details on this reduction procedure); from this $X_g \equiv X_s =: X$ follows. (The same can be deduced if one assumes reactions to occur only on the external surface of the catalytic surface [3], [8].)

Taking into account the above considerations, the simplified model follows [10]:

$$0 = u_m(t,x) \rho_g(t,x) c_g \frac{\partial T_g}{\partial x} + hG_A(T_g - T_s) \quad (12.3a)$$

$$(1-\varepsilon)\rho_s c_s \frac{\partial T_s}{\partial t} = (1-\varepsilon)\lambda_s \frac{\partial^2 T_s}{\partial x^2} + hG_A(T_g - T_s)$$
$$- h_{\text{ext}} S_{\text{ext}}(T_s - T_{\text{ext}}) - \Delta H^T R'(X, T_s) \quad (12.3b)$$

$$(1-\varepsilon)\frac{\partial X}{\partial t} = -u_m(t,x)\frac{\partial X}{\partial x} - R(X, T_s) \quad (12.3c)$$

The boundary and initial conditions are ($t \geq 0, x \in [0, L]$)

$$\frac{\partial T_s}{\partial t}(t, L) = 0 \quad \text{(adiabatic constraint)} \quad (12.4a)$$

$$T_g(t, 0) = T_g^*(t), \quad X(t, 0) = X_g^*(t) \quad (12.4b)$$

$$T_s(0, x) = T_s^*(x), \quad X(0, x) = X^*(x) \quad (12.4c)$$

where $T_g^*(t)$ and $X_g^*(t) = (X_{g1}^*(t), \ldots, X_{g6}^*(t))^T$ are, respectively, the temperature of the exhaust gas and the concentrations of the chemical components at the inlet of the TWC, $T_s^*(x)$ is the initial temperature of the substrate, $X^*(x) = (X_1^*(x), \ldots, X_6^*(x))^T$ are the initial concentrations and L is the TWC length.

Integration Algorithm

Before developing the algorithm of numerical integration, the set of equations (12.3) has been cast into a dimensionless form; from now on, for the sake of simplicity, the same symbols will refer to dimensionless quantities.

Solving Equation (12.3c) For the reader's convenience, we rewrite here the concentration equations (12.3c) as follows

$$\frac{\partial X(t,x)}{\partial t} + \frac{u_m(t,x)}{(1-\varepsilon)}\frac{\partial X(t,x)}{\partial x} = -\frac{R(X(t,x), T_s(t,x))}{(1-\varepsilon)}. \quad (12.5)$$

The initial conditions are prescribed along the axes of the (t, x)-plane

$$X(t, 0) = X_g^*(t), \quad t \geq 0, \qquad X(0, x) = X_s^*(x), \quad x \in [0, 1] \quad (12.6)$$

If the temperature pattern $T_\mathrm{s} = T_\mathrm{s}(t,x)$ *were known*, (12.5) would be a system of 'quasi-linear' hyperbolic PDEs and could be solved using the characteristics method [7], [26], [27] yielding the ODEs

$$\frac{dt}{ds} = 1 \qquad (12.7\mathrm{a})$$

$$\frac{dx}{ds} = \frac{u_\mathrm{m}(t(s),x(s))}{(1-\varepsilon)} \qquad (12.7\mathrm{b})$$

$$\frac{dX}{ds} = -\frac{\widetilde{R}(t(s),x(s),X(s))}{(1-\varepsilon)} \qquad (12.7\mathrm{c})$$

where $\widetilde{R}(t,x,X) \triangleq R(X,T_\mathrm{s}(t,x))$.

In particular equation (12.7a) relates the parameter s of each characteristic curve to the time t; Equation (12.7b) describes the motion of the particle along the characteristic curve from $x=0$ (inlet of TWC) to $x=1$ (outlet of TWC); Equation (12.7c) describes the change of concentrations as the particle moves along the characteristic and, hence, along the converter. It is interesting to notice that the characteristics method allows a 'Lagrangian' approach to the problem, since it describes the motion and the effects of chemical reactions on the single gas particle.

Since u_m is always positive, $x(s)$ is invertible in view of (12.7b), and we can define new variables $\widetilde{X}(x)$, $\widetilde{t}(x)$ as

$$\widetilde{X}(x) \triangleq X(s(x)), \qquad \widetilde{t}(x) \triangleq t(s(x)) \qquad (12.8)$$

Thus, along a characteristic curve, system (12.7) simply reduces to

$$\frac{d\widetilde{t}}{dx} = \frac{(1-\varepsilon)}{u_\mathrm{m}(\widetilde{t},x)} \qquad (12.9\mathrm{a})$$

$$\frac{d\widetilde{X}}{dx} = -\frac{\widetilde{R}(\widetilde{t}(x),x,\widetilde{X}(x))}{u_\mathrm{m}(\widetilde{t}(x),x)} \qquad (12.9\mathrm{b})$$

where $x \in [0,1]$ and the initial conditions are prescribed as

$$\widetilde{t}(0) = \hat{t}, \qquad \widetilde{X}(0) = X_\mathrm{g}^*(\hat{t}) \qquad (12.10)$$

Numerical integration. Actually, Equation (12.3c) in the set of PDEs (12.3) cannot be solved as above because the function $T_\mathrm{s}(t,x)$ is not *a priori* known but it is a solution of (12.3a)–(12.3b). Conversely, the concentrations X in Equation (12.3b) are the solution of (12.3c). In other words, the system (12.3) is coupled (Figure 12.4).

However, we can by-pass this difficulty by noticing that, in practice, the thermal dynamics described by Equations (12.3a)–(12.3b) are much slower than the chemical dynamics of (12.3c). In particular, the time spent by each gas particle inside the converter is very short compared to the time scale of the thermal phenomenon. It is, thus, possible to set up an approximate integration scheme (Figure 12.5) inspired by the singular perturbation approach [19], [23], [13]:

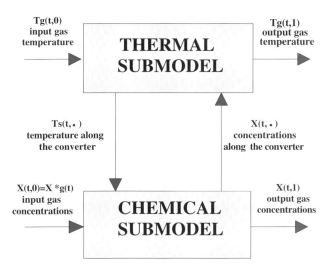

Figure 12.4 Structure of the system.

- Consider equation (12.9a) and notice that the time spent by a gas particle in crossing the converter, i.e. $\tilde{t}(1) - \tilde{t}(0)$, is small compared to the characteristic time of thermal phenomena. This enables us to set $\tilde{t}(x) = \hat{t}$ for all $x \in [0, 1]$ and Equation (12.9b) becomes

$$\frac{d\widetilde{X}}{dx} = -\frac{\widetilde{R}(\hat{t}, x, \widetilde{X}(x))}{u_m(\hat{t}, x)} = -\frac{R(\widetilde{X}(x), T_s(\hat{t}, x))}{u_m(\hat{t}, x)} \qquad (12.11)$$

In other words, we integrate (12.3c) separately from (12.3a)–(12.3b) by considering the temperature T_s to depend only on the space variable x. The solution of Equation (12.11) depends on the initial conditions $\hat{t}$, $X_g^*(\hat{t})$ and the temperature pattern $T_s(\hat{t}, \cdot)$. We will simply indicate this by writing $\widetilde{X}(x; \hat{t})$.

- Now, to solve the thermal equations (12.3a)–(12.3b), we define the effectiveness of catalysis along the TWC as

$$\eta_i(\hat{t}, x) := \frac{\widetilde{X}_i(x; \hat{t})}{X_{g,i}^*(\hat{t})}, \qquad i = 1, \ldots, 6 \qquad (12.12)$$

then, at the time interval $[\hat{t}, \hat{t} + \Delta t]$, we use the linear approximation

$$X_i(t, x) \approx \eta_i(\hat{t}, x) X_{g,i}^*(t), \qquad i = 1, \ldots, 6 \qquad (12.13)$$

The width Δt must be chosen so that for $t \in [\hat{t}, \hat{t} + \Delta t]$

$$\Delta T(t) := \max_{x \in [0,1]} |T_s(t, x) - T_s(\hat{t}, x)| \leq \Delta_T \qquad (12.14a)$$

$$\Delta X_i(t) := |X_{g,i}^*(t) - X_{g,i}^*(\hat{t})| \leq \Delta_{X,i} \quad i = 1, \ldots, 6 \qquad (12.14b)$$

where Δ_T and $\Delta_{X,i}$, $i = 1,\ldots,6$ are, respectively, fixed temperature and concentration thresholds.

The distributed parameter model (12.3a)–(12.3b) has been converted into a lumped one by a finite difference scheme ('method of lines') [30], where a discrete number of spatial elements (i.e., slices of the converter) is considered, each described by time-varying variables. In this way, from the temperature PDEs a system of ODEs is obtained that can be solved through usual integration packages.

The outline of the integration algorithm on the interval $[0, t_{\text{fin}}]$ follows:

$t \leftarrow 0$;
$T_s^*(x) \leftarrow T_s(0, x)$;   (initial pattern of substrate temperature)
while $t \leq t_{\text{fin}}$
    solve the chemical equations with a temperature pattern T_s^*;
    evaluate $\eta_i(x)$ $x \in [0, 1]$ $i = 1,\ldots,6$;
    $\Delta T \leftarrow 0$;
    $\Delta X_i \leftarrow 0$ $i = 1,\ldots,6$;
    while $(\Delta T \leq \Delta_T)$ and $(\Delta X_i \leq \Delta_{X,i} \text{for} i = 1,\ldots,6)$ and $(t \leq t_{\text{fin}})$
        $X_i(t, x) = \eta_i(x) X_{g,i}^*(t)$, $i = 1,\ldots,6$;
        solve the thermal equations;
        $T_s^*(x) \leftarrow T_s(t, x)$;   (update the pattern of temperature)
        evaluate ΔT;
        evaluate ΔX_i $i = 1,\ldots,6$;
        increment t;
    end while
end while

Identification Procedure and Results

The simplified model (12.3) has been identified in order to fit the experimental data. In particular the model structure is characterised by 12 independent parameters: seven for the temperature and five for the concentrations dynamics.

Our experience taught us that gradient-based optimization algorithms are quite ineffective in this case suggested the use of a genetic algorithm (GA), a kind of stochastic optimisation algorithm [14], in order to explore the whole space of solutions and avoid trapping in local minimum points.

In this parametric estimation problem, we used a purposely designed GA (see [10] for details) to adjust the model parameters in order to minimise the sum-squared error between model outputs and measured experimental

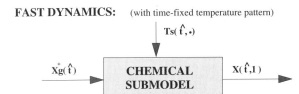

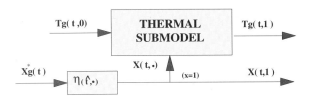

Figure 12.5 Approximate system in the time interval $[\hat{t}, \hat{t} + \Delta t]$

data

$$\sum_h \frac{1}{v_h} \int_0^{t_{\text{fin}}} (y_h - \hat{y}_h)^2 dt \qquad (12.15)$$

where y_h is the measured output, $\hat{y}_h$ is the simulation result, h refers to the output channel, v_h are scaling factors, 0 and t_{fin} are, respectively, the initial and final time instant of the simulation.

The model behaviour has been compared to experimental data for a four-stroke four-cylinder 16-valve two-litres displacement FIAT engine, gasoline-powered passenger vehicle equipped with fresh Pt/Rh converter monolith.

In Figures 12.6–12.7, we show, as an example, some of the simulation results from the TWC model along an FTP cycle comparing real data and model output. The parameter identification phase covers the first 160sec and the interval between 400–550 s. The window 160–400 s is for validation. Figures also show the experimental data at the TWC inlet used as model inputs.

Results show a good fit between the model and the real behaviour of the TWC for the temperature dynamics, also in the validation phase. As regards the concentrations, we can see that in all cases the pre-lightoff phase is well approximated, even if the quality of the training data in that region is not particularly good. Most important, the model always captures correctly the transition between the pre-light-off and the post-light-off phase. After the light-off phase the catalysis effect is reproduced reasonably well.

The algorithm has been developed in Matlab^{TM} 5.3/Simulink 3.0 environment with the support of a C-compiled S-function to shorten the compu-

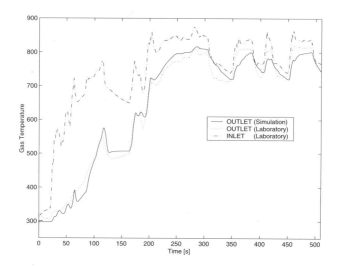

Figure 12.6 T_g time history. Dotted line: experimental data referred to the TWC outlet. Solid line: model output referred to the TWC outlet. Dash-dotted line: experimental data at the TWC inlet

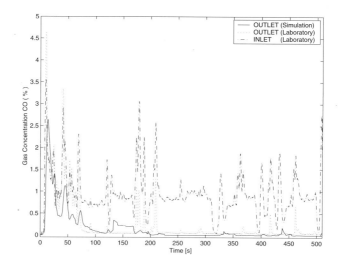

Figure 12.7 CO time history. Dotted line: experimental data referred to the TWC outlet. Solid line: model output referred to the TWC outlet. Dash-dotted line: experimental data at the TWC inlet

tational time, and the simulation of 510 real seconds of the warm-up along an FTP cycle takes 5 s on a PC, Intel Pentium III Processor, 128 Mb RAM.

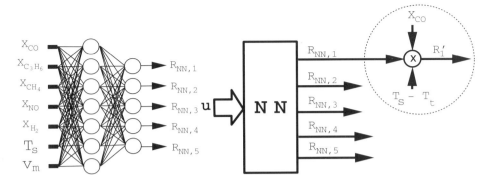

Figure 12.8 A two-layer network architecture (left) applied to structure of the reaction kinetics submodel.

12.4 Machine Learning for Reaction Kinetics

The lack of precise kinetic measurements, especially during the transient warm-up phase, makes the choice of the suitable expressions and the *tuning* of their parameters a crucial aspect for model reliability.

Currently, NNs are used for modelling static nonlinear maps with satisfactory results. Recently, novel interconnections of NNs have been tested within dynamic contexts: namely for the identification of unknown dynamical systems described by a set of ordinary differential equations [33], or as a part of lumped parameter model [21], [29]. The training is realised through standard gradient methods, such as backpropagation.

Here we present a novel approach that considers the NN as a subsystem in a PDE model and uses GA to train it. More precisely, the NN submodel computes the reaction rates of the catalyst during warm-up and is embedded into the TWC partial differential equations model. This modelling technique allows us to preserve the most important features of an accurate distributed parameter TWC model, it circumvents both the structural and the parameter uncertainty of standard reaction kinetics models, and saves computational time. The GA bypasses the above-mentioned problems with the data and the difficulties to apply gradient-based methods in this complex modelling schema. The whole procedure uses for integration the fast *ad hoc* PDE integration procedure introduced in Sec. 12.3.

The NN for the Reaction Kinetics

In order to describe the 'neural' reaction rates that have to be used in model (12.3), we refer to the following structure (Figure 12.8) [11]:

$$R' = \begin{cases} \mathbf{D}\, R_{NN}\,(T_s - T_t), & \text{for} \quad T_s \geq T_t \\ 0, & \text{for} \quad T_s < T_t \end{cases} \quad (12.16)$$

where $R' = (R'_1, \ldots, R'_5)^T$ is the specific reaction rates vector (Eq. (12.3b)); $\mathbf{D} \triangleq \mathrm{diag}(X_{\mathrm{CO}}, X_{\mathrm{C_3H_6}}, X_{\mathrm{CH_4}}, X_{\mathrm{NO}}, X_{\mathrm{H_2}}) \in \mathbb{R}^{5 \times 5}$; $R_{NN} \in \mathbb{R}^5$ is the output vector of a fully connected multilayer feedforward NN; T_s is the temperature of the reactive surface (substrate), and T_t is a prefixed temperature threshold. Notice the structure (12.16) ensures that the kinetic rates are zero in absence of reactants or when the substrate is at ambient temperature. On the basis of the reaction kinetics, it is straightforward to obtain the vector of the rates of the chemical components $R = (R_1, \ldots, R_6)^T$ (Equations (12.3c)) since it is linearly related to R'.

The mathematical model of the NN is completely defined by the number of layers, the specification of the activation function for the neurons in each layer, and the weight matrices for each layer. The NN size has to be chosen according to the trade-off between model accuracy and computational burden. In our case, a two-layers feedforward neural network (see Figure 12.8) is used in the kinetic model and it can be specified as follows:

$$z = \tanh(W_1 u + b_1), \qquad (12.17a)$$
$$R_{NN} = W_2 z + b_2, \qquad (12.17b)$$

where $z \in \mathbb{R}^7$ is the output vector of the first layer; $W_1 \in \mathbb{R}^{7 \times 7}$ and $W_2 \in \mathbb{R}^{5 \times 7}$ are the network weight matrices; $b_1 \in \mathbb{R}^7$ and $b_2 \in \mathbb{R}^5$ are bias column vectors; u is the input vector of the neural net defined as $u = (X_{\mathrm{CO}}, X_{\mathrm{C_3H_6}}, X_{\mathrm{CH_4}}, X_{\mathrm{NO}}, X_{\mathrm{H_2}}, T_s, v_m)^T$ and the $\tanh(\cdot)$ operator is intended component-wise. The mean gas velocity in monolith v_m is included in the input vector u because previous identification experiments of the system have shown that this further piece of information improves the performance of the whole model.

Identification Procedure and Results

Since the kinetic data, *i.e.*, inputs and outputs of the net, are not directly available, we have to base the training on the TWC inlet-outlet concentration and temperature profiles. Towards this goal, here we use a GA that minimises the functional with respect to the NN parameters W and b.

The model behaviour was compared to experimental data along the transient thermal phase; the results refer to a four-stroke four-cylinder 1400 cc FIAT engine, equipped with fresh Pt/Rh converter monolith. As an example of model behaviour, some simulation results are shown in Figures 12.9- 12.10.

In all figures, real TWC input and output data and simulations output along an FTP cycle are plotted. The model captures the most important features of the TWC warm-up; in particular, the thermal behaviour is very well reproduced. As regards the concentration, the model clearly detects when, due to the low temperature, the catalyst is not properly working. Once the device is sufficiently warm, the conversion is reproduced reasonably well. The simulation of 500 real seconds of the warm-up along an FTP cycle takes 3 on a PC, Intel Pentium III Processor, 128 Mb RAM. Our algorithms

Three-way Catalytic Converters

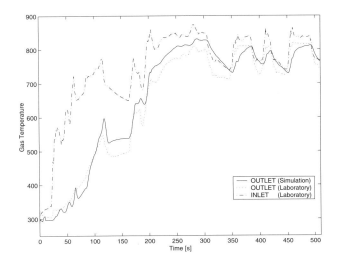

Figure 12.9 T_g time history. Dotted line: experimental data referred to the TWC outlet. Solid line: model output referred to the TWC outlet. Dash-dotted line: experimental data at the TWC inlet

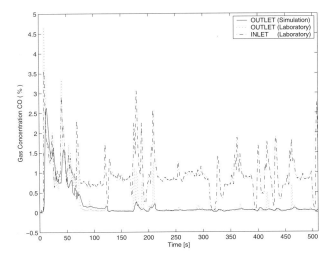

Figure 12.10 C_3H_6 time history. Dotted line: experimental data referred to the TWC outlet. Solid line: model output referred to the TWC outlet. Dash-dotted line: experimental data at the TWC inlet

have been developed on Matlab^{TM} 5.3/Simulink 3.0 environment with the support of C compiled S-function. Details on the whole identification procedure of the NN sub-model can be found in [11].

12.5 A Phenomenological Model of TWC

Many detailed chemistry studies (see, for example, [16], [32]), show that modulating the composition of the feedgas with different amplitudes and frequencies strongly influence the efficiency of a 'warmed-up' TWC. This behaviour can be attributed to the ability of some component of the washcoat to be periodically oxidised and reduced depending on the gas redox environment. The component that seems to play a key role in these dynamic oxidation-reduction phenomena is Cerium [22]. The presence of Cerium improves the TWC performance by allowing the *oxygen storage* phenomenon [31], [35]: during transients, in presence of oxygen excess, there is an oxygen chemiadsorption on the catalyst, while in conditions of defect, there is a release. In other words, if an excess of O_2 participates in the combustion, it will be chemically stored (up to a certain capacity); conversely, if a deficit of O_2 exists then the catalyst will give up oxygen (as long as some is available) to allow the reactions to happen. Thus, the oxygen storage is a key mechanism that enhances the catalyst activity helping the catalysed oxidation-reduction reactions.

The phenomenological model we present here (see the block diagram in Figure 12.11) is based on the hypothesis that the catalyst dynamics are dominated by the oxygen storage phenomenon. This allows description of the catalyst activity only in terms of the oxygen buffer dynamics using the pre- and post-catalyst and a transport delay. Since, during warm-up, unburned hydrocarbons allow capture of the device light-off, a chemical model describes only THC oxidation. Other TWC control-oriented models can be found in the literature, but only for warmed-up catalysts [6], [12], [28].

Oxygen Storage

The oxygen storage phenomenon is modelled with two actions aimed at correcting the quantity of oxygen in the gas at the inlet of the catalyst, trying to maintain the air-fuel mixture at the stoichiometric point.

Letting C[g] denote the oxygen capacity of the catalyst, we denote by Θ_{FG} and Θ_{St} the fraction of such a capacity present in the actual feedgas and the stoichiometric feedgas. Hence

$$\dot{\Theta}_{FG} = \frac{\dot{m}_f \cdot \lambda_{FG} \cdot S \cdot 0.23}{C} \quad (12.18a)$$

$$\dot{\Theta}_{St} = \frac{\dot{m}_f \cdot S \cdot 0.23}{C} \quad (12.18b)$$

where $\dot{m}_f$ [g/s] is the fuel mass flow rate, λ_{FG}[/] is the air-fuel ratio at the feedgas, S is the stoichiometric value (~ 14.6); and 0.23 represents the

Three-way Catalytic Converters

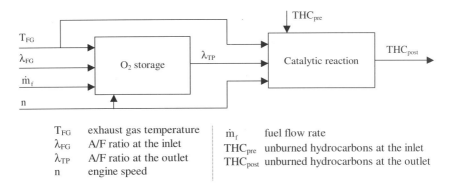

T_{FG}	exhaust gas temperature	$\dot{m}_f$	fuel flow rate
λ_{FG}	A/F ratio at the inlet	THC_{pre}	unburned hydrocarbons at the inlet
λ_{TP}	A/F ratio at the outlet	THC_{post}	unburned hydrocarbons at the outlet
n	engine speed		

Figure 12.11 Block diagram of TWC model

percentage of oxygen present in the air.

At each instant of time, the quantity of oxygen that could be stored in lean condition or released in rich condition is computed as

$$\dot{\Theta}_{cor} = (\dot{\Theta}_{FG} - \dot{\Theta}_{St}) \cdot g(T_{FG}) \cdot \begin{cases} f_L(\Theta), & \lambda_{FG} \geq 1 \\ f_R(\Theta), & \lambda_{FG} < 1 \end{cases} \quad (12.19)$$

where $(\dot{\Theta}_{FG} - \dot{\Theta}_{St})$ is the quantity of oxygen available for the storage or requested by the catalytic reactions; Θ is the relative oxygen level stored in the TWC computed later, and $T_{FG}[\r{r}C]$ is the feedgas temperature affecting the TWC efficiency by means of a function $g(T_{FG})$, switching around a threshold value T_{th}, according to

$$g(T_{FG}) = \frac{\tanh[(T_{FG} - T_{th})\gamma] + 1}{2} \quad (12.20)$$

where γ is a parameter to be identified and represents the slope of the function, and $f_L(\Theta)$ and $f_R(\Theta)$ calculate the percentage of the surplus or deficit oxygen, respectively, stored or released, according to

$$f_L(\Theta) = (1 - \Theta^8) \quad (12.21a)$$
$$f_R(\Theta) = [1 - (1 - \Theta)^8] \quad (12.21b)$$

In particular, in lean conditions, if the catalyst is full ($\Theta = 1$) the oxygen is not stored ($f_L = 0$); conversely, if the catalyst is empty ($\Theta = 0$) all the oxygen available is stored ($f_L = 1$). Conversely, in rich conditions, when the catalyst is full ($\Theta = 1$) all the deficit oxygen is released ($f_R = 1$) and if it is empty ($\Theta = 0$) it is not possible to release oxygen ($f_R = 0$).

The quantity of oxygen in the gas after this first correction, $\Theta_{FG_{cor}}$, and the corresponding air-fuel ratio, $\lambda_{FG_{cor}}$ are now given by

$$\dot{\Theta}_{FG_{cor}} = \dot{\Theta}_{FG} - \dot{\Theta}_{cor} \quad (12.22a)$$
$$\lambda_{FG_{cor}} = \frac{\dot{\Theta}_{FG_{cor}} \cdot C}{\dot{m}_f \cdot S \cdot 0.23} \quad (12.22b)$$

The second correction is modelled with a first order system with unit gain and time-varying time constant

$$\dot{\lambda}_{\text{aux}} = -\frac{\lambda_{\text{aux}}}{\tau(\lambda_{\text{FG}}, \lambda_{\text{aux}}, T_{\text{FG}})} + \frac{\lambda_{\text{FG}_{\text{cor}}}}{\tau(\lambda_{\text{FG}}, \lambda_{\text{aux}}, T_{\text{FG}})} \quad (12.23)$$

where $\tau(\lambda_{\text{FG}}, \lambda_{\text{aux}}, T_{\text{FG}})$ is the time constant, depending on the air/fuel ratio of the gas at the inlet and the outlet of the catalyst, and the feedgas temperature. The corresponding relative quantity of oxygen present in the gas at the outlet of the catalyst, Θ_{aux}, is computed as

$$\dot{\Theta}_{\text{aux}} = \frac{\dot{m}_{\text{f}} \cdot \lambda_{\text{aux}} \cdot S \cdot 0.23}{C}. \quad (12.24)$$

The quantity of oxygen that is altogether stored or released from the TWC is simply the difference between the oxygen at the feedgas, $\dot{\Theta}_{\text{FG}}$, and the oxygen present in the gas after the two actions, $\dot{\Theta}_{\text{aux}}$. In this way, it is possible to upgrade the relative oxygen level stored in the TWC, according to:

$$\dot{\Theta} = \begin{cases} \dot{\Theta}_{\text{FG}} - \dot{\Theta}_{\text{aux}} & 0 \leq \Theta \leq 1 \\ 0 & \text{otherwise} \end{cases} \quad (12.25)$$

Finally the air-fuel ratio at the tailpipe, λ_{TP}, is computed as follows:

$$\lambda_{\text{TP}}(t) = \lambda_{\text{aux}}(t - \Delta(\dot{m}_{\text{a}})) \quad (12.26)$$

where $\dot{m}_{\text{a}}$ [g/sec] is the air mass flow rate and $\Delta(\dot{m}_{\text{a}})$ is the transport delay of the gas

$$\Delta(\dot{m}_{\text{a}}) = \frac{\delta_1}{\dot{m}_{\text{a}}} + \delta_2 \quad (12.27)$$

The parameters C, T_{th}, γ, δ_1, δ_2, $\tau_{\min}$, and $\tau_{\max}$ have to be identified for each specific catalytic converter.

Catalytic Reactions

As before, the total hydrocarbons (THC) in the exhaust gas are divided into two groups: *fast oxidizing hydrocarbons* (propylene, C_3H_6, 86%) and *slow oxidizing hydrocarbons* (methane, CH_4, 14%). They participate in the reactions

$$C_3H_6 + 4.5O_2 \rightarrow 3CO_2 + 3H_2O$$
$$CH_4 + 2O_2 \rightarrow CO_2 + 2H_2O$$

The kinetic model [3] has one output (the hydrocarbons at the outlet of the TWC, THC_{post} [g]) and five inputs (the hydrocarbons at the inlet of the

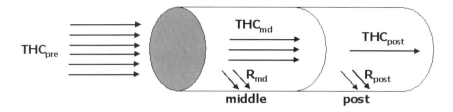

Figure 12.12 Catalytic reaction schema

TWC, THC_{pre} [g]; the engine speed, n [rpm]; the fuel flow rate, $\dot{m}_f$ [g/sec]; the air-fuel ratio of the gas corrected by the oxygen storage phenomenon, λ_{TP} [/]; the feedgas temperature, T_{FG} [°C]). The TWC is supposed to be divided into two parts connected in series, as shown in Figure 12.12.

The unburned hydrocarbons produced by the engine combustion enter in the first part of the TWC, and there the first oxidation reactions take place. The remaining unburned hydrocarbons, called THC_{md}, pass through the first part in which is divided the TWC and arrive in the second part, where other oxidation reaction take place. The unburned hydrocarbons that are not oxidised in the two parts of the TWC, called THC_{post}, are the output of the system. In conclusion, the model is composed of two states (the unburned hydrocarbons in the middle of the catalyst, THC_{md} [g] and at the outlet, THC_{post} [g]) as follows:

$$\dot{\text{THC}}_{\text{md}} = -K_1 n (\text{THC}_{\text{md}} - \text{THC}_{\text{pre}})$$
$$\quad\quad\quad -R_{\text{md}}(T_{\text{FG}}, \text{THC}_{\text{md}}, \dot{m}_f, \lambda_{\text{TP}})$$
$$\dot{\text{THC}}_{\text{post}} = -K_2 n (\text{THC}_{\text{post}} - \text{THC}_{\text{md}})$$
$$\quad\quad\quad -R_{\text{post}}(T_{\text{FG}}, \text{THC}_{\text{post}}, \dot{m}_f, \lambda_{\text{TP}})$$

where

- $R_i(T_{\text{FG}}, \text{THC}_i, \dot{m}_f, \lambda_{\text{TP}})$ $i = \text{md, post}$, are the reaction rates described by the equation

$$R_i(T_{\text{FG}}, \text{THC}_i, \dot{m}_f, \lambda_{\text{TP}}) = K_i \text{THC}_i \sqrt{\dot{m}_f \lambda_{\text{TP}}} \frac{\tanh[(T_{\text{FG}} - T_{\text{th}_i})\gamma_i] + 1}{2}$$

- K_1, K_2, K_{md}, K_{post}, $T_{\text{th}_{\text{md}}}$, $T_{\text{th}_{\text{post}}}$, γ_{md}, γ_{post} are the coefficients to be identified.

Identification Procedure and Results

A hybrid identification procedure, using both a genetic algorithm and a least square optimization, has been designed to identify the coefficients of the catalytic converter model.

In Figure 12.13, the air-fuel ratio of the exhaust gas at the input (dashed line) and the output (solid line) of the catalyst are plotted; Figure 12.14 shows the unburned hydrocarbons at the input of the TWC (dashed

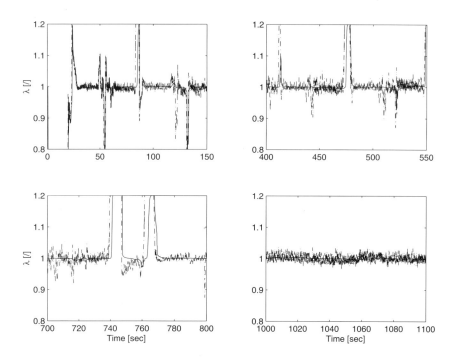

Figure 12.13 Air/fuel ratio at the inlet and outlet of the TWC λ_{FG}

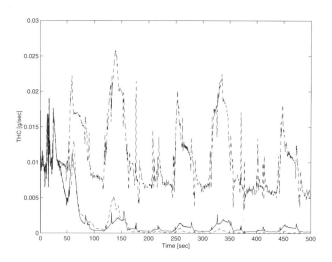

Figure 12.14 Unburned hydrocarbons at the inlet and outlet of the TWC (THC). The identification range is [0 s, 300 s]; the remaining period is the validation range

line), the real output (dash-dot line) and simulated output (solid line). The experimental data utilised for the identification and validation process refer to a Golf 1.6 performing an ECE cycle.

12.6 Conclusions

Mathematical models of the dynamic behaviour of automotive catalysts were presented. The TWC models differ in their accuracy and computational complexity and are aimed to provide a basis for the design and testing of warm-up control strategies and for board diagnosis.

The minimisation of harmful emissions during the transient warm-up phase requires dynamical thermochemical TWC models simple enough for on-line computation. In this work, a two-timescale infinite-adsorption TWC model is a first step toward this goal.

The lack of precise kinetic measurements, especially during the transient, make the kinetic reaction sub-model a key point in TWC modelling. These reasons suggested a way to ulteriorly simplify the two time scale infinite-adsorption TWC model and, thus, shorten the computational time. The use of machine learning techniques to include black box reaction kinetics sub-models into the TWC model is proposed, showing the validity of this approach using a simple static NN model as a case of study.

At last, a simplified, control-oriented phenomenological TWC model, purposely designed for real-time control application, is presented. All the models are identified and validated with experimental data.

12.A Appendix—Mathematical Reduction Procedure

Asymptotic Expansion of Singularly Perturbed Quasi-linear PDEs

In Section 12.3, we showed, quite naively, that letting the adsorption parameters $k_{D,i}$ go to infinity leads to a simpler PDE structure. Here, we briefly illustrate how a more formal approach can justify the approximation, improve its quality if needed, and provide further insight into the physical phenomenon at hand. Further readings on the subjects are, for example, in [4], [17].

For the sake of brevity, we focus just on the chemical equations in (12.1), *i.e.*, (12.1b) and (12.1d), assuming that the reaction rates R do not depend on the substrate temperature T_s; further we assume we are dealing with only one chemical components and that the void factor $\varepsilon = 0$. We leave to the interested reader the tedious but straightforward effort of generalization.

The equations reduce to[1]

$$\mu u_m(t,x) \frac{\partial X_g}{\partial x} = -(X_g - X_s) \qquad (12.31a)$$

$$\mu \frac{\partial X_s}{\partial t} = (X_g - X_s) - \mu R(X_s) \qquad (12.31b)$$

where $\mu = (k_D G_A)^{-1}$ is a "small" positive number. We adjoin the boundary conditions $X_g(t,0) = m(t)$, $X_s(0,x) = n(x)$.

The above equations are a system of quasi-linear PDEs in canonical form, since in each equation only one unknown appears differentiated along a characteristic direction. In particular, X_g is differentiated along the direction $(0,1)$ of the (t,x)-domain, X_s is differentiated along the direction $(1,0)$. An equivalent but non-canonical form of the above equations, suitable for our purposes, is obtained by replacing (12.31a) into (12.31b):

$$\mu u_m \frac{\partial X_g}{\partial x} = -(X_g - X_s) \qquad (12.32a)$$

$$\frac{\partial X_s}{\partial t} = -u_m \frac{\partial X_g}{\partial x} - R(X_s) \qquad (12.32b)$$

From the above equation, one easily sees that we are dealing with a singularly perturbed problem, in that for $\mu = 0$ the first differential equation becomes an algebraic equation. In order to obtain an asymptotic expansion of the solution of (12.32) of order $O(\mu^2)$, we let

$$\begin{aligned} X_g(t,x) &\cong X_{g0}(t,x) + \mu X_{g1}(t,x) \\ X_s(t,x) &\cong X_{s0}(t,x) + \mu X_{s1}(t,x) \end{aligned} \qquad (12.33)$$

We now replace the above expression into (12.32), we consider that $R(X_s) \cong R(X_{s0}) + \mu X_{s1} R'(X_{s0})$ (in this context and differently from Section 12.2 we denote by R' the derivative of the function R in (12.32)), we compare equal powers of μ and obtain two systems of PDEs to construct the approximate solution. In particular, the zero-order system is

$$X_{g0} = X_{s0} \qquad (12.34a)$$

$$\frac{\partial X_{s0}}{\partial t} = -u_m \frac{\partial X_{s0}}{\partial x} - R(X_{s0}) \qquad (12.34b)$$

with the boundary condition $X_{s0}(t,0) = m(t)$, $X_{s0}(0,x) = n(x)$. This is exactly the approximate solution we propose in section 12.3. It should be apparent that, due to the loss of one degree of freedom necessary to satisfy the boundary condition, this is an "outer" approximation of the solution, i.e. it holds for $\mu \to 0$, $t > 0$, and $x > 0$, yielding reasonable results away from the domain boundary but giving errors around the axes, and particularly around the origin where the solution surface has a step discontinuity.

---

[1] We drop the dependence on the variables when this does not lead to ambiguity.

If μ is "small", but not "very small", a better approximation requires the solution of a further system

$$u_m \frac{\partial X_{g0}}{\partial x} = -(X_{g1} - X_{s1}) \tag{12.35a}$$

$$\frac{\partial X_{s1}}{\partial t} = -u_m \frac{\partial X_{g1}}{\partial x} - R'(X_{s0}) X_{s1} \tag{12.35b}$$

which reduces to the single quasilinear PDE in the unknown X_{s1}

$$\frac{\partial X_{s1}}{\partial t} = -u_m \left[\frac{\partial X_{s1}}{\partial x} - \frac{\partial}{\partial x}\left(u_m \frac{\partial X_{g0}}{\partial x} \right) \right] - R'(X_{s0}) X_{s1} \tag{12.36}$$

with the boundary condition $X_{s1}(t,0) = u_m(t,0) \frac{\partial X_{g0}}{\partial x}(t,0)$, $X_{s1}(0,x) = 0$.

A similar procedure can be used to obtain an "inner" approximation, valid for $\mu \to 0$, and around $t = 0$ or $x = 0$. To do this, we first change the timescale and the length scale defining, for $\mu \neq 0$, $\tau = t/\mu$, $\sigma = x/\mu$; we introduce the new unknown functions $\widehat{X}_g(\tau,\sigma) = X_g(\mu\tau, \mu\sigma)$, $\widehat{X}_s(\tau,\sigma) = X_s(\mu\tau, \mu\sigma)$, and rewrite (12.31) as follows

$$\hat{u}_m \frac{\partial \widehat{X}_g}{\partial \sigma} = -(\widehat{X}_g - \widehat{X}_s), \tag{12.37a}$$

$$\frac{\partial \widehat{X}_s}{\partial \tau} = (\widehat{X}_g - \widehat{X}_s) - \mu R(\widehat{X}_s), \tag{12.37b}$$

where $\hat{u}_m(\tau,\sigma) = u_m(\mu\tau, \mu\sigma)$, with the boundary condition $\widehat{X}_g(\tau,0) = m(\mu\tau)$, $\widehat{X}_g(0,\sigma) = n(\mu\sigma)$. One can see that the change of variables yields a regularly perturbed system, but now the domain of integration in the plane (τ,σ) grows as μ^{-1} for $\mu \to 0$.

Again, to develop the approximation, we let

$$\begin{aligned}\widehat{X}_g(\tau,\sigma) &\cong \widehat{X}_{g0}(\tau,\sigma) + \mu \widehat{X}_{g1}(\tau,\sigma) \\ \widehat{X}_s(\tau,\sigma) &\cong \widehat{X}_{s0}(\tau,\sigma) + \mu X_{s1}(\tau,\sigma)\end{aligned} \tag{12.38}$$

The zero-order system is

$$\hat{u}_m \frac{\partial \widehat{X}_{g0}}{\partial \sigma} = -(\widehat{X}_{g0} - \widehat{X}_{s0}) \tag{12.39a}$$

$$\frac{\partial \widehat{X}_{s0}}{\partial \tau} = (\widehat{X}_{g0} - \widehat{X}_{s0}) \tag{12.39b}$$

Expanding $m(\mu\tau) \cong m(0) + \mu\tau m'(0)$, $n(\mu\sigma) \cong n(0) + \mu\sigma n'(0)$, we obtain the boundary conditions $\widehat{X}_{g0}(\tau,0) = m(0)$, $\widehat{X}_{s0}(0,\sigma) = n(0)$.

To improve the approximation, we solve a further system

$$\hat{u}_m \frac{\partial \widehat{X}_{g1}}{\partial \sigma} = -(\widehat{X}_{g1} - \widehat{X}_{s1}) \tag{12.40a}$$

$$\frac{\partial \widehat{X}_{s1}}{\partial \tau} = (\widehat{X}_{g1} - \widehat{X}_{s1}) - R(X_{s0}) \tag{12.40b}$$

with the boundary conditions $\widehat{X}_{g1}(\tau,0) = \tau m'(0)$, $\widehat{X}_{s1}(0,\sigma) = \sigma n'(0)$.

It seems interesting to notice that in the outer approximation of the system (namely, far from the axes) the phenomenon is essentially described by the reaction rate, while in the inner approximation (close to the axes) what counts is the adsorption phenomenon. A toy example, illustrating these results, can be found in [10].

Nomenclature

c_g	J/kg K	specific heat capacity of gas
c_s	J/kg K	specific heat capacity of substrate
R_i	mol/m²s	specific reaction rate for component i
R'_l	mol/m²s	specific reaction rate for the chemical reaction l
G_A	m²/m³	active area/volume ratio of the monolith
h	W/m²K	convective heat transfer coefficient (from gas to substrate)
h_{ext}	W/m²K	heat transfer coefficient
$k_{D,i}$	m/s	mass transfer coefficient for component i
S_{ext}	m²/m³	external area/volume ratio
T_{ext}	K	external temperature
u_m	m/s	mean gas velocity in monolith
ΔH_i	J/mol	heat of ith-reaction
ε		void fraction
ρ_g	kg/m³	gas density
ρ_s	kg/m³	substrate density
λ_s	W/m K	substrate thermal conductivity (from substrate to ambient)
$X_{g,i}$	mol/m³	gas phase concentration of ith component
$X_{s,i}$	mol/m³	solid phase concentration of ith component

References

[1] *Automotive Handbook*, Robert Bosch GmbH distributed by SAE, 4th Edition, 1996.

[2] G. Abate, L. Glielmo, M. Milano, P. Rinaldi, S. Santini and G. Serra. Numerical Simulation and Identification of the Dynamic Behaviour of Three Way Catalytic Converters, *ICE97-3rd International Conference on Internal Combustion Engines: Experiments and Modelling*, 409–412, Capri, Italy, September, 1997

[3] F. Aimard, S. Li and M. Sorine, Mathematical Modeling of Automotive Three-Way Catalytic Converters with Oxygen Storage Capacity, *Control Eng. Practice*, 4(8), 1119–1124, 1996.

[4] C. M. Bender and S. A. Orszag. *Advanced Mathematical Methods for Scientists and Engineers: Asymptotic Methods and Perturbation Theory*, Springer-Verlag, New York, 1999.

[5] N. Baba, K. Ohsawa, and S. Sugiura. Numerical Approach for Improving the Conversion Characteristics of Exhaust Catalysts under Warming-Up Condition, *SAE paper 962076*, 1996.

[6] E. P. Brandt, Y. Wang and J. W. Grizzle. Dynamic Modeling of a Three-Way Catalyst for SI Engine Exhaust Emission Control, *IEEE Trans. on Control System Technology*, 85, 767–776, 2000.

[7] C. R. Chester. *Techniques in Partial Differential Equations*, McGraw-Hill, 1971.

[8] C. Cussenot, M. Basseville, and F. Aimard. Monitoring the Vehicle Emission System Components, *Proc. 13th IFAC World Congress*, San Francisco, USA, 1996.

[9] C. Dubien and D. Schweich. Three-way Catalytic Converter Modelling. Numerical Determination of Kinetic Data, *Catalysis and Automotive Pollution Control IV, Studies in Surface Science and Catalysis*, 116, 399–408, 1998.

[10] L. Glielmo and S. Santini. A Two-Time-Scale Infinite-Adsorption Model of Three Way Catalytic Converters during the Warm-up Phase, *ASME Journal of Dynamic Systems, Measurement and Control*, 123(1), March 62–70, 2001.

[11] L. Glielmo, S. Santini, and M. Milano. A Machine Learning Approach to Modeling and Identification of Automotive Three-Way Catalytic Converters, *IEEE/ASME Transaction on Mechatronics*, 2(5), 132–141, June 2000.

[12] G. Fiengo, L. Glielmo and S. Santini. On Board Diagnosis for Three-Way Catalytic Converters, *International Journal of Robust Nonlinear Control*, 11, 1073–1094, 2001.

[13] L. Glielmo and S. Santini. A Fast Integration Algorithm for Three-Way Catalytic Converters PDE Models, *3rd MathMode-IMACS Symposium on Mathematical Modelling*, 335–338, Vienna, Austria, 2-4 February 2000.

[14] D. E. Goldberg. *Genetic Algorithms in Search, Optimization, and Machine Learning*, Addison-Wesley, 1989.

[15] R. D. Hawthorn. Afterburner Catalysis: Effect of Heat and Mass Transfer between Gas and Catalyst Surface, *AIChE Symposium Series*, 70(137), 428–438, 1974.

[16] R. K. Herz, J. B. Kiela, and J. A. Sell. Dynamic behavior of automotive catalyst 2. Carbonmonoxide conversion under transient air/fuel conditions, *Ind. Chem. Prod. Res. Dev.*, 22, 387–396, 1983.

[17] J. Kevorkian and J. D. Cole. *Multiple Scale and Singular Perturbation Methods*, Springer-Verlag, New York, 1996.

[18] O. Levenspiel. *Chemical Reaction Engineering*, 2nd Edition, Wiley, New York, Chap. 14, 1972.

[19] C. C. Lin and L. A. Segel. *Mathematics Applied to Deterministic Problems in the Natural Sciences*, Society for Industrial & Applied Mathematics, 1988.

[20] C. N. Montreuil, S. C. Williams, and A. A. Adamczyk. Modeling Current Generation Catalytic Converters: Laboratory Experiments and Kinetic Parameter Optimization—Steady State Kinetics, *SAE paper 920096*, 1992.

[21] K. Narendra and K. Parthasarathy. Identification and Control of Dynamic Systems using Neural Networks, *IEEE Transaction on Neural Networks*, 1, 4–27, 1990.

[22] J. G. Nunan, H. J. Roberta, M. J. Cohn, S. A. Bradley. Physicochemical properties of Ce-containing three-way catalysts and the effect of Ce on catalyst activity, *J. Catal.*, 133, 309–324, 1992.

[23] P. V. Kokotović, H. K. Khalil and J. O'Reilly, *Singular Perturbation Methods in Control: Analysis and Design*. Academic Press, London, 1986.

[24] G. C. Koltsakis, P. A. Kostantinidis, and A. M. Stamatelos, 'Development and Application Range of Mathematical Models for 3-way Catalytic Converters,' *Applied Catalysis B: Environmental*, 12, 161–191, 1997.

[25] G. C. Koltsakis, I. P. Kandylas, and A. M. Stamatelos. Three-Way Catalytic Converter Modeling and Applications, *Chem. Eng. Comm.*, 164, 153–189, 1998.

[26] A. Jeffrey. *Quasilinear Hyperbolic Systems and Waves*, Pitman Publishing, London, 1976.

[27] F. John. *Partial Differential Equations*, Springer-Verlag, New York, 1975.

[28] J. C. Jones and R. A. Jackson and J. B. Roberts and P. Bernard. A Simplified Model for the Dynamics a Three-Way Catalytic Converter, *SAE Paper 2000-01-0652*, 2000.

[29] I. H. J. Ploemen and M. J. G. van de Molengraft. Hybrid Modeling for Mechanical Systems: Methodologies and Applications, *Journal of Dynamical Systems Measurements and Control*, 121, 270–277, 1999.

[30] W. E. Schiesser. *The Numerical Method of Lines: Integration of Partial Differential Equations*, Academic Press, San Diego, 1991.

[31] E. C. Su and C. N. Montreuil and W. G. Rothschild. Oxygen Storage Capacity of Monolith Three Way Catalysts, *Applied Catalysis*, 17, 75–86, 1985.

[32] K. C. Taylor and R. M. Sinkevitch. Behaviour of Automobile Exhaust Catalysts with Cycled Feedstreams, *Ind. Eng. Chem. Prod. Res. dev.*, 22, 45–51, 1983.

[33] W. Wang and C. Lin. Runge-Kutta Neural Networks for Identification of Dynamical Systems in High Accuracy, *IEEE Trans. on Neural Networks*, 9, 294–307, 1998.

[34] S. E. Voltz, C. R. Morgan, D. Liederman, and S. M. Jacob. Kinetic Study of Carbon Monoxide and Propilene Oxidation on Platinum Catalysts, *Ind. Eng. Chem. Prod. Res. Dev.*, 12(4), 1994.

[35] H. C. Yao and Y. F. Yu Yao, Ceria in Automotive Exhaust Catalysts. Oxygen Storage, *Journal Catalysis*, 86, 254–265, 1984.

13

Control of Gasoline Direct Injection Engines using Torque Feedback: A Simulation Study

M. Gäfvert   K.-E. Årzén   B. Bernhardsson   L. M. Pedersen

Abstract

A novel approach to the control of a GDI engine is presented. The controller consists of a combination of sub-controllers, where torque feedback is a central part. The sub-controllers are, with a few exceptions, designed using simple linear feedback and feedforward control-design methods, in contrast to traditional table-based engine control. A silent extremum-controller is presented. It is used to minimize the fuel consumption in stratified mode. The controller has been evaluated with good results on the European driving cycle using a dynamic simulation model.

13.1 Introduction

Modern internal combustion engines constitute very complex systems in need of precise and robust control. The control objectives in engine control fall into two categories: tracking of reference values such as engine speed, engine torque or air-fuel ratio, and optimizing operating conditions such that fuel consumption and emissions are kept at a minimum. The *gasoline direct injection* (GDI) engine is more complex than an ordinary port fuel injection (PFI) engine and therefore requires a more advanced control system. Conventional control of car engines is normally based on extensive use of engine maps, *i.e.*, multi-dimensional look-up tables that have been derived through extensive engine test-bench experiments [8]. This is an open-loop approach which can be sensitive to engine-to-engine variations and variations due to aging and wear. In this paper, it is shown, using simulations based on a calibrated GDI engine model, that it is possible to instead employ simple feedback control strategies with a low number of tuning parameters and yet achieve good control performance, provided the effective torque is available for feedback.

A GDI engine can operate in two main modes: homogeneous mode and stratified mode. The mode that should be used is decided by engine load and engine speed. Smooth mode-switching is important for driver comfort. In this paper, a mode-switching strategy is presented that minimizes torque bumps and reduces fuel consumption.

The stratified charge of the cylinder is used in combination with extremely lean air fuel mixtures to reduce fuel consumption. The air fuel ratio can then be chosen in an optimal manner with respect to engine efficiency and emissions. In this paper, a silent extremum-controller is presented. It is used to find the air-fuel ratio that generates a desired torque with the smallest possible fuel consumption. Engine control is a well known application area for extremum control, see *e.g.*, [4], [20]. A problem with the lean-burn approach is the increased formation of nitrogen oxides. To reduce the pollutant emissions, it is possible to introduce exhaust gas recirculation (EGR) to the engine. Then there is need for optimizing the amount of recirculated gas with respect to engine efficiency and emissions. The extremum-control strategy is also extended to the case where EGR is present.

The work presented in this paper has been performed within the EU/ESPRIT Long Term Research project FAMIMO. Within FAMIMO a Matlab^{TM}/Simulink dynamic simulation model of a two-liters GDI engine provided by Siemens Automotive has served as a common benchmark for evaluation of different control strategies.

The paper is organized as follows. Section 13.2 discusses the key characteristics of GDI engines. The overall structure of the benchmark used in the simulation studies is described in Section 13.3. The GDI engine model used as the basis for the control design is overviewed in Section 13.4. The control strategy and the overall structure of the control system are presented in Section 13.5, followed by a detailed description of the design of the different

Control of Gasoline Direct Injection Engines using Torque Feedback   291

sub-controllers, including the extremum controller, and the switching logic in Section 13.6. The results of the controller are presented in Section 13.7. In Section 13.8, a complete engine management system based on the core controller from Section 13.6 with additional idle-speed control is described. The results of the complete engine management system on the European driving cycle are presented in Section 13.9. Comparisons are made with other control designs. Section 13.10 includes a discussion on the feasibility of having the torque signal available for feedback. Finally, conclusions are given in Section 13.11.

This paper is based on the conference papers [9], [10]. Other related results are presented in [25] and [23].

13.2 GDI Engines

Increased environmental requirements and requirements of decreased fuel consumption have, during recent years, led to the development of new types of combustion engines within the car industry. One of these engine types is the GDI engine, also known as the DISI (direct injection, spark ignition) engine. The main difference between a GDI engine and a conventional PFI engine concerns where air and fuel are mixed, and when the mixing takes place.

In a PFI engine, the fuel is mixed with the air in the intake ports outside the cylinder. The air-fuel mixture enters the cylinders during the intake stroke. This gives rise to a homogeneous mixture in the cylinders that, to minimize emissions, should be kept at a stoichiometric ratio ($\lambda = 1$). In a GDI engine, instead, the fuel is injected directly into the cylinder late in the compression stroke. Direct injection in the compression stroke gives a stratified charge that only needs to be sufficiently rich locally, close to the spark. This gives rise to an extremely lean air fuel mixture, which reduces fuel consumption. The difference between PFI and GDI engines is shown in Figure 13.1. GDI engines put high demands on the timing of the injection. Good performance requires that exactly the right amount of fuel is injected at the correct time.

The stratified combustion mode in GDI engines can only be used at low or moderate engine loads and for low or moderate engine speeds, *e.g.*, in low-speed urban traffic. For high loads, the amount of fuel that needs to be injected is too large to be timely injected. For high engine speeds, the time window when the fuel must be injected is too small. Therefore, GDI engines also use an alternative combustion mode, the homogeneous mode. Here, the fuel is still injected directly into the cylinder, but now it is done in the intake stroke rather than in the compression stroke. In homogeneous mode, the GDI engine behaves similarly to a PFI engine.

GDI engines were commercially pioneered by Mitsubishi. However, currently most major car manufacturers have GDI engines in their production, *e.g.*, Ford, GM, and VW. More information on the operation of GDI engines can be found in [34].

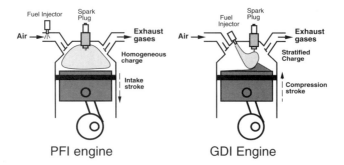

Figure 13.1 Differences between a PFI and a GDI engine. In a GDI engine the top of the piston head is shaped to create a fuel swirl that gives maximum fuel concentration at the spark.

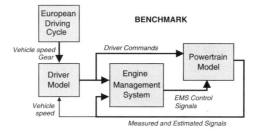

Figure 13.2 Structure of the GDI benchmark.

13.3 The GDI Benchmark

The overall structure of the GDI benchmark is shown in Figure 13.2. The goal is to design an engine management system (EMS), or controller, that follows the command signals from the driver while minimizing fuel consumption and maintaining driving comfort. Additionally, it is required that the air-fuel ratio is kept at $R_{a/f} = 1$ during homogeneous combustion mode. The engine-speed shall always be larger than a specified idle-speed limit to avoid stalling, also in the presence of load disturbances. Minimizing fuel consumption is, in principle, equivalent to using the stratified combustion mode whenever possible. Maintaining driver comfort involves minimizing torque bumps due to mode changes and avoiding too large torque gradients, especially at low gears.

The EMS control signals are the desired position of the fresh-air throttle (ϕ_{MTC}), the ignition-advance angle (IGA), the start of injection (SOI), and the fuel-injection duration (T_{inj}), i.e., the amount of fuel injected. The start of injection essentially determines the combustion mode that the engine operates in. The problem of finding optimal IGA and SOI is not part of the benchmark, and suitable values are assumed to be known for all operating

Control of Gasoline Direct Injection Engines using Torque Feedback

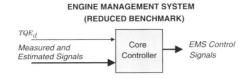

Figure 13.3 Structure of the Engine Management System in the reduced benchmark.

conditions. The manifold pressure P_{man}, throttle position ϕ_{MTCPOS}, engine speed N, and air-fuel ratio $R_{a/f}$ are measured signals available for feedback. The air inlet temperature and the atmospheric pressure are measured, but regarded as constant in this study. It is also assumed that the effective torque (TQE) is available as a feedback signal. The feasibility of this assumption is discussed in Section 13.10.

The first stage of the benchmark is to design a core controller that handles torque-tracking and mode-switching, see Figure 13.3. During this stage a reduced benchmark model is used, where the driver is excluded. The reference inputs to the EMS are instead effective torque trajectories generated by calculating the corresponding demanded torque from three small excerpts, or scenarios, of the full European driving cycle. The scenarios are chosen such that the gear is constant.

13.4 The GDI Engine Model

The engine model is derived and implemented in Simulink^{TM} by Siemens Automotive. The model is a continuous-variable mean-value engine model (MVEM) for one cylinder. MVEMs are generally accepted as the modeling paradigm for engine control, and are extensively described in the literature [13], [14], [31], [1], [15]. Therefore, this section will be kept brief, and the reader is referred to the cited references for further details.

The model is built using equations derived from first principles combined with engine maps derived through test-bench experiments on a real GDI engine, see [25] and [26]. Hence, some of the analytical physical relations found in standard MVEMs are parameterized with look-up tables. The basic version of the model does not include EGR. However, in Section 13.6 it is extended with a simple EGR model. The power train model can be divided into two parts: the engine model and the engine-load model. The engine part can be decomposed into three subsystems: air intake, intake manifold, and cylinder. The engine-load part includes clutch, gearbox, differential, brakes, and the chassis, see Figure 13.4. The dynamics of the model can be described using three state variables, and the model has a large number of nonlinearities.

The engine hosts two of the states: the throttle position (ϕ_{MTCPOS}) and the intake-manifold pressure (P_{man}). The engine is equipped with a

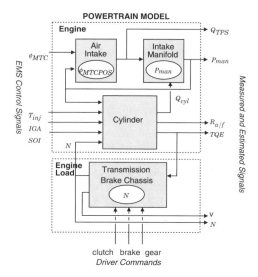

Figure 13.4 The power train model is composed of the engine model and the engine load model. The engine model hosts the throttle position state ϕ_{MTCPOS} and the intake-manifold pressure state P_{man}. The engine load model hosts the engine speed state N.

controlled electronic air throttle with dynamics that may be approximated by a first order LTI system. The throttle position ϕ_{MTCPOS} is thus described by

$$T_{MTC}(d/dt)\phi_{MTCPOS} + \phi_{MTCPOS} = \phi_{MTC} \quad (13.1)$$

The flow of intake air is controlled with the desired throttle-position signal ϕ_{MTC}. The air mass-flow rate through the throttle into the intake manifold (Q_{TPS}) is described by a calibrated look-up table parameterization of the standard orifice equation as

$$Q_{TPS} = \Psi_0(\phi_{MTCPOS}, P_{man}/P_{atm}) \quad (13.2)$$

The build-up of the manifold pressure P_{man} is described by the ideal gas law together with a mass-flow balance equation

$$(d/dt)P_{man} = K_P(Q_{TPS} - Q_{cyl}) \quad (13.3)$$

The air-flow into the cylinder (Q_{cyl}) depends on the engine speed, manifold pressure and the volumetric efficiency, and is described by a look-up table

$$Q_{cyl} = \Psi_7(P_{man}, N) \quad (13.4)$$

The torque production in the cylinder is described by a nonlinear mapping that depends on air-flows, engine speed, and the engine control variables

T_{inj}, SOI and IGA. The effective torque is the indicated torque generated by the cylinder (TQI), subtracted by the torque losses due to pumping work and friction (TQL).

$$TQE = TQI - TQL \tag{13.5}$$

The torque losses depend on the engine speed and the air-flow and are described by a look-up table:

$$TQL = \Psi_8(Q_{cyl}, N) \tag{13.6}$$

In stratified mode, torque production is modeled by the following nonlinear maps.

$$\begin{aligned} TQI &= TQI_{S,ref} G_{S,phase} G_{S,air} \\ TQI_{S,ref} &= \Psi_{10}(M_{fuel}, N) \\ G_{S,phase} &= \Psi_{11}\Big(IGA - \Psi_{12}(TQI_{S,ref}, N), EOI - IGA - \Psi_{13}(TQI_{S,ref}, N)\Big) \\ G_{S,air} &= \Psi_{34}(\Psi_{33}(TQI_{S,ref}, N) - M_{air}) \end{aligned} \tag{13.7}$$

The indicated torque TQI is modeled as the ideal torque $TQI_{S,ref}$ for a certain fuel-flow and engine speed, penalized by $G_{S,phase}$ for inappropriate ignition timing, and by $G_{S,air}$ for inappropriate air flow. In homogeneous mode, the corresponding maps are

$$\begin{aligned} TQI &= TQI_{H,ref} G_{H,IGA} G_{H,air/fuel} \\ TQI_{H,ref} &= \Psi_{32}(M_{air}, N) \\ G_{H,IGA} &= \Psi_{15}\Big(IGA - \Psi_{16}(M_{air}, N)\Big) \\ G_{H,air/fuel} &= \Psi_{17}(R_{a/f} - \Psi_{18}(M_{air}, N)) \end{aligned} \tag{13.8}$$

The indicated torque TQI is modeled as the ideal torque $TQI_{H,ref}$ for a certain air-flow and engine speed, penalized by $G_{H,IGA}$ for inappropriate ignition timing, and by $G_{H,air/fuel}$ for inappropriate air-fuel ratio.

Together with the engine speed N, the manifold pressure affect the amount of air intake of the cylinder, Q_{air}, while T_{inj} and SOI determine the amount of fuel injected and the injection timing, respectively. Note that the cylinder subsystem contains no dynamics. The outputs TQE and $R_{a/f}$ are, thus, momentarily determined by the inputs IGA, SOI, T_{inj}, P_{man}, and N. The third state, the engine speed N, is hosted by the transmission-brake-chassis subsystem, and varies slowly compared to the dynamics of the engine subsystem. It is described by

$$J(d/dt)N = TQE - TQR \tag{13.9}$$

where TQR depends on the gear, and includes rolling-resistance, aerodynamic drag, climbing-resistance, and brake resistance. Since it is not an objective to control N, it will be regarded as an external variable on the engine subsystem in the core-controller design.

13.5 Core Control Strategies

The engine is a highly nonlinear multi-input multi-output system. To handle the complexity a bottom-up approach has been adopted. A number of specialized simple controllers for the different subsystems of the motor have been designed, and then integrated into a complete control system.

Considerations on model uncertainties and tuning issues led to a controller structure based on extensive use of feedback loops with simple feedforward paths. In combination with the bottom-up approach called for by the complexity, this led to a core controller based on several sub-controllers. The subcontrollers can be tuned individually, with only rough process knowledge, and with only a few parameters to calibrate.

Choice of Control Variables

In the context of torque control, the cylinder operation can be described as nonlinear static maps, constructed by aggregating the look-up tables Ψ_k of the previous section

$$(T_{inj}, P_{man}, N, IGA, SOI) \overset{\text{Homogeneous}}{\mapsto} TQE$$
$$(T_{inj}, P_{man}, N, IGA, SOI) \overset{\text{Stratified}}{\mapsto} TQE \tag{13.10}$$

The variables IGA and SOI are chosen to yield optimal efficiency. Since N varies slowly, it is then sensible to consider a family of maps

$$(P_{man}, T_{inj}) \overset{\text{Homogeneous}}{\underset{N}{\mapsto}} TQE$$
$$(P_{man}, T_{inj}) \overset{\text{Stratified}}{\underset{N}{\mapsto}} TQE \tag{13.11}$$

that are parameterized in N. These maps describe how the generated effective torque TQE depends on the manifold pressure P_{man} and the fuel injection time T_{inj} at given engine speeds N. Experiments on the Simulink^{TM} model of the cylinder were used to derive these maps. Figure 13.5 shows the maps for $N = 1500$ rpm.

Understanding the characteristics of these maps is the essential prerequisite for successful torque control. For example, from the maps it is easy to see that the sensitivity of TQE to variations in P_{man} is large in homogeneous mode, while it is large to variations in T_{inj} in stratified mode.

Recall from the previous section that T_{inj} can be set instantaneously, while P_{man} is the result of a dynamic response to MTC. In homogeneous

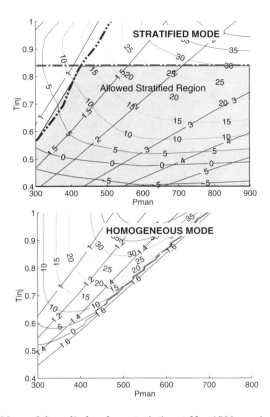

Figure 13.5 Maps of the cylinder characteristics at $N = 1500$ rpm in stratified mode (left) and homogeneous mode (right). The plot shows constant-torque level-curves as a function of the manifold pressure P_{man} and the injection duration T_{inj}, with IGA and SOI chosen optimally. The diagonal straight lines show constant $R_{a/f}$. The shaded area shows the allowed region for stratified mode. The dash-dotted line marks the constraint on $TQI_{S,ref}$, and the dash-dot-dotted line the constraint on M_{air}

mode, the $R_{a/f}$ constitutes the most critical controlled variable. Therefore, T_{inj} is chosen as control variable for the $R_{a/f}$ in this mode. The P_{man} is then used to control TQE. In stratified mode, the only requirement on the air-fuel ratio is $R_{a/f} \geq 1$. Thus, T_{inj} is the natural choice for controlling TQE, and P_{man} can be chosen such as to yield the $R_{a/f}$ that minimizes the fuel consumption. (It is also evident from Figure 13.5 that the torque sensitivity to changes in P_{man} is small for lean air-fuel ratios.) It can be seen in Figure 13.5 that there exists an optimal combination of P_{man} and T_{inj} for obtaining a specific torque TQE from the engine in stratified mode. This choice minimizes T_{inj}, which is equivalent to minimizing instant fuel consumption.

Mode Switch Strategies

An elaborate switching strategy is used for switching between homogeneous and stratified mode. The mode-switches should ideally be performed without large torque-bumps. The strategy tries to switch mode where the torque level curves in homogeneous mode and stratified mode meet. To achieve this the $R_{a/f} = 1$ constraint in homogeneous mode is temporarily relaxed during the mode-switch.

The engine should run in stratified combustion mode whenever possible. The controller thus has to detect when conditions for stratified operation are fulfilled, and then perform a mode switch. When the constraints are no longer fulfilled, the combustion mode immediately has to be switched back to homogeneous. The constraints for running in stratified mode have in the benchmarks been defined as:

- $TQI_{S,ref} < 50$ (essentially a bound on T_{inj});
- M_{air} is large enough (essentially a bound on P_{man});
- $N < 3000$.

The constraint limits of the physical engine are confidential, therefore the model constraints are set to typical values.

Only the first constraint poses a real challenge. Since T_{inj} varies quickly at fast torque-reference changes, the constraint may be violated with short notice. To handle such situations, a simple $TQI_{S,ref}$ predictor was implemented as

$$\widehat{TQI}_{S,ref}(kh+t) = TQI_{S,ref}(kh) + (TQE_d(kh) - TQE_d(kh-h))T/h \quad (13.12)$$

With a prediction horizon $T = 0.2\,\text{s}$ the need for mode switches can be detected in advance, and constraint violations can be avoided.

The mode switches should ideally be performed without large torque bumps. By superimposing the torque level curves of Figure 13.5 the plot in Figure 13.6 is obtained. In this plot, it is observed that there exist points in the (P_{man}, T_{inj})-space where the torque difference between homogeneous and stratified mode is zero. A key observation is that these points lie close to a line described by a constant air-fuel ratio close to the lean limit, $R_{a/f} \approx 1.5$. The mode switch strategy adopted in this work is to try to perform all combustion-mode switches on this line. For descriptions of similar mode switch strategies please refer to [17], [22], [34], [5].

Core Controller Structure

The resulting controller structure is shown in Figure 13.7. The engine management system consists of five sub-controllers. The P_{man} controller is central in the total controller design. In stratified mode, the extremum-controller minimizes the fuel consumption by producing a suitable reference to the P_{man}-controller. The TQE-controller controls the effective torque using

Control of Gasoline Direct Injection Engines using Torque Feedback

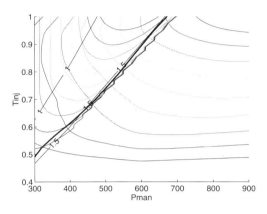

Figure 13.6 Static cylinder map $(P_{man}, T_{inj}) \mapsto TQE$ for stratified and homogeneous mode. The bold line indicates a set of points where the torques in homogeneous and stratified mode are equal

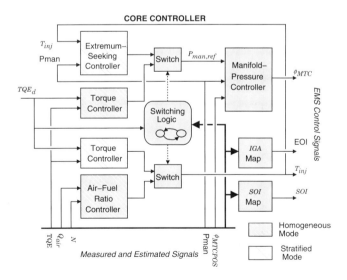

Figure 13.7 Overall structure of the sub-controllers in the core controller

T_{inj}. In homogeneous mode, the TQE-controller controls the effective torque by producing a suitable reference to the P_{man}-controller. A simple controller regulates the $R_{a/f}$ in homogeneous mode using T_{inj}. The control signals IGA and SOI are obtained from maps. The mode switches are controlled by a discrete-event logic controller.

13.6 Controller Designs

This section presents the design of the different sub-controllers and explains the switching-scheme. The extensive use of feedback in the controllers makes it possible to use only rough engine-models in the design, as will be evident below.

Manifold Pressure Controller

In [22], it is stated that a quick and accurate control of the air flow is necessary to ensure smooth transitions between stratified and homogeneous mode. Since the air-flow is closely related to the manifold pressure, it was therefore decided to control the air-flow indirectly by controlling P_{man}.

The manifold pressure dynamics (13.3) can be approximated by the differential equation

$$(d/dt)P_{man} \approx k_1 \phi_{MTCPOS} f(\eta) - k_2 N P_{man} \tag{13.13}$$

where $\eta = P_{man}/P_{atmospheric}$. From the look-up table relating P_{man} and Q_{TPS} it was heuristically found that this nonlinearity can be approximated with $\hat{f}(\eta) = 1 - \eta^8$.

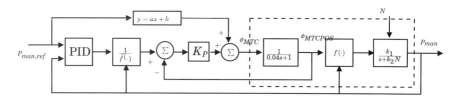

Figure 13.8 Block diagram showing throttle and manifold-pressure controllers and the corresponding dynamics. Blocks inside the dashed box are in continuous time

A proportional control of ϕ_{MTCPOS} was first applied to obtain a desired time constant of the fresh-air throttle.

$$\phi_{MTC}(k) = K_P \left[\phi_{MTCPOS_d}(k) - \phi_{MTCPOS}(k) \right] \tag{13.14}$$

where ϕ_{MTCPOS_d} is a desired throttle-position. A PID-controller was then introduced to control P_{man} to follow the desired trajectory $P_{man,ref}$. To eliminate the gain variations for the P_{man} dynamics, the control signal is scaled as

$$\phi_{MTC} = \frac{u}{\hat{f}(\eta)} \qquad (13.15)$$

where u is the output from the PID-controller for controlling P_{man}. The $\hat{f}(\eta)$ term can be viewed as a local feedback linearization term. Assuming that $\phi_{MTCPOS} \approx \phi_{MTC}$, the resulting differential equation, seen from the PID-controller, is

$$(d/dt)P_{man}(t) \approx k_1 u(t) - k_2 N(t) P_{man}(t) \qquad (13.16)$$

Note that the open-loop time-constant from P_{man} still varies with N. Since N is measured, a possibility would be to use gain scheduling. Instead the PID-controller was robustly tuned using the Nonlinear Control Design Blockset in Simulink^{TM}, to handle the variations in N. The resulting closed-loop system has a time constant of around 60 ms, which appears constant in the entire work space. The resulting P_{man} controller, hence, is insensitive to variations in N. A feedforward term corresponding to an affine approximation of the static relation from P_{man} to ϕ_{MTCPOS} was added to speed up the response to changes in $P_{man,ref}$. The result is the cascade feedback controller-structure with feedback linearization and feedforward in Figure 13.8.

Torque Controller (Stratified Mode)

As stated above, T_{inj} is used for controlling TQE in stratified mode. Since the references for TQE contain ramp variations, it is necessary to have two integrators in the controller to achieve good tracking. Therefore, a PII^2-controller was used:

$$C(q) = K_p + \frac{K_i h}{q-1} + \frac{K_{i^2} h^2}{(q-1)^2} \qquad (13.17)$$

In stratified mode, a rough linear approximation of the torque generation map is $\delta TQE = (60 \pm 30)\delta T_{inj} + (0.02 \pm 0.02)\delta P_{man}$. With the unit delay introduced by the computational delay in the controller, a rough model is

$$\delta TQE \approx \frac{60 \pm 30}{q} \delta T_{inj} \qquad (13.18)$$

The resulting open-loop system is of third order. A pole-placement design assuming a nominal gain of 60 gives satisfactory performance and robustness.

A feedforward term corresponding to an affine approximation of the static relation from T_{inj} to TQE was added to speed up the response to reference changes.

Torque Controller (Homogeneous Mode)

In homogeneous mode, a rough linear approximation of the torque generation map is $\delta TQE \approx (0.1 \pm 0.05)\delta P_{man}$ (with P_{man} and T_{inj} constrained by the requirement that $R_{a/f} = 1$). Combining this with a linear first-order system approximation of the controlled manifold pressure dynamics gives

$$\delta TQE \approx \frac{(0.10 \pm 0.05)(1 - e^{-h/0.06})}{q - e^{-h/0.06}} \delta P_{man,ref} \qquad (13.19)$$

A PII^2-controller was tuned using pole placement, assuming a nominal gain of 0.1. It was necessary to also introduce a lead compensator to obtain a phase gain around the cross-over frequency. The torque control is considerably slower in homogeneous mode than in stratified mode, due to the controlled manifold-pressure dynamics present in the loop.

A feedforward term corresponding to an affine approximation of the static relation from P_{man} to TQE was added to speed up the response to reference changes.

Air-fuel Ratio Controller (Homogeneous Mode)

The air-fuel ratio control problem in homogeneous mode is not the main focus of this work. Therefore, it is assumed that a well working controller exists, approximated by controlling $R_{a/f}$ using T_{inj}, with a pure feedforward strategy based on the stoichiometric ratio. This control yields perfect control of $R_{a/f}$ with a delay of one sample.

Extremum Controller

In stratified mode, the torque TQE is regulated with the control variable T_{inj}. The $R_{a/f}$ can be chosen arbitrarily. In [17], it is suggested to use a constant ultra-lean air-fuel ratio in stratified mode, while in [18] light throttling is suggested to reduce emissions. In the present work, the $R_{a/f}$ is instead chosen as to maximize engine efficiency, which is equivalent to choose P_{man} to minimize fuel consumption. Thus, for constant TQE and N the engine efficiency can be described by the fuel consumption $T_{inj} = f(P_{man})$, see Figure 13.5. The minimum of this function is dependent on operating conditions such as N and TQE, but also on e.g., aging effects on the engine. An extremum-controller can be used to find and track this minimum during engine operation, without any other prior knowledge on the shape of f than an assumption of convexity. The controller applies a perturbation δP_{man} to the base manifold pressure $P_{man,base}$. The effect of the perturbation on the engine output-torque TQE is eliminated by closed-loop control using T_{inj}. In this way, the effects of the perturbations are only visible internally in the controller as correcting variations ΔT_{inj} in the fuel mass flow. The suggested structure with the interconnection of an extremum-controller and a controller that eliminates the effects of the perturbation signals on the process output is therefore denoted a *silent*

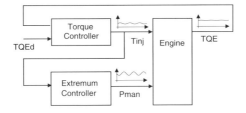

Figure 13.9 Structure of the silent extremum control strategy.

extremum-controller. To the authors knowledge this is a new construction. The gradient direction of the engine efficiency with respect to the manifold pressure is computed from the perturbation and the corresponding variation in the fuel mass-flow. The base manifold pressure is then changed in the negative gradient direction. The control structure is shown in Figure 13.9, and may be implemented as

$$P_{man,base}(k) = P_{man,base}(k-1) - \gamma(|\Delta T_{inj}(k)|)\,\text{sgn}\left(\frac{\Delta T_{inj}(k)}{\Delta P_{man}(k)}\right) + \Delta \delta P_{man}(k)$$

where

$$\Delta P_{man}(k) = P_{man}(k) - P_{man}(k-1)$$
$$\Delta T_{inj}(k) = T_{inj}(k) - T_{inj}(k-1)$$
$$\Delta \delta P_{man}(k) = \delta P_{man}(k) - \delta P_{man}(k-1)$$

where $\gamma(\cdot)$ is a function chosen to give an appropriate step length. The perturbation signal is chosen as $\delta P_{man} = A \sin \omega t$. The consecutive corrections make the base manifold pressure converge to a value corresponding to the optimal $R_{a/f}$ for the present operating conditions.

Note that the extremum-controller does not use any information about the engine, but only relies on measurements collected during normal engine operation. The silence property of the controller is critically dependent on the closed-loop torque control. It would be possible to use control of other engine outputs such as engine speed instead. A review of extremum control can be found in [30]. Some approaches to off-line optimization of GDI engines are presented in [21].

Extremum Control of $R_{a/f}$ and EGR In [28], it is reported how EGR reduces fuel consumption and NO_x emissions in a GDI engine running in stratified mode. An interesting feature of the results is that the fuel consumption attains its minimum at the desired EGR operation point regarding pollutant emissions. Therefore, extremum control of $EGR = Q_{EGR}/(Q_{TPS} + Q_{EGR})$ with minimization of T_{inj} as a control objective may be feasible. The recirculated exhaust-gas mass-flow Q_{EGR} is controlled with the desired EGR valve position ETC. If it is desirable to optimize fuel consumption with respect to both P_{man} and EGR simultaneously, it is necessary to modify the previous approach such that the extremum-controller can separate the effects of the perturbation signals on P_{man} and EGR on the engine output, i.e., in effect on T_{inj}. This may be achieved by choosing orthogonal perturbation signals. The effect of small deviations δP_{man} and δETC from a base manifold-pressure $P_{man,base}$ and a base EGR valve position ETC_{base} can then be described by the first order approximation

$$T_{inj} \approx T_{inj}|_{base} + \left.\frac{\partial T_{inj}}{\partial P_{man}}\right|_{base} \delta P_{man} + \left.\frac{\partial T_{inj}}{\partial ETC}\right|_{base} \delta ETC$$

With $\delta P_{man} = a_P \sin(m_P \omega t)$ and $\delta ETC = a_E \sin(m_E \omega t)$ with integers $m_P \neq m_E$, it follows after multiplication of the expression above with a perturbation signal, and integration over k periods, that the gradient can be computed as

$$\left.\frac{\partial T_{inj}}{\partial P_{man}}\right|_{base} = \frac{m_P \omega}{k a_P^2 \pi} \int_{t-k2\pi/\omega}^{t} \delta P_{man}(s) T_{inj}(s)\, ds$$

$$\left.\frac{\partial T_{inj}}{\partial ETC}\right|_{base} = \frac{m_E \omega}{k a_E^2 \pi} \int_{t-k2\pi/\omega}^{t} \delta ETC(s) T_{inj}(s)\, ds$$

After an integration period, $P_{man,base}$ and ETC_{base} are updated with a step in the negative gradient direction. This corresponds to a steepest-descent optimization, which is known to have slow convergence. A possible refinement would be to estimate also the Hessian matrix, and use more elaborate methods to compute the step.

To evaluate the extremum-controller on combined EGR and P_{man} optimization, the calibrated engine model was extended with a simple isothermal EGR model [15], where the EGR flow is described by the standard orifice equation. A more accurate EGR model for transient behavior would be obtained by using an adiabatic manifold pressure model as in [7], but for the purposes of this study the isothermal model is considered to be sufficient. An efficiency map derived from the results in [28] was added to the torque production model. For a certain TQE, N and EGR an efficiency factor on the indicated torque was computed by multi-linear interpolation of the experimental data. This is believed to yield at least a crude approximation of real engine behavior. Moreover, the results of the extremum-controller are valid even though the real efficiency dependence would deviate from the model, since the lack of *a priori* knowledge of the exact shape of the target function is one of the distinguished features of extremum control. The resulting efficiency maps are shown in Figure 13.10.

Injection Start Controller

The start of injection control signal SOI is directly calculated from a given end of injection EOI, by $SOI = EOI + N \cdot 0.006 \cdot T_{inj}$. In stratified mode, EOI

Control of Gasoline Direct Injection Engines using Torque Feedback   305

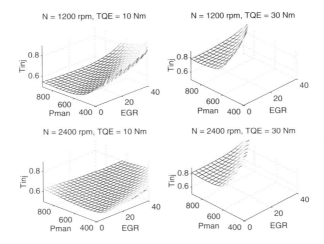

Figure 13.10 Maps of cylinder characteristics regarding fuel consumption T_{inj}, showing how P_{man} and EGR can be chosen optimally to yield a certain desired TQE for a given N

is obtained from a look-up table. In homogeneous mode, EOI is fixed at the value 190.

Ignition Advance Controller

The ignition advance IGA is obtained from a look-up table based on the engine maps.

Switching Logic

The mode switching strategy is best illustrated by paths in a torque-level diagram as in Figure 13.11. In this diagram, lines of constant torque in homogeneous and stratified mode are plotted, together with lines of constant air-fuel ratio, stratified-mode operation constraints, and the line representing zero torque difference.

In the following subsections, the *stratified* → *homogeneous* and *homogeneous* → *stratified* strategies are described in some detail. The mode switching logic is modeled as a state machine that is implemented using Simulink^{TM} Stateflow. The letters used in the descriptions below refer to the paths in the figure.

Stratified → homogeneous mode switch. This mode change poses the greatest challenge because of the constraint on $TQI_{S,ref}$ being upper bounded in stratified mode. A mode-switch is initiated when $TQI_{S,ref}$ is predicted to violate the stratified constraints within 0.2 s. The mode switch consists of a number of submodes:

A ⟶ B: In this step, P_{man} is decreased as quickly as possible to reach the zero torque-difference line before the $TQI_{S,ref}$ limit is reached (torque

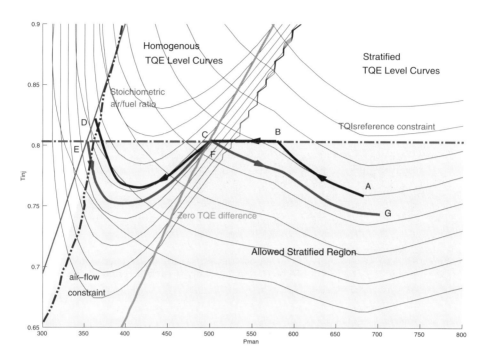

Figure 13.11 Illustration of the switch strategy between homogeneous and stratified mode. The path $A \to B \to C \to D$ illustrates a switch from stratified to homogeneous mode, where the torque limit of the stratified mode is reached before the switch is ended. The mode switch occurs at C and the switch of control strategy occurs at D. The path $E \to F \to G$ illustrates a switch from homogeneous to stratified mode. Here, the control strategy switch occurs at E and the mode switch occurs at F

is probably increasing rapidly). The extremum-controller is switched off, and a reference value $P_{man,ref}$ corresponding to $R_{a/f} = 1.5$ is passed to the P_{man}-controller.

B $\longrightarrow$ C: If the $TQI_{S,ref}$ limit is reached, T_{inj} has to be kept constant until $R_{a/f}$ has reached the zero torque-difference line. At this point, the TQE -controller is switched off, and T_{inj} is frozen at its present value. Therefore as P_{man}, and in effect $R_{a/f}$, decreases, the torque error increases.

C: The actual mode switch takes place close to the zero torque-difference line $R_{a/f} \approx 1.5$.

C $\longrightarrow$ D: After the mode switch, P_{man} is used to regulate $R_{a/f}$ to its nominal reference in homogeneous mode. The torque is controlled with T_{inj}.

D: When $R_{a/f}$ is close to the nominal reference the homogeneous-mode controllers are switched on, which use T_{inj} to control the air-fuel ratio, and P_{man} to control the torque.

Control of Gasoline Direct Injection Engines using Torque Feedback 307

Homogeneous → stratified mode switch. A mode switch from homogeneous to stratified mode is triggered when the constraints for running in stratified mode are fulfilled. This switch does not impose as critical timing conditions as the switch in the opposite direction, since the torque is decreasing, and $TQI_{S,ref}$ is moving away from its bound in stratified mode. A certain margin is required regarding $TQI_{S,ref}$ to avoid constraint violations during the mode switch, and to avoid a situation with instant chattering mode switches.

E ⟶ F: A controller that regulates the torque with T_{inj} is switched on, while P_{man} is ramped up to the zero torque difference line.

F ⟶ G: The actual mode switch takes place close to the zero torque-difference line. P_{man} is ramped up to a suitable starting position for the extremum-controller.

G: The extremum-seeking controller is switched on.

During the switching, the mode-switching strategy described above temporarily relaxes the constraint that the air-fuel ratio should be equal to 1 in homogeneous mode. This increases the amount of emissions slightly. Since the benchmark does not include an emissions model, it is difficult to evaluate the severity of these additional emissions.

Tracking

The final controller consists of a number of sub-controllers that are used in either homogeneous mode, stratified mode, or during mode switches. These controllers have to be switched in and out in a graceful manner. To ensure this, an extensive *tracking-scheme* is included in the core controller. The same tracking scheme is used to prevent integrator-windup when the actuators saturate.

Sampling and filtering

The core controller is a discrete-time controller with sampling period 5 ms. Anti-aliasing filters were added to the measured signals. Please refer to [2] for more information on the design and implementation of discrete-time controllers.

13.7 Core Controller Results

The core controller was tested in Simulink^{TM} simulations with the reduced-benchmark model. The tests include three different scenarios of different durations and torque references, and six different engine configurations:

1. The nominal model without measurement noise.
2. The nominal model with a unmodeled 10% torque loss to simulate aging phenomena and engine-to-engine variations. Without measurement noise.

3. The nominal model with a unmodeled 10 % torque gain to simulate engine-to-engine variations. Without measurement noise.

4. The nominal model with measurement noise.

5. Configuration 2 with measurement noise.

6. Configuration 3 with measurement noise.

The main demands of the control system are that the torque should be within ±2 Nm from the reference during normal operation, that torque-bumps at the mode-switches should be less than 10 Nm and that the air-fuel ratio $R_{a/f}$ should be controlled with an accuracy within 2 %.

The core controller presented here gave good performance in simulations, well within the specifications. It was compared with other more traditional controllers that used feedforward from detailed inverse engine maps, and competed well in tracking performance. The extensive use of feedback also made the present controller more robust to engine variations. The torque-gain variations did not affect tracking or mode switch performance significantly. The controller also worked well in the presence of noise, even though noisy signals were not explicitly taken into account in the design procedure. Figure 13.12 shows the noise-free simulations of the torque response in a torque scenario at gear 1.

The mode switch from homogeneous to stratified mode takes place at $t \approx 2.6$ s. The simulations show that the torque is followed with good accuracy and that the mode-switch is carried out practically without bumps. The price paid for the small torque-bump is that the $R_{a/f}$ deviates more from the reference than specified during the transition phase in the mode switch.

The oscillations that are present in the MTC, P_{man}, and T_{inj} signals in stratified mode, are due to the perturbation signal applied by the extremum controller. The perturbation signal was chosen as $\delta P_{man} = 30\sin(10\pi t)$ mbar, and $P_{man,base}$ was updated every 50 ms. Note that the effect of the perturbation signal is not visible in TQE. The controller correctly finds the optimal P_{man} with a convergence time of about 1.5 s. In a real engine the presence of periodic pumping fluctuations in the manifold pressure [16] that interferes with the perturbation signals may have a deteriorating effect on the extremum-controller. A remedy would be to use the controller of Section 13.6, where the perturbation signal may be chosen orthogonal to the pumping fluctuations.

Results with EGR

The extremum control of EGR was evaluated on simulations on the full benchmark extended with the EGR model as described above.

In Figure 13.13 is shown how the extremum controller performs when optimizing P_{man} and EGR simultaneously. The engine enters stratified mode at $t = 10.5$ s. The P_{man} extremum controller is started directly when entering stratified mode, while the EGR extremum controller is started at $t = 17$ s. Convergence is somewhat slower than in the previous case when only

Control of Gasoline Direct Injection Engines using Torque Feedback

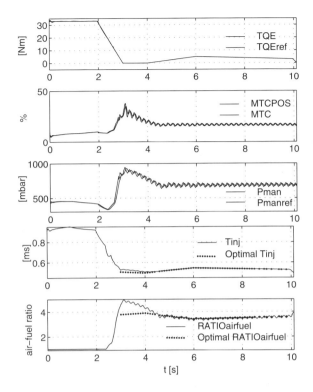

Figure 13.12 Results for torque scenario at low gear. Stratified mode is entered and the extremum-controller is started at $t = 2.6$ s. Note how the perturbation signal is visible as a periodic variation in MTC and P_{man}, but invisible in TQE. The extremum controller converges to minimal T_{inj} in approximately 1.5 s, and is then able to track the optimum. The optimal values are computed by means of numerical optimization on the simulation model

P_{man} is optimized. This can partly be explained by the inefficient steepest-gradient step in the optimization. The perturbation signals were chosen as $\delta ETC = 0.5 \sin(2\pi t)$ and $\delta P_{man} = 10 \sin(3\pi t)$. Integration was done over one period, and ETC_{base} and $P_{man,base}$ were updated every 1 s.

Notice how the effect of the perturbation signals is completely invisible in TQE. The perturbation signals were chosen more or less arbitrarily, and the indicated performance of the controllers may be improved further with more careful choices.

13.8 A Complete Engine Management System

In the second stage of the benchmark study, the goal was to design an engine management system for the complete benchmark shown in Figure 13.2. Instead of a computed torque reference, the command signals to the controllers are now the accelerator, brake, and clutch pedal positions, and the

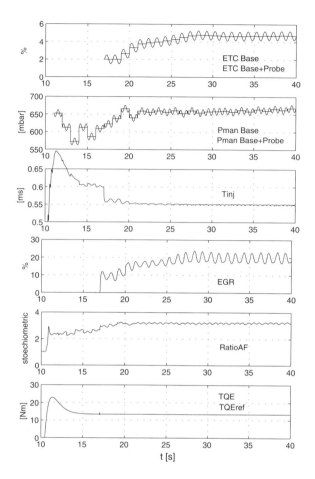

Figure 13.13 Results from driving-cycle scenario at gear 3 and velocity 40 km/h with extremum control of $R_{a/f}$ and EGR ($N \approx 1900$ rpm). Note how P_{man} and ETC simultaneously converge to optimal base values.

gear. These signals come from a model that mimics the behavior of a human driver. The inputs to the driver model are the reference signals (vehicle speed and gear) from the European driving cycle. The control problem is extended to also include idle-speed control with the presence of electric load disturbances. The idle speed has been defined as 1000 rpm and the engine speed at which the engine starts misfiring has been chosen as 900 rpm.

The controller structure for the complete engine management system is shown in Figure 13.14. The accelerator pedal position is mapped to a corresponding torque reference using a simple linear transformation where the accelerator pedal range $0-100\%$ is mapped to the range $TQE_{\min}(N) - TQE_{\max}(N)$, where the limits depend on the engine speed. The idle-speed controller is switched in when the accelerator pedal is fully released, or

Control of Gasoline Direct Injection Engines using Torque Feedback

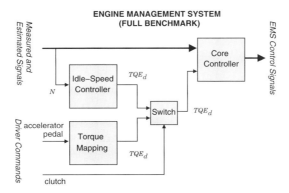

Figure 13.14 Complete Engine management system structure. The core controller is extended with the idle-speed controller, the accelerator pedal interpreter, and switching logic that determines which of these blocks that provides the torque reference.

when the clutch is fully disengaged and the torque reference generated by the idle-speed controller is larger than the torque reference generated by the driver. The latter part of the condition is used to prevent engine stalling during the switch from idle-speed control to driver control during gear switches. A rate limiter is used on the torque reference to improve the performance during gear switches. To avoid unnecessary mode changes, it is not possible to switch from homogeneous mode to stratified mode during a gear switch (when the clutch is disengaged).

The resulting control structure is of multi-level cascade type. The outermost level is the driver or the idle speed controller. The next level is the torque controller in the core controller. In homogeneous mode the torque controller generates the reference for the P_{man}-controller, which is also a cascade controller. Hence, in the extreme case, the system contains four controller levels.

Idle Speed Control

To evaluate the idle-speed controller, load disturbances l corresponding to electrical loads are introduced during the idle-speed control parts of the driving cycle.

The scenario contains two types of loads. Small loads are modeled as non-measurable constant loads of size 3.5 Nm. Large loads, corresponding, *e.g.*, to the air-conditioning system, are also modeled as non-measurable constant loads, now of size 15 Nm. However, in the latter case the idle-speed controller has access to a warning signal that signals that a large load is pending, and may delay the arrival of the large load for a short while. This can be used to prepare the idle-speed controller for the forthcoming load, and is a common solution in modern vehicles.

In line with the simple controller structures used in the core controller, the idle-speed controller has also been kept simple. It consists of a simple PI-controller, where the proportional gain depends slightly on the engine speed. The input to the idle-speed controller is the idle-speed error, *i.e.*, $1000 - N$ and the output is the torque reference. The warning signal for large loads is used to add a constant value to the integrator state of the PI-controller. This is a simple solution that prevents any dangerous undershoots in the engine speed when the large load arrives.

13.9 Full Benchmark Results and Comparisons

In the full benchmark, the evaluation scenario consists of the full European driving cycle including gear switches and idle-speed phases. The core controller without EGR control is used together with the idle-speed controller described in the previous section. The reason for not incorporating EGR in these evaluations is that the EGR was introduced to the benchmark at a late stage. Therefore the EGR was not present in the evaluation criteria, and was not part of the solutions of all participating research groups. In the reduced benchmark, it was important to achieve good torque-tracking performance. This resulted in the additional I^2 terms in some of the PI-controllers. In the full benchmark, torque tracking is no longer of major importance, since the torque reference is now an internal signal in the EMS. As a result of this, it is probably possible to reduce the complexity of the core controller even further.

The engine speed and the vehicle speed for the full driving cycle are shown in Figure 13.15.

A typical gear-shift phase is shown in Figure 13.16. The accelerator pedal is released during the gear switch. A combustion-mode switch from homogeneous to stratified mode takes place immediately before the gear switch. After the gear-switch the system switches back to homogeneous mode. Small irregularities in the torque signal occur during the mode changes. During a short interval, the idle-speed controller becomes active. The controller was evaluated with the same configurations as for the core controller. Small and large load disturbances are applied during the idle-speed control phases. The simulations are evaluated using three different criteria values that were defined in the benchmark:

- $C_{Throttle} = T_{samp} \cdot \sum |\phi_{MTC}(k) - \phi_{MTC}(k-1)|/T_{scenario}$ is a measure of the variations in ϕ_{MTCPOS}.

- $C_{Cons} = \int Q_{fuel} dt / (\rho_{fuel} \cdot \int v dt)$ is a measure of the fuel consumption (liters/100 km)

- C_{Global} is the average value of five criteria that measure the performance of the air-fuel ratio control in homogeneous mode, penalize large torque gradients, penalize torque bumps caused by mode changes, measure the performance of the vehicle-speed control, and measure the performance of the idle-speed control.

Control of Gasoline Direct Injection Engines using Torque Feedback   313

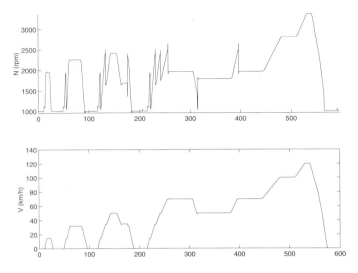

Figure 13.15 Engine speed and vehicle speed with the complete engine management system in a simulation of the European driving cycle. Gear switches and idle-speed control engagement and disengagement can be seen in the engine speed signal.

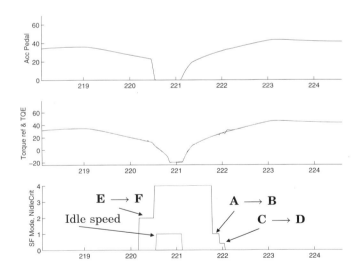

Figure 13.16 Accelerator pedal (top), torque and torque reference (middle), and combustion mode and idle-speed mode during a gear-switch phase. The mode-switching states as defined above are indicated in the figure.

All three criteria should be minimized.

During the full driving cycle, 26 combustion mode changes are performed. The engine spends about the same amount of time in homogeneous mode as in stratified mode. The perturbation signal of the extremum controller was chosen as $\delta P_{man} = 5\sin(2.5\pi t)$ mbar, which is significantly smaller and slower than in the core controller simulations. This is to reduce the cost of throttle actuation that is part of the evaluation criteria. The mean fuel consumption was 6.78 l/100 km, compared to 6.88 l/100 km for a controller running with fully-open throttle in stratified mode. The results of the present work thus indicate that throttling may be used to improve fuel consumption. It should be emphasized that these results apply to the benchmark engine model, and that it is uncertain how they translate to a real engine.

The controller has been compared with two other designs: *(MPC)*, a model-predictive controller based on a piecewise-linear model of the GDI engine, developed using clustering-based identification on the Simulink^{TM} engine model [23], and *(PWL)*, a design based on feedforward using piecewise-linear inverse approximations of the different engine maps, combined with feedback [25]. The latter control design is relatively close in spirit to a production engine management system, with the exception that piecewise linear models are used instead of look-up tables. A summary of the results is shown in Figure 13.17. The approach taken in the present work proved to yield better fuel economy and tracking performance for all configurations. The throttle command signal cost $C_{throttle}$ is somewhat higher in configurations 1–3, due to the probing actions of the extremum-controller. This cost may be reduced by a more clever choice of probing signals. There is no significant deterioration in throttle cost in the noisy configurations 4–6. The better result with respect to fuel consumption can be explained by the on-line optimization using the extremum-controller in combination with the torque control algorithms. In the PWL design, ignition timing is used for fast torque control, which results in lower efficiency of the engine and as a result higher fuel consumption. In the global criteria, which is essentially a measure of tracking performance and the ability to switch mode without large torque bumps, it is interesting to see that the present design is significantly more robust to gain variations from the nominal design in configuration 1. This is due to the extensive use of feedback, which results in robustness to engine variations. Also, here the introduction of noise does not significantly deteriorate performance.

13.10 Torque Estimation and Sensing

The issue of real-time estimation of the engine torque is not trivial. Earlier work has mainly focused on the estimation of indicated torque for diagnosis purposes. With the introduction of electronic throttle control, the interest has shifted to construct real-time torque estimators that can act as virtual sensors in closed-loop engine control. In [27], effective torque estimates are

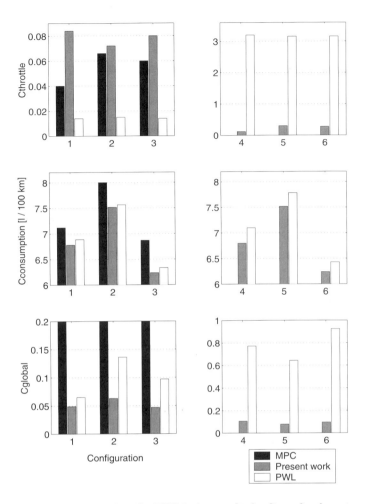

Figure 13.17 Criteria results. The MPC design results for C_{global} has been truncated to keep the results from the present work and the PWL design distinguishable. There are no results available for the MPC design for configurations 4–6.

obtained from a second-order passive circuit model using crankshaft speed fluctuations. [3] presents a summary of results concerning the estimation of indicated, effective, and load torques in IC engines using nonlinear second-order sliding mode observers using driveline and crankshaft speed measurements. Other related work is presented in [24], [33], [11], [12]. Torque estimation is, in general, dependent on sophisticated and accurate engine models. Hence, the possibility to use simplified models for controller design and implementation are weighted against the additional complexity of the estimators.

Another alternative is to measure the effective torque directly. Magnetoelastic or magnetostrictive sensors are a realistic solution to this problem as they provide low-cost contact-free measurements suitable for harsh environments. Magnetoelasticity is a property of ferromagnetic materials that causes the magnetic permeability of the material to change when it is subjected to mechanical stress. The basic principle of a magnetoelastic sensor is to let a magnetic field penetrate the shaft, and detect changes in flux caused by the torsional stress in the shaft. In [29], [19], [32], practical application of the ABB Torductor sensor is demonstrated. More work on magnetostrictive torque measurements is found in *e.g.*, [6]. Even if these sensors are generally not found on current production vehicles, they definitively constitute a realistic and mature solution to torque measurements in automotive vehicles.

13.11 Conclusions

Conventional engine management systems are dominated by feedforward control based on look-up tables containing inverse approximations of different engine maps. The reason for this is the lack of inexpensive sensors for, *e.g.*, engine torque and in-cylinder pressure. Such open-loop approaches are sensitive to engine-to-engine variations and aging and wear phenomena. Since feedback is not possible without a sensor, the look-up tables must be very accurate. Development of the tables is time-intensive and costly. This is one of the reason for the large interest in the engine-control community for different data-based nonlinear approximation schemes, *e.g.*, neural networks and fuzzy systems.

The work presented in this paper indicates that much simpler control designs may be used with very good results, provided that the effective torque produced by the engine is available for feedback. The work has aimed at the design of a robust, simple, and efficient EMS for a GDI engine. A bottom-up approach has been adopted, where standard building blocks like linear PID-controllers and static affine feedforward have been used extensively for building a controller for torque, idle-speed, and combustion-mode control. The controller also includes extremum control of the air-fuel ratio and EGR to minimize fuel consumption and emissions. The feedback controllers rely on the availability of the effective torque signal. The silent extremum control strategy that is developed also requires torque feedback to eliminate effects of perturbation signals on the engine output. The suggested silent extremum-control structure may find applications in other similar control problems. It has the useful property of optimizing a target function with respect to a control variable by adding disturbances that are invisible in the system output.

The EMS was evaluated in computer simulations on the full European driving cycle with good results that compare well with traditional table-based engine control with respect to tracking performance, ride comfort, and fuel consumption. The strategy was proven to be robust to engine variations and disturbances.

The results show that it is possible to design engine management systems with little prior knowledge of engine model parameters if feedback control and on-line optimization are used extensively. The reasons why the control design works as well as it does are the inherent robustness of feedback to uncertainties, the linearizing nature of feedback, the extremum-based model-free on-line optimization, and the switching strategy that minimizes fuel consumption and torque bumps.

Acknowledgments

This work has been performed within the EU ESPRIT FAMIMO project. The work of the first author has been supported by the DICOSMOS project within the NUTEK Complex Technical Systems Program, and the work of the third author has been supported financially by The Danish Steel Works Ltd. We would like to thank Dominique Passaquay, Pascal Bortolet, and Serge Boverie at Siemens Automotive, Toulouse, and Bengt Johansson at Department of Heat and Power Engineering, Division of Combustion Engines, Lund Institute of Technology.

13.12 References

[1] P. Andersson, L. Eriksson, and L. Nielsen. Modeling and architecture examples of model based engine control. In *Proceedings of the Second Conference on Computer Science and Systems Engineering*, Linköping, Sweden, 1999.

[2] K. J. Åström and B. Wittenmark. *Computer-Controlled Systems*. Information and System Sciences Series. Prentice-Hall, third edition, 1997.

[3] P. Azzoni, D. Moro, F. Ponti, and G. Rizzoni. Engine and load torque estimation with application to electronic throttle control, *SAE Technical Paper 980795*, 1998.

[4] C. Draper and Y. Li. Principles of optimalizing control systems and an application to the internal combustions engine. *ASME*, 1951.

[5] M. Druzhinina, I. Kolmanovsky, and J. Sun. Hybrid control of a gasoline direct injection engine. In *Proceedings of the 38th conference on Decision & Control*, 2667–2672, Phoenix, Arizona USA, 1999.

[6] W. J. Fleming. Magnetostrictive torque sensor performance — nonlinear analysis. *IEEE Trans. Vehicular Technology*, 38(3):159–167, 1989.

[7] M. Fons, M. Muller, A. Chevalier, C. Vigild, E. Hendricks, and S. C. Sorenson. Mean value engine modelling of an SI engine with EGR, *SAE Technical Paper 1999-01-0909*, 1999.

[8] J. Gerhardt, H. Honninger, and H. Bischof. A new approach to functional and software structure for engine management systems, *SAE Technical Paper 980801*, 1998.

[9] M. Gäfvert, L. M. Pedersen, K.-E. Årzén, and B. Bernhardsson. Simple feedback control and mode switching strategies for GDI engines. In *SAE 2000 World Congress*, Detroit, Michigan, USA, *SAE paper 2000-01-0263*, 2000.

[10] M. Gäfvert, K.-E. Årzén, and L. M. Pedersen. Simple linear feedback and extremum seeking control of GDI engines. In *Proceedings of Seoul 2000 FISITA World Automotive Congress*, Seoul, Korea, June 2000.

[11] P. Gyan, S. Ginoux, J. C. Champoussin, and Y. Guezennec. Crankangle based torque estimation: Mechanistic/stochastic, *SAE Technical Paper 2000-01-0559*, 2000.

[12] I. Haskara and L. Mianzo. Real-time cylinder pressure and indicated torque estimation via second order sliding modes. In *Proc. American Control Conference*, 3324–3328, Arlington, 2001.

[13] E. Hendricks. Mean value modelling of spark ignition engines, 1990. *SAE Technical Paper 900616*, 1990.

[14] E. Hendricks. Engine modelling for control applications: A critical survey. *Meccanica*, 32:pp. 387–396, 1997.

[15] E. Hendricks, D. Engler, and M. Fam. A generic mean value engine model for spark ignition engines. In *41st Simulation Conference, SIMS 2000*, Lyngby, Denmark. Technical University of Denmark, 2000.

[16] E. Hendricks, M. Jensen, A. Chevalier, and T. Vesterholm. Problems in event based engine control. In *Proc. American Control Conference*, 1585–1587, Baltimore, MD, 1994.

[17] Y. Iwamoto, K. Noma, O. Nakayama, T. Yamauchi, and H. Ando. Development of gasoline direct injection engine, *SAE Technical Paper 970541*, 1997.

[18] N. S. Jackson, J. Stokes, P. A. Whitaker, and T. H. Lake. Stratified and homogeneous charge operation for the direct injection gasoline engine — high power with low fuel consumption and emissions, *SAE Paper 970543*, 1997.

[19] J. Jeremiasson and C. Wallin. Balancing of individual cylinders in a V8 diesel engine based on crankshaft torque measurements, 1998. SAE Technical Paper 981063.

[20] U. Kiencke. The role of automatic control in automotive systems. In *Proceedings of the IFAC 10th Triennal World Congress*, Munich, FRG, 1987.

[21] I. Kolmanovsky, M. van Nieuwstadt, and J. Sun. Optimization of complex powertrain systems for fuel economy and emission. In *Proceedings of the 1999 IEEE International Conference on Control Applications*, 833–839, Hawaii, USA, 1999.

[22] T. Kume, Y. Iwamoto, K. Lida, M. Murakami, K. Akishino, and H. Ando. Combustion control technologies for direct injection SI engine, 1996. SAE Technical Paper 960600.

[23] S. Mollov, P. J. van der Veen, and R. Babuska. Fuzzy model-based predictive control of a GDI engine. In *European Control Conference*, Porto, Portugal, 2001. To appear.

[24] J. J. Moskwa and C. H. Pan. Engine load torque estimation using nonlinear observers. In *Proc. 34th Conf. Decision & Control*, 3397–3402, New Orleans, LA, 1995.

[25] D. Passaquay. *Modélisation et Commande de processus multivariables à base de logique floue: Application à la commande de moteurs thermiques.* Ph.D. thesis, Laboratoire d'Analyse et d'Architecture des Systèmes du CNRS, 2000. (Referred sections in English.).

[26] D. Passaquay, P. Bortolet, and S. Boverie. Engine benchmark architecture decomposition, reduced benchmark, controller design rules. Technical report, Siemens, Toulouse, Sep 1998. Release 1.2 (Confidential).

[27] G. Rizzoni. Estimate of indicated torque from crankshaft speed fluctuations: A model for the dynamics of the IC engine. *IEEE Trans. Vehicular Technology*, 38(3), 168–179, 1989.

[28] S. Sasaki, D. Sawada, T. Udeda, and H. Sami. Effects of EGR on direct injection gasoline engine. *JSAE Review*, (19):223 – 228, 1998.

[29] J.R. Sobel, J. Jeremiasson, and C. Wallin. Instantaneous crankshaft torque measurements in cars, *SAE Technical Report 960040*, 1996.

[30] J. Sternby. A review of extremum control. Technical Report LUTFD2/(TFRT–7161)/1-47/(1979), Department of Automatic Control, Lund Institute of Technology, April, 1979.

[31] Jing Sun, I. Kolmanovsky, D. Brehob, J.A. Cook, Julie Buckland, and Mo Haghgooie. Modeling and control of gasoline direct injection stratified charge (DISC) engines. In *Proc. 1999 IEEE Int. Conf. Control Applications*, 471–477, 1999.

[32] C. Wallin, L. Gustavsson, and M. Donovan. Engine monitoring of a formula 1 racing car based on direct torque measurements, *SAE Technical Paper 02P-195*, 2002.

[33] Y.Y. Wang, V. Krishnaswami, and G. Rizzoni. Event-based estimation of indicated torque for IC engines using sliding-mode observers. *Control Engineering Practice*, 5(8), 1123–1129, 1997.

[34] Fu-Quan Zao, Ming-Chia Lai, and D.L. Harrington. A review of mixture preparation and combustion control strategies for spark-ignited direct-injection gasoline engines, *SAE Paper 970627*, 1997.

14

Closed-loop Combustion Control of HCCI Engines

P. Tunestål   J.-O. Olsson   B. Johansson

Abstract

The HCCI engine, with its excellent potential for high efficiency and low NO_X emissions, is investigated from a control perspective. Combustion timing, *i.e.*, where in the thermodynamic cycle combustion takes place, is identified as the most challenging problem with HCCI engine control. A number of different means for controlling combustion timing are suggested, and results using a dual-fuel solution are presented. This solution uses two fuels with different ignition characteristics to control the time of auto-ignition. Cylinder pressure measurement is suggested for feedback of combustion timing. A simple net-heat release algorithm is applied to the measurements, and the crank angle of 50% burnt is extracted. Open-loop instability is detected in some high-load regions of the operating range. This phenomenon is explained by positive feedback between the cylinder wall heating and ignition timing processes. Closed-loop performance is hampered by time delays and model uncertainties. This problem is particularly pronounced at operating points that are open-loop unstable.

14.1 Homogeneous Charge Compression Ignition (HCCI)

The Homogeneous Charge Compression Ignition (HCCI) engine with its excellent potential for combining low exhaust emissions with high efficiency gained substantial interest towards the end of the 20th century. The HCCI engine combines features of the traditional spark ignited (SI) Otto-cycle and compression ignited (CI) Diesel-cycle engines into something that must be characterized as a separate engine concept.

The HCCI Engine Concept

The HCCI engine combines features of the traditional SI and CI engines into a new engine concept. The HCCI engine features homogeneous charge like the SI engine, and compression ignition like the CI engine. HCCI operation can be two-stroke or four-stroke, and the first studies [19], [15] were performed on two-stroke engines. Later studies [23], [3] on four-stroke engines show that high efficiency can be combined with low NO_x emissions for HCCI engines running with a high compression ratio and lean operation. This text will focus exclusively on the four-stroke version of the HCCI engine.

The HCCI Cycle

The four-stroke HCCI cycle can be described by its four strokes: *intake*, *compression*, *expansion*, and *exhaust*. During the *intake stroke*, a more or less homogeneous mix of fuel and air is inducted into the cylinder. During the *compression stroke*, this charge is compressed by the upward motion of the piston. Towards the end of the *compression stroke*, temperature and pressure have reached levels where pre-combustion reactions start to take place. Somewhere near the TDC (top dead center), actual combustion starts. During the initial part of the *expansion stroke*, the bulk of combustion takes place during the course of a few crank-angle degrees. During the rest of the *expansion stroke*, the high pressure caused by combustion forces the piston down towards BDC (bottom dead center). During the *exhaust stroke*, the upward motion of the piston forces the exhaust gas to leave the cylinder through the exhaust valve.

Ignition. An HCCI engine, contrary to SI and diesel-cycle CI engines, has no direct means for controlling ignition timing. The SI engine has spark timing, and the diesel-cycle CI engine has the start of fuel injection, which both directly control the onset of combustion. However, for an HCCI engine, ignition timing is dictated by the conditions of the charge and the cylinder walls at the time when the intake valve closes. This is one of the biggest challenges with practical implementation of HCCI engine technology. Ignition timing can only be controlled indirectly through adjustments in the cylinder charge preparation. The following paragraphs will describe the most important parameters that control ignition timing for an HCCI engine.

The temperature of the air when it enters the cylinder has a large influence on the charge temperature towards the end of the compression stroke. With a compression ratio of 18:1, a change in intake temperature by 30 K will result in a change in temperature at TDC by almost 100 K. Since temperature is a very important factor in auto-ignition, an increase in intake temperature will have a very strong advancing influence on ignition timing.

The portion of the exhaust gas that is not expelled during the exhaust stroke, the residual gas, is particularly important for HCCI operation. The thermal energy provided by the residual gas contributes in heating the charge of the following cycle, and affects the crank angle at which ignition takes place. On an engine with variable valve timing, the residual-gas fraction can be controlled—e.g., by early closing of the exhaust valve, which will trap a larger amount of exhaust gas in the cylinder for the following cycle. It is necessary to remember though that exhaust gas also acts as a diluent, and thereby slows down the combustion chemistry. This will tend to retard ignition timing, and with a very high residual-gas fraction this effect will dominate.

Closely related to residual gases is EGR (exhaust gas recirculation). This refers to exhaust gas that is routed back from the exhaust manifold to the intake manifold. Combined with an EGR cooler, this can be used for diluting the charge and thus lowering the reaction rate. An increase in EGR rate will retard ignition timing.

Another important factor is the cylinder-wall temperature. Hot cylinder walls will heat the charge throughout the intake and compression strokes, and will advance ignition timing.

The fuel-air equivalence ratio affects both fuel concentration and oxygen concentration. However, since HCCI engines operate lean, the equivalence ratio has a stronger influence on fuel concentration than on oxygen concentration. The dominating effect of increasing the equivalence ratio, thus, is an increase in fuel concentration, which will result in a higher reaction rate. Thus, increasing the equivalence ratio serves to advance ignition timing.

Another possible way to control ignition timing is by changing the fuel composition. Addition of a second fuel with higher reactivity will serve to advance ignition timing. Examples are the addition of hydrogen to natural gas and n-heptane to iso-octane.

A variable compression ratio provides an effective means of controlling the temperature towards the end of the compression stroke. A higher compression ratio increases the charge temperature near the TDC, and tends to advance ignition timing.

Charge stratification—i.e., inhomogeneous charge distribution—can be used to locally increase the equivalence ratio, and thus the reaction rate, in order to advance the ignition timing. Charge stratification can be achieved through late fuel injection. The drawback is locally high temperatures, causing an increase in NO_X production.

Evidently, there are many parameters that affect ignition timing, but they all do so in non-trivial ways, and furthermore, many of the parameters affect each other as well. Some of the parameters are even affected by ignition timing itself. The cylinder wall temperature, *e.g.*, increases with advanced ignition timing. When ignition timing is advanced, the peak cylinder temperature increases which, in turn, causes an increase in cylinder wall temperature. It follows that ignition timing is very sensitive to operating conditions

14.2 Closed-loop Control of Ignition Timing

It is evident from above that ignition control is much more of a challenge for an HCCI engine than for an SI or diesel-cycle CI engine. The most readily available means of controlling ignition timing is by adjusting the fuel composition. This does not require any novel mechanical design like variable valve timing or variable compression ratio. It merely requires a doubling of the port fuel injection system.

Selection of feedback. The sensitivity of ignition timing to operating conditions does not allow an open-loop solution in the form of *e.g.*, a look-up table. Furthermore, the system becomes unstable for some operating conditions at high load. Thus, closed-loop control is an absolute necessity, which poses the question of what to use for feedback. Cylinder pressure is the natural choice, since ignition is an in-cylinder phenomenon. What characteristic of the cylinder pressure trace reflects when combustion takes place, though?

The crank angle of maximum pressure gives some information about when the bulk of combustion is taking place, but for combustion timing before or near the TDC, this angle tends to gravitate towards TDC due to the dependence on volume in the ideal gas law. Furthermore, for very late combustion timing, the pressure maximum from compression dominates the one from combustion. Another problem is that a maximum has a certain flatness to it, which makes it non-unique.

The crank angle of maximum pressure derivative suffers from the same problems as the crank angle of maximum pressure in addition to the inherent noise problems with numerical differentiation. Another possibility is to search for the inflection point, where the pressure trace transitions are from negative to a positive second derivative due to the onset of combustion. This also suffers from the problems with numerical differentiation.

It turns out that a first-law analysis based on the pressure measurements and heat release analysis provides a very robust source of feedback. Heat release analysis applies the first law of thermodynamics to the combustion chamber during the entire combustion event in order to estimate the rate at which chemical energy is converted to thermal energy. If no adjustments are made for heat transfer or flow into and out of crevices, the net heat release is obtained. Integration with respect to the crank angle yields the cumulative heat release, which roughly reflects the mass fraction burned.

Combustion in an HCCI engine is usually very fast. The mass fraction burned usually goes from 10% to 90% in about 5 crank angle degrees, which means that the crank angle of 50% heat release, CA50, provides a very accurate measure of when combustion is taking place. In the following, CA50, combustion timing, and ignition timing will be used interchangeably to denote the crank angle of 50% heat release.

Processing cylinder pressure measurements. Cylinder pressure measurements are normally performed with either piezoelectric elements combined with charge amplifiers or with fiber-optical sensors. Both methods fail to measure the DC component of the cylinder pressure. A thermodynamically-based method of estimating the DC component is detailed in [26], and amounts to estimating an initial pressure and a measurement offset based on pressure measurements during the compression stroke. It is essential to select the crank-angle interval for estimation between the intake-valve closing and the start of combustion for the thermodynamic assumptions to hold.

The cylinder pressure measurements, p_m, can be decomposed into the actual pressure, p, and a sensor offset, Δp, according to (14.1).

$$p_m = p + \Delta p \tag{14.1}$$

The real pressure can be modeled with polytropic compression:

$$p = CV^{-\kappa} \tag{14.2}$$

where V is the combustion chamber volume, κ is the polytropic exponent, and C depends on the initial pressure according to:

$$C = p_0 V_0^\kappa \tag{14.3}$$

If the polytropic exponent is assumed to be known (normally between 1.3 and 1.4), two parameters need to be estimated from the measurements: C and Δp. Thus, the vector of estimates is given by:

$$\theta = \begin{pmatrix} \Delta p \\ C \end{pmatrix} \tag{14.4}$$

The output vector, assuming n measurements, is given by:

$$Y = \begin{pmatrix} y_1 \\ \vdots \\ y_n \end{pmatrix} = \begin{pmatrix} (p_m)_1 \\ \vdots \\ (p_m)_n \end{pmatrix} \tag{14.5}$$

and the regression matrix is given by:

$$\Phi = \begin{pmatrix} \varphi_1 \\ \vdots \\ \varphi_n \end{pmatrix} = \begin{pmatrix} 1 & V_1^{-\kappa} \\ \vdots & \vdots \\ 1 & V_n^{-\kappa} \end{pmatrix} \tag{14.6}$$

The least squares estimate of θ is then given by (14.7).

$$\hat{\theta} = \left(\Phi^T \Phi\right)^{-1} \Phi^T Y = \Phi^+ Y \tag{14.7}$$

where Φ^+ is the Moore-Penrose pseudo inverse of Φ.

In [26], a method of estimating the polytropic exponent is also provided. In cases where the polytropic exponent is thought to vary significantly from cycle to cycle, this method can be used. Intake temperature and fuel composition as well as the equivalence ratio affect the polytropic exponent.

Heat release analysis. An analysis of the combustion chamber, based on the first law of thermodynamics, relates the rate at which chemical energy is converted to thermal energy to the pressure in the combustion chamber. This type of analysis is conventionally called *heat release analysis*, and can be used to determine when combustion is taking place. This term stems from the simplification that is normally done, in which the charge composition is assumed to be constant, and that the increase in internal energy is interpreted as heat. If the actual heat transfer to the cylinder walls as well as crevice flow is neglected, equation (14.8) relates heat release to cylinder pressure:

$$\delta Q_{ch} = \frac{c_v}{R} V \mathrm{d}p + \frac{c_p}{R} p \mathrm{d}V \tag{14.8}$$

This equation is integrated over a crank angle interval which includes the whole combustion event. The parameters c_v and c_p are the specific heats at constant volume and pressure, respectively, and technically depend on temperature. However, if the only objective is to determine combustion timing, they can be assumed to be constant. R is the universal gas constant.

The result of the integration of the heat release equation is the cumulative net heat release as a function of crank angle, $Q_{ch}(\alpha)$. A typical heat release trace is plotted in Figure 14.1 together with some definitions. The most important definition in this context is CA50, the crank angle of 50% heat release. Since combustion is very fast, CA50 can be used as a robust source of feedback for combustion timing.

14.3 Closed-Loop Combustion Control of HCCI Engines

Feedback structure. Two fuel injectors and one cylinder pressure sensor per cylinder allows separate control loops for each cylinder. Thus, the

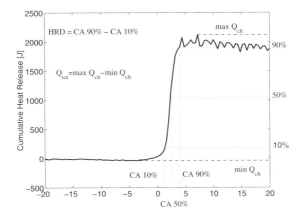

Figure 14.1 Definitions of some heat-release based cycle parameters

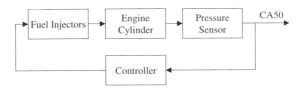

Figure 14.2 Control structure for combustion timing control. The CA50 based on cylinder pressure measurements provides feedback about combustion timing, and octane ratio provides a control input.

control structure indicated in Figure 14.2 can be used for the combustion timing control of each cylinder. Fuel octane number is a measure of a fuel's resistance to auto-ignition, and can be used as the control input for combustion timing control of an HCCI engine cylinder. When using a mixture of iso-octane and n-heptane, the octane number is, by definition, the percentage of iso-octane.

The plant. The sensitivity of CA50 to changes in fuel octane number varies by orders of magnitude for different operating points, see Figure 14.3. Each line in the plot represents a specific intake temperature and load. Within each line, the fuel octane number has been varied to achieve an interval of combustion timings. The strong nonlinearity of the plant makes a linear controller unsuitable for the task. The situation can be remedied, however, if the sensitivity is mapped over the multi-dimensional space of operating conditions. This map can be used for gain-scheduling the otherwise linear controller.

In [18], a multivariable function is fitted to measurements of the sensitivity of CA50 to changes in octane number for a multitude of operating conditions. In order to get a simple, computationally inexpensive model, the

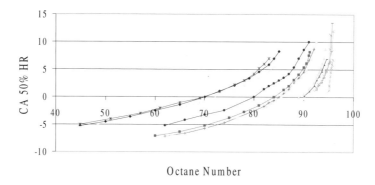

Figure 14.3 Combustion timing versus fuel octane number for various operating points. Measurements on a Scania D12 6-cylinder engine converted for HCCI operation. Octane number varied through fueling with a variable mixture of iso-octane and n-heptane

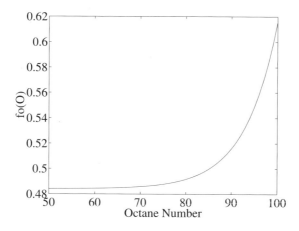

Figure 14.4 The octane number component of the sensitivity function

sensitivity is modeled as a product of functions of one variable each. This approach is entirely empirical, but yields a sensitivity model with acceptable residuals (within 3%). The variables that are included in this model are engine speed, inlet air temperature, fuel octane number, fuel mass per cycle, and CA50. A later model revision includes inlet pressure as well. Figure 14.4 shows the octane number component of the sensitivity function. The sensitivity model is used for scaling the controller gains, which implicitly assumes that the dynamic behavior of the plant is independent of the operating point. Only the DC gain of the plant changes. For the case of hydrogen enrichment of natural gas, the sensitivity change with respect to operating condition is more modest. Figure 14.5 shows the hydrogen requirement as a function of CA50 for two different engine loads, and the

Closed-loop Combustion Control of HCCI Engines

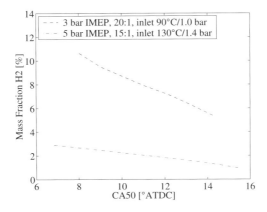

Figure 14.5 Dependence of CA50 on H_2 fraction for light load (3 bar IMEP) and medium load (5 bar IMEP)

dependence is roughly linear within each engine load.

Open-loop stability. An interesting phenomenon that appears in some regions of the operating space of an HCCI engine is open-loop instability. This phenomenon results when wall temperature effects dominate the ignition dynamics. Figure 14.6 shows the open-loop behavior at a stable and an unstable operating point, respectively. All control inputs are held constant in both cases. The effect of open-loop instability under closed-loop operation is non-minimum-phase behavior, *i.e.*, the control input starts out in the "wrong" direction after a setpoint change (compare Figures 14.7 and 14.8).

The cause of instability is the positive thermal feedback provided through the interaction between ignition timing and cylinder wall temperature. A small increase in cylinder wall temperature results in a hotter cylinder charge, which advances ignition timing. Advanced ignition timing, however, results in higher gas temperature and more heat transfer to the walls, thus higher wall temperature. The reversed case is a small drop in cylinder wall temperature, which results in cooler cylinder charge. Ignition timing is retarded, which results in lower gas temperature, which in turn reduces the heat transfer to the walls, and thus the wall temperature. It is evident that operating points where this effect dominates are unstable.

The positive feedback mentioned above is always present, but not all operating points are unstable. The stabilizing negative feedback responsible is closely related to the destabilizing positive feedback. Early ignition leads to high peak temperature and heat transfer, but this results in lower gas temperature towards the end of the cycle, which means both colder residual gas and more of it. This reduces the reactivity of the charge for the next cycle, and retards ignition timing. The opposite holds for late ignition. Thus,

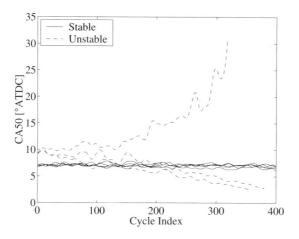

Figure 14.6 Repeated open-loop operation at one *stable* and one *unstable* operating point

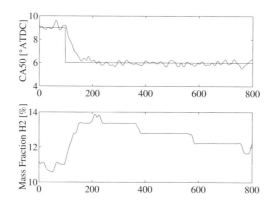

Figure 14.7 Change of CA50 setpoint at a stable operating point. The independent axis indicates number of engine cycles

the residual gas provides the stabilizing negative feedback.

Closed-loop Performance

The bandwidth of the closed-loop system is limited by time delays. In the case of pulse width modulated fuel injection using solenoid injectors, the time between fuel injection command and actual fuel injection is very small. For hydrogen enrichment of natural gas however, a mass flow controller with its own very conservative control system is used, which increases the lag from command to actual change in fuel injection.

The most severe time delay is, however, in the control system itself.

Closed-loop Combustion Control of HCCI Engines

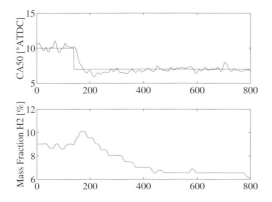

Figure 14.8 Change of CA50 setpoint at an unstable operating point. The independent axis indicates number of engine cycles

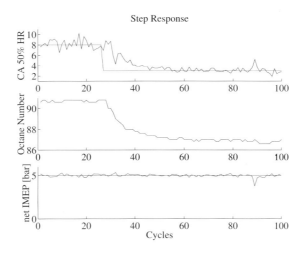

Figure 14.9 Change of combustion timing setpoint. Operation with iso-octane and n-heptane. The time delay is clearly seen in the response.

Data handling and communication with fuel-injector drivers causes a delay between a change in measurement and a corresponding change in control input of approximately four engine cycles, see Figure 14.9. With this delay present, it is, of course, impossible to achieve closed-loop bandwidth better than a few multiples of the delay.

14.4 Conclusion and Discussion

The field of closed-loop control of HCCI engines is a very new one, and a lot remains to be done. This text aims to show the potential and to highlight the difficulties. It is shown that closed-loop control of ignition timing can be achieved using a set-up with a secondary fuel as ignition improver. Closing the loop does, however, require a measurement also. The preferred measurement is the crank angle of 50 % heat release, CA50, which offers a robust measure of when the rapid HCCI combustion is taking place. In lieu of a CA50 sensor, CA50 has to be computed from crank-angle based cylinder pressure measurements.

Both the choice of control input and the choice of measurement can be questioned. A secondary fuel is impractical on a vehicle, unless it can be produced on-board. For a stationary application, it may be acceptable, however. Using cylinder pressure as a measurement is expensive, but cheaper cylinder pressure sensors can be expected in the future. For control input, other solutions exist; e.g., variable compression ratio and variable valve timing. For feedback, however, there is no good alternative to cylinder pressure measurements.

Transient performance is limited at present. It is the authors' opinion, however, that most of this problem is due to time delays in the control system. A physical change in CA50 takes in the order of four engine cycles to propagate through the control system to a physical change in fuel injection. This delay is mostly due to inefficient communication and data handling, and should be possible to cut down to around one engine cycle.

14.5 References

[1] T. Aoyama, Y. Hattori, J. Mizuta, and Y. Sato. An experimental study on premixed-charge compression ignition of Diesel fuel. *SAE Technical Paper 960081*, 1996.

[2] I. Arsie, C. Pianese, and G. Rizzo. A non linear observer for fuel film dynamics into the intake manifold of a spark ignition engine, *Proc. 1999 IEEE/ASME Int. Conf. Advanced Intelligent Mechatronics*, 251–256, Atlanta, GA, Sept 19-23, 1999,

[3] M. Christensen, P. Einewall, and B. Johansson. Homogeneous charge compression ignition (HCCI) using isooctane, ethanol and natural gas — a comparison to spark ignition operation. *SAE Technical Paper 972874*, 1997.

[4] M. Christensen and B. Johansson. Influence of mixture quality on homogeneous charge compression ignition. *SAE Technical Paper 982454*, 1998.

[5] M. Christensen and B. Johansson. Homogeneous charge compression ignition with water injection. *SAE Technical Paper 1999-01-0182*, 1999.

[6] M. Christensen, B. Johansson, P. Amnéus, and F. Mauss. Supercharged homogeneous charge compression ignition (HCCI). *SAE Technical Paper 980787*, 1998.

[7] P. Duret and S. Venturi. Automotive calibration of the IAPAC fluid dynamically controlled two-stroke combustion process. *SAE Technical Paper 960363*, 1996.

[8] R. Gentili, S. Frigo, L. Tognotti, P. Hapert, and J. Lavy. Experimental study of ATAC (active thermo-atmosphere combustion) in a two-stroke gasoline engine. *SAE Technical Paper 970363*, 1997.

[9] A. W. Gray and T. W. Ryan. Homogeneous charge compression ignition (HCCI) of Diesel fuel. *SAE Technical Paper 971676*, 1997.

[10] A. Harada, N. Shimazaki, S. Sasaki, T. Miyamoto, H. Akagawa, and K. Tsujimura. The effect of mixture formation on premixed lean Diesel combustion engine. *SAE Technical Paper 980533*, 1998.

[11] A. Hultqvist, M. Christensen, B. Johansson, A. Franke, M. Richter, and M. Aldén. A study of the homogeneous charge compression ignition combustion process by chemiluminescence imaging. *SAE Technical Paper 1999-01-3680*, 1999.

[12] Y. Ishibashi and M. Asai. Improving the exhaust emissions of two-stroke engines by applying the activated radical combustion. *SAE Technical Paper 960742*, 1996.

[13] N. Lida. Combustion analysis of methanol-fueled active thermo-atmosphere combustion (ATAC) engine using a spectroscopic observation. *SAE Technical Paper 940684*, 1994.

[14] P. Najt and D. E. Foster. Compression-ignited homogeneous charge combustion. *SAE Technical Paper 830264*, 1983.

[15] M. Noguchi, Y. Tanaka, T. Tanaka, and Y. Takeuchi. A study on gasoline engine combustion by observation of intermediate reactive products during combustion. *SAE Technical Paper 790840*, 1979.

[16] J.-O. Olsson, O. Erlandsson, and B. Johansson. Experiments and simulations of a six-cylinder homogeneous charge compression ignition (HCCI) engine. *SAE Technical Paper 2000-01-2867*, 2000.

[17] J.-O. Olsson, P. Tunestål, G. Haraldsson, and B. Johansson. A turbo charged dual fuel HCCI engine. *SAE Technical Paper 2001-01-1896*, 2001.

[18] J.-O. Olsson, P. Tunestål, and B. Johansson. Closed-loop control of an HCCI engine. *SAE Technical Paper 2001-01-1031*, 2001.

[19] S. Onishi, S. H. Jo, K. Shoda, P. Do Jo, and S. Kato. Active thermo-atmosphere combustion (ATAC)—A new combustion process for internal combustion engines. *SAE Technical Paper 790501*, 1979.

[20] T. W. Ryan and T. J. Callahan. Homogeneous charge compression ignition of Diesel fuel. *SAE Technical Paper 961160*, 1996.

[21] T. Seko and E. Kuroda. Methanol lean burn in auto ignition DI engine. *SAE Technical Paper 980531*, 1998.

[22] M. Stockinger, H. Schäpertöns, and P. Kuhlmann. Versuche an einem gemischansaugenden mit selbstzündung, *Motortechnische Zeitschrift*, 53, 1992.

[23] H. Suzuki, N. Koike, H. Ishii, and M. Odaka. Exhaust purification of Diesel engines by homogeneous charge with compression ignition. *SAE Technical Paper 970315*, 1997.

[24] H. Suzuki, N. Koike, and M. Odaka. Combustion control method of homogeneous charge Diesel engines. *SAE Technical Paper 980509*, 1998.

[25] R. H. Thring. Homogeneous-charge compression-ignition (HCCI) engines. *SAE Technical Paper 892068*, 1989.

[26] P. Tunestål. The Use of Cylinder Pressure for Estimation of the In-Cylinder Air/Fuel Ratio of an Internal Combustion Engine, Ph.D. thesis, University of California, Berkeley, Dec 2000. Department of Mechanical Engineering, 2000.

15

Approximations of Maximal Controlled Safe Sets for Hybrid Systems

L. Berardi   E. De Santis   M. D. Di Benedetto   G. Pola

Abstract

In the determination of the *"maximal safe set"* for a hybrid system, the core problem lies in the computation of a maximal controlled invariant set contained in a constraint set for a continuous-time dynamical system. In the case of a linear system, we propose a procedure that, on the basis of a controlled invariant set for the exponential discretization of the continuous-time system, leads to an arbitrarily good approximation of the maximal controlled invariant set for the continuous-time system. The approximating set has the interesting property that the constraints can be satisfied by means of a piecewise constant control. An example of an application of the proposed procedure to idle control is illustrated.

15.1 Introduction

In the area of hybrid system control, emphasis has been placed on solving problems with safety specifications, which are described by giving a set of good states within which the controlled hybrid system should evolve. The problem of finding the *"maximal safe set"*, *i.e.*, the set of all initial states guaranteeing that the evolution of the system remains in the good set, can be decomposed into a number of sub-problems, each consisting basically in the computation of a maximal controlled invariant set contained in a set of constraints [5].

For discrete-time linear systems, iterative procedures for the computation of maximal controlled invariant sets are well known in the literature (*e.g.*, [9], [12]). However, these procedures may not converge in a finite number of steps. Inner-approximation algorithms were proposed in [3] that yield tight bounds for an error originating from stopping the procedure after a finite number of steps. For continuous-time linear systems, this problem is still open. A result that could, in principle, be used to determine an approximation of the maximal controlled invariant set for a continuous-time linear system is that if a set S is controlled invariant for the Euler approximating system, then it is also controlled invariant for the continuous-time system [10]. However, if the control law that makes the set S invariant is sampled (for example in digital implementation), there is no guarantee that the trajectory of the continuous-time system remains in S between consecutive sampling times. Hence, we need to resort to a different approximation scheme.

The key idea is to compute a controlled safe set instead of a controlled invariant set. A controlled safe set is a set such that, using an appropriate piecewise constant control, the evolution of the controlled continuous-time system satisfies the constraints at all times. In this paper, we show how the exponential discretization of the continuous time system can be used for the derivation of a controlled safe set. In Section 15.2, we summarize some properties of controlled safe sets. In Section 15.3, we show that, by an appropriate choice of the sampling time, an arbitrarily good approximation of the maximal controlled invariant set can be obtained. Then, we prove that there is no loss of generality in assuming polyhedral state and input constraints, thus simplifying all the numerical calculations, and we give some tools for the computation of controlled safe sets. An example of an application of the proposed procedure to the idle speed control problem is illustrated in Section 15.4. Section 15.5 contains some concluding remarks.

15.2 Definition and Properties of Controlled Safe Sets

Consider a continuous-time linear system:

$$\dot{x}(t) = Ax(t) + Bu(t) \tag{15.1}$$

with state and input constraints

$$x(t) \in \Lambda, \quad u(t) \in U, \quad \forall t \geq 0 \tag{15.2}$$

and a discrete-time linear system:

$$x(k+1) = A_d x(k) + B_d u(k) \tag{15.3}$$

with state and input constraints

$$x(k) \in \Lambda, \quad u(k) \in U, \quad \forall k \in \mathbb{N} \tag{15.4}$$

where $\Lambda \subset \mathbb{R}^n$, $U \subset \mathbb{R}^p$. We denote by $x(t, x_0, u)$ and $x(k, x_0, u)$ the state evolution of the continuous-time system (15.1) at time t and of the discrete-time system (15.3) at time k, respectively, with initial state x_0 and control law $u|_{[0,t)}$ (resp. $u|_{[0,k]}$). We abuse notation somwwhat by writing either $x(t, x_0, u)$ or $x(t, x_0, u(0))$ if $u(t) = u(0)$, $\forall t \geq 0$.

A control law $u(.)$ is said to be feasible with respect to continuous-time system (15.1) if $u(t) \in U$, $\forall t \geq 0$ (or with respect to the discrete-time system (15.3) if $u(k) \in U$, $\forall k \in \mathbb{N}$).

We first give the standard definition of controlled invariant set and the definition of controlled safe set:

DEFINITION 15.1
A set $\Omega \subset \Lambda$ is the controlled invariant for system (15.1) (and (15.3)) if $\forall x_0 \in \Omega$ there exists a feasible control law $u(.)$ such that the evolution $x(t, x_0, u) \in \Omega$, $\forall t \geq 0$ (and $x(k, x_0, u) \in \Omega$, $\forall k \in \mathbb{N}$, respectively) and we say that $u(.)$ makes the set Ω invariant for (15.1) (and (15.3), respectively). A set $\Omega \subset \Lambda$ is a controlled safe set for system (15.1) (and (15.3), respectively) if $\forall x_0 \in \Omega$ there exists a feasible control law $u(.)$ such that the evolution $x(t, x_0, u) \in \Lambda$, $\forall t \geq 0$ (and $x(k, x_0, u) \in \Lambda$, $\forall k \in \mathbb{N}$ respectively), and we say that $u(.)$ makes the set Ω safe for (15.1) (and (15.3), repectively). □

Throughout the paper, we assume that:

- ASSUMPTION 1: Λ and U are C-sets, *i.e.*, convex, compact sets with the origin in their interior;

- ASSUMPTION 2: (A, B) is asymptotically stabilizable

Let I_c be the maximal controlled invariant subset of Λ for the continuous-time system (15.1). By Assumptions 1 and 2, I_c is a C-set. By definition, any controlled safe set $\Omega \subset \Lambda$ is contained in I_c. Since the union of controlled safe subsets of Λ is a controlled safe set, there exists a unique maximal controlled safe subset of Λ. The maximal controlled safe set in Λ coincides with I_c.

Consider now a particular discrete-time system of the form (15.3), the exponential discretization of the continuous-time system (15.1) with sampling time $T > 0$:

$$x(k+1) = A_d(T)x(k) + B_d(T)u(k) \qquad (15.5)$$

where $A_d(T) = e^{AT}$ and $B_d(T) = \int_0^T e^{A(T-t)}B \, dt$. If Assumption 2 holds, then $(A_d(T), B_d(T))$ is asymptotically stabilizable with respect to system (15.5), $\forall T \in \check{\mathbb{R}}$, where $\check{\mathbb{R}} = (0, \infty)$ a.e. (see [13]).

In the following, we give the tools for the computation of a set that is controlled invariant for (15.5), controlled safe for (15.1) and is an arbitrarily good approximation of I_c. Moreover, this set has the property that the constraints can be satisfied by means of an appropriately chosen piecewise-constant control law. More precisely, the class of controls we focus on is

$$U_T = \{u_T(.) : u_T(t) = u(kT), \forall t \in [kT, (k+1)T),$$
$$k \in \mathbb{N}, \text{ and } u(kT) \in U, \forall k \in \mathbb{N}\}$$

DEFINITION 15.2
A set $\Omega \subset \Lambda$ is a T-controlled safe set for the continuous-time system (15.1) if $\forall x_0 \in \Omega$ there exists a feasible piecewise constant control law $u_T(.) \in U_T$ such that the evolution $x(t, x_0, u_T) \in \Lambda$, $\forall t \geq 0$. □

Let $S(T)$ be a controlled invariant C-subset of Λ for the exponential discretization system (15.5). Given $x \in S(T)$, set

$$U(x) = \{u \in U : A_d(T)x + B_d(T)u \in S(T)\} \qquad (15.6)$$

By definition, $\forall x_0 \in S(T)$ there exists a control law $u_T(.) \in U_T$ such that $x(t, x_0, u_T) \in \Lambda, \forall t = kT, k \in \mathbb{N}$. However, since in general $S(T)$ is not controlled invariant for the continuous-time system (15.1), this is not true for $t \neq kT$. We therefore introduce a precise measure of how much the state evolution of the continuous-time system deviates from the constraint.

DEFINITION 15.3
Let $S(T)$ be a controlled invariant C-subset of Λ for system (15.5). The expanding factor $\mu(S(T))$ is the minimum scalar value greater than or equal to one, such that $\forall x_0 \in S(T)$, there exists $u_T(.) \in U_T$ such that the state evolution of system (15.1) $x(t, x_0, u_T) \in \mu(S(T))\Lambda, \forall t \geq 0$, and $x(t, x_0, u_T) \in S(T), \forall t = kT, k \in \mathbb{N}$. □

Given a controlled invariant set for the exponential discretization system (15.5), a T-controlled safe set can be computed for the continuous-time system (15.1) using the following:

PROPOSITION 15.4
Suppose Assumptions 1 and 2 hold and $T \in \check{\mathbb{R}}$. If $S(T)$ is a controlled invariant C-subset of Λ for system (15.5), then the set

$$S(T)/\mu(S(T))$$

is a T-controlled safe set for the continuous-time system (15.1). Moreover, if $u'(.)$ is the control law that makes $S(T)/\mu(S(T))$ invariant for the discrete-time system (15.5), the piecewise constant control $u(t) = u'(kT), \forall t \in [kT, (k+1)T], k \in \mathbb{N}$ makes $S(T)/\mu(S(T))$ safe for the continuous-time system (15.1). □

15.3 Inner Approximations of the Maximal Controlled Invariant Set

By Proposition 15.4, under Assumptions 1 and 2 and $T \in \check{\mathbb{R}}$, if $S(T)$ is a controlled invariant C-subset of Λ for system (15.5), then $S(T)/\mu(S(T))$ is T-controlled safe set for (15.1) and is therefore contained in I_c. This suggests to approximation of I_c with $I(T)/\mu(I(T))$, where $I(T)$ is the maximal controlled invariant subset of Λ for the discrete-time system (15.5) under constraints (15.4).

The next result shows that, by an appropriate choice of the sampling time T, $I(T)/\mu(I(T))$ can be an arbitrarily good approximation of I_c. The following definition will be used.

DEFINITION 15.5
Let $\mathcal{A}$ and $\mathcal{B}$ be two C-sets. Given $\varepsilon > 0$, $\mathcal{B}$ is an ε-approximation of $\mathcal{A}$ if:

$$\mathcal{B} \subset \mathcal{A} \subset (1+\varepsilon)\mathcal{B}$$

□

THEOREM 15.6—[8]
If Assumptions 1 and 2 hold, then $\forall \varepsilon > 0 \; \exists \mathbf{T} > 0$ such that

$$\frac{I(T)}{\mu(I(T))}$$

is an ε-approximation of $I_c, \forall T \in (0, \mathbf{T}] \cap \check{\mathbb{R}}$. □

We now show that there is no loss of generality in assuming that the C-sets Λ and U are polyhedral.

Given the C-sets Λ and U, consider the sets Λ_p and U_p, respectively polyhedral ε-approximations of Λ and U, i.e., for any given $\varepsilon > 0$,

$$\Lambda_p \subset \Lambda \subset (1+\varepsilon)\Lambda_p \qquad (15.7)$$
$$U_p \subset U \subset (1+\varepsilon)U_p$$

Consider the discrete-time system (15.5) with state and input constraints Λ_p and U_p, i.e.,

$$x(k) \in \Lambda_p, \quad u(k) \in U_p, \quad \forall k \in \mathbb{N} \tag{15.8}$$

Let $I_p(T)$ be the maximal controlled invariant set for discrete-time system (15.5) under the constraints of (15.8).

By definition, if (15.7) holds, we have

$$I_p(T) \subset I(T) \subset (1+\varepsilon)I_p(T)$$

In [3], it is shown that under Assumptions 1 and 2 and $T \in \check{\mathbb{R}}$, $I(T)$ and $I_p(T)$ are C-sets. However, $I_p(T)$ is not necessarily polyhedral, but for any given $\varepsilon > 0$ there exists a polyhedral controlled invariant set $\mathcal{P}(T)$, for system (15.5) under constraints (15.8), which satisfies

$$\mathcal{P}(T) \subset I_p(T) \subset (1+\varepsilon)\mathcal{P}(T) \tag{15.9}$$

As a consequence of relation (15.7), the sets $I_p(T)$ and $\mathcal{P}(T)$ are controlled invariant with respect to system (15.5) under the constraints of (15.4). Therefore the sets $I_p(T)/\mu(I_p(T))$ and $\mathcal{P}(T)/\mu(\mathcal{P}(T))$ are T-controlled safe sets for continuous-time system (15.1) under constraints (15.2). Moreover, the following holds:

THEOREM 15.7—[4]
Suppose Assumptions 1 and 2 hold and $T \in \check{\mathbb{R}}$. $\forall \delta > 0$, $\exists \varepsilon > 0$, such that, if $\mathcal{P}(T)$ is a polyhedral controlled invariant set for the discrete-time system (15.5) satisfying

$$\mathcal{P}(T) \subset I(T) \subset (1+\varepsilon)\mathcal{P}(T) \tag{15.10}$$

then

$$\mu(\mathcal{P}(T)) - \delta \leq \mu(I(T)) \leq \mu(\mathcal{P}(T)) + \delta \tag{15.11}$$

and

$$\frac{\mathcal{P}(T)}{\mu(\mathcal{P}(T)) + \delta} \subset \frac{I(T)}{\mu(I(T))} \subset (1+\varepsilon)\frac{\mathcal{P}(T)}{\mu(\mathcal{P}(T)) - \delta} \tag{15.12}$$

□

As a consequence of Theorem 15.7, we can assume hereafter that

- ASSUMPTION 3: $\Lambda = \Lambda_p$ and $U = U_p$

In the following subsection a procedure is presented in order to compute a polyhedral ε-approximation of $I(T)$ which, under Assumption 3, coincides with $I_p(T)$. Some tools are then given for the computation of the expanding factor.

Computation of a Polytopic Approximating Set of $I(T)$

We illustrate here some results, established in [3], which give the means for the computation of an ε-approximation of $I(T)$.

PROPOSITION 15.8—[3]
Suppose Assumptions 1 and 2 hold and $T \in \check{\mathbb{R}}$. Define the sequence of sets $\{\Lambda_i, \ i = 0, 1, 2 \ldots\}$

$$\Lambda_i = \Lambda \quad i = 0 \tag{15.13}$$
$$\Lambda_i = \{x : \exists u \in U : A_d(T)x + B_d(T)u \subset \Lambda_{i-1}\} \cap \Lambda, \quad i > 0$$

This sequence converges, in general asymptotically, to the set $I(T)$. $\forall \varepsilon > 0$ $\exists \mathbf{i}$ such that $I(T)$ is an ε-approximation of Λ_i, $\forall i \geq \mathbf{i}$. □

The recursion (15.13) is not guaranteed to terminate in a finite number of steps. Each of the sets Λ_i contains the set $I(T)$ and hence can be viewed as an *outer approximation* of $I(T)$. Obviously, this approximation becomes better and better as i increases. However, the sets Λ_i are not controlled invariant, and hence they are not useful from a control synthesis point of view.

We then resorted in [3] to give algorithms for the computation of *inner controlled invariant approximations* of the set $I(T)$, as illustrated below.

Suppose Assumptions 1, 2 and 3 hold and $T \in \check{\mathbb{R}}$. Let S be a controlled invariant C-subset of Λ for system (15.5). Define the sequence

$$\{\Omega^i(S), \quad i = 0, 1, 2 \ldots\}$$

where

$\Omega_0(S) = S$

$$\Omega_i(S) = \left\{ \begin{array}{l} x : x_0 = x, x(i, x_0, u) \in S, \ x(k, x_0, u) \in \Lambda, \ \forall k \ \ 0 \leq k \leq i \\ \text{for some } u = \begin{pmatrix} u(0) \\ u(1) \\ \vdots \\ u(i-1) \end{pmatrix}, \text{ where } u(j) \in U, 0 \leq j \leq i-1 \end{array} \right\}$$

The sets $\Omega_i(S)$ are controlled invariant $\forall i \geq 0$ for system (15.5). Moreover, they can be computed by means of the following procedure:

$$\Omega_i(S) = S \quad i = 0 \tag{15.14}$$
$$\Omega_i(S) = \{x : \exists u \in U : A_d(T)x + B_d(T)u \subset \Omega_{i-1}(S)\} \cap \Lambda \quad i > 0$$

The following procedure computes an ε-approximation of $I(T)$.

PROCEDURE 15.9—[3]
Suppose Assumptions 1, 2 and 3 hold, and $T \in \check{R}$. Let $\varepsilon > 0$ be given.

1. Compute a controlled invariant C-set $S \subset \Lambda$ for system (15.5) (as shown e.g., in [11] and [6]).
2. $i = 0$;
repeat $i = i + 1$; $\mathcal{P}(T) = \Omega_i(S)$ until $\mathcal{P}(T)$ is an ε-approximation of Λ_i, and hence an ε-approximation of $I(T)$.

□

THEOREM 15.10—[3]
Suppose Assumptions 1, 2 and 3 hold, and $T \in \check{\mathbb{R}}$. For any given $\varepsilon > 0$, Procedure 15.9 terminates in a finite number of steps. The computed set $\mathcal{P}(T)$ is controlled invariant for system (15.5) and is an $\varepsilon-$approximation of $I(T)$.

□

Computation of the Expanding Factor

We summarize here some results from [4] which establish some properties of the expanding factor and give the means for its computation.

Given a C-set Ω and $x \in \mathbb{R}^n$, consider the so-called Minkowski functional [7]:

$$\Psi_\Omega(x) = \inf\{\mu \in \mathbb{R}, \mu \geq 0 : x \in \mu\Omega\}$$

From this definition, Ω can be thought as a unit ball and $x \in \Omega$ if and only if $\Psi_\Omega(x) \leq 1$. The Minkowski functional satisfies the following properties:

$$\forall x, y \in \mathbb{R}^n \qquad (15.15)$$
$$\Psi_\Omega(x) \geq 0$$
$$\Psi_\Omega(\lambda x) = \lambda \Psi_\Omega(x), \quad \forall \lambda \geq 0$$
$$\Psi_\Omega(x + y) \leq \Psi_\Omega(x) + \Psi_\Omega(y).$$

The following Theorem holds.

THEOREM 15.11—[4]
Suppose Assumptions 1 and 2 hold, and $T \in \check{\mathbb{R}}$. Let $S(T)$ be a controlled invariant C-subset of Λ for system (15.5), then

$$\mu(S(T)) = \max\left\{1, \max_{x_0 \in S(T)} \min_{u_T(0) \in U(x_0)} \max_{t \in [0,T]} \Psi_\Lambda(x(t, x_0, u_T(0)))\right\}$$

where $u_T(\cdot) \in U_T$, $u_T(t) = u_T(0)$, $\forall t \geq 0$, and

$$U(x) = \{u \in U : A_d(T)x + B_d(T)u \in S(T)\}$$

□

The expanding factor can be rendered arbitrarily close to 1 by appropriately choosing the sampling time T, as stated below.

THEOREM 15.12—[4]
Suppose Assumptions 1 and 2 hold and $T \in \check{\mathbb{R}}$. $\forall \varepsilon > 0$, $\exists \widehat{T} > 0$, such that if $S(T)$ is a controlled invariant C-subset of Λ for (15.5), then

$$\mu(S(T)) \leq 1 + \varepsilon, \forall T \leq \widehat{T}.$$

$\square$

The following theorem specializes the result of Theorem 15.11 to the polytopic controlled invariant set $\mathcal{P}(T)$. In particular, last part of the proof of Theorem 15.13 illustrates a step-by-step procedure for the calculation of the expanding factor of $\mathcal{P}(T)$.

THEOREM 15.13—[4]
Suppose Assumptions 1, 2 and 3 hold and $T \in \check{\mathbb{R}}$. Write

$$\mathcal{P}(T) = Conv\{v_1, v_2, ..., v_q\}$$

where $v_1, v_2, ..., v_q$ are the vertices of $\mathcal{P}(T)$. Let the convex sets $U(v_i)$ be represented by

$$U(v_i) = Conv\left\{u_i^1, u_i^2, ..., u_i^{N_i}\right\}, i = 1, 2, ..., q$$

Then

$$\mu(\mathcal{P}(T)) = \max\left\{1, \max_{i=1,2,...,q} \min_{k=1,2,...,N_i} \max_{t \in [0,T]} \Psi_\Lambda(x(t, v_i, u_i^k))\right\}$$

where $u_i^k \in U_T$ and $u_i^k(t) = u_i^k(0)$, $\forall t \geq 0$.

Proof. First we show that

$$\mu(\mathcal{P}(T)) = \max\left\{1, \max_{i=1,2,...,q} \min_{u_T(0) \in U(v_i)} \max_{t \in [0,T]} \Psi_\Lambda(x(t, v_i, u_T(0)))\right\} \quad (15.16)$$

where $u_T \in U_T$ and $u_T(t) = u_T(0)$, $\forall t \geq 0$.
For each $x_0 \in \mathcal{P}(T)$, consider $x_0 = \sum_{i=1}^q \lambda_i v_i$, with $\lambda_i \geq 0, \forall i = 1, 2, ..., q$ and $\sum_{i=1}^q \lambda_i = 1$. In [4], it is shown that

$$\sum_{i=1}^q \lambda_i U(v_i) \subset U(x_0) \quad (15.17)$$

Then, the following chain of inequalities can be obtained using relation (15.17), the Minkowski functional properties (15.15), the operator max and the definition of the coefficients λ_i:

$$\min_{u_T(0)\in U(x_0)} \left(\max_{t\in[0,T]} \Psi_\Lambda(x(t,x_0,u_T(0)))\right) \tag{15.18}$$

$$\leq \min_{u_T(0)\in\sum_{i=1}^q \lambda_i U(v_i)} \left(\max_{t\in[0,T]} \Psi_\Lambda(x(t,x_0,u_T(0)))\right)$$

$$= \min_{u_T(0)=\sum_{i=1}^q \lambda_i u_i \in \sum_{i=1}^q \lambda_i U(v_i)} \left(\max_{t\in[0,T]} \Psi_\Lambda\left(\sum_{i=1}^q \lambda_i x(t,v_i,u_i)\right)\right)$$

$$\leq \min_{u_T(0)=\sum_{i=1}^q \lambda_i u_i \in \sum_{i=1}^q \lambda_i U(v_i)} \left(\sum_{i=1}^q \lambda_i \max_{t\in[0,T]} \left(\Psi_\Lambda\left(x(t,v_i,u_i)\right)\right)\right)$$

$$= \sum_{i=1}^q \lambda_i \left(\min_{u_i\in U(v_i)} \left(\max_{t\in[0,T]} \left(\Psi_\Lambda\left(x(t,v_i,u_i)\right)\right)\right)\right)$$

$$\leq \max_{i=1,2,...,q} \left(\min_{u_i\in U(v_i)} \left(\max_{t\in[0,T]} \left(\Psi_\Lambda\left(x(t,v_i,u_i)\right)\right)\right)\right) \sum_{i=1}^q \lambda_i$$

$$= \max_{i=1,2,...,q} \min_{u_T(0)\in U(v_i)} \max_{t\in[0,T]} \Psi_\Lambda(x(t,v_i,u_T(0)))$$

where $u_i \in U(v_i)$ $\forall i = 1, 2, ..., q$. Therefore

$$\max_{x_0\in\mathcal{P}(T)} \min_{u_T(0)\in U(x_0)} \left(\max_{t\in[0,T]} \Psi_\Lambda(x(t,x_0,u_T(0)))\right) \tag{15.19}$$

$$\leq \max_{i=1,2,...,q} \min_{u_T(0)\in U(v_i)} \max_{t\in[0,T]} \Psi_\Lambda(x(t,v_i,u_T(0)))$$

On the other hand, by definition,

$$\max_{x_0\in\mathcal{P}(T)} \left(\min_{u\in U(x_0)} \left(\max_{t\in[0,T]} \Psi_\Lambda(x(t,x_0,u)))\right)\right) \tag{15.20}$$

$$\geq \max_{x_0\in\{v_1,v_2,...,v_q\}} \min_{u_T(0)\in U(x_0)} \max_{t\in[0,T]} \Psi_\Lambda(x(t,x_0,u_T(0)))$$

Combining (15.20) with (15.19), (15.16) holds. We now show that the minimum in Relation (15.16) can be computed with respect to the vertices of the polytopes $U(v_i)$.

Given some vector v and some matrix M, the symbols v_i and M_i denote, respectively, the i-th component of the vector v and the i-th row of the matrix M. Let the external representation of the polytope Λ be as follows:

$$\Lambda = \{x \in R^n : Lx \leq l\}$$

where $L \in \mathbb{R}^{s\times n}$, $l \in \mathbb{R}^s$, and $l_i > 0$, $\forall i = 1, 2, ..., s$, since Λ is a polyhedral C-set.

For a fixed v_i and $u \in U(v_i)$, set

$$\eta(v_i, u) = \max_{t \in [0,T]} \Psi_\Lambda(x(t, v_i, u))$$

It is easy to show that

$$\eta(v_i, u) = \max_{k=1,\dots,s} \theta_k(v_i, u)$$

where

$$\theta_k(v_i, u) = \max_{t \in [0,T]} \Phi_k(t, v_i, u), \text{ with}$$

$$\Phi_i(t, v_i, u) = \frac{L_k}{l_k}(e^{At} v_i + e^{At} F(t) B u) \text{ where } F(t) = \int_0^t e^{-A\alpha} d\alpha$$

This optimization problem is solvable by numerical methods, since it is the search of the maximum of a scalar function on a scalar compact domain. Let $t_i^* \in [0,T]$ be the time at which $\Phi_k(t, v_i, u)$ attains the maximum and let $\theta_j(v_i, u)$ be the maximal value among $\theta_k(v_i, u)$, $k = 1, \dots, s$.
Then,

$$\eta(v_i, u) = \frac{L_j}{l_j} e^{At_j^*} v_i + \frac{L_j}{l_j} e^{At_j^*} F(t_j^*) B u$$

We have now to find the minimum with respect to $u \in U(v_i)$, i.e.:

$$\min_{u \in U(v_i)} \eta(v_i, u) = \frac{L_j}{l_j} e^{At_j^*} v_i + \min_{u \in U(v_i)} \frac{L_j}{l_j} e^{At_j^*} F(t_j^*) B u$$

Since $U(v_i)$ is a polytope, the above optimization is a constrained optimization of a linear function on a polytopic domain. From Weierstrass Theorem this minimum exists, and from linear programming theory it is on one of the vertexes of $U(v_i)$. Hence we have

$$\mu(\mathcal{P}(T)) = \max_{i=1,2,\dots,q} \min_{u_T(0) \in U(x_0)} \max_{t \in [0,T]} \Psi_\Lambda(x(t, v_i, u_T(0)))$$

$$= \max_{i=1,2,\dots,q} \min_{k=1,2,\dots,N_i} \max_{t \in [0,T]} \Psi_\Lambda(x(t, v_i, u_i^k))$$

where $u_i^k \in U(v_i) = Conv\left\{u_i^1, u_i^2, \dots, u_i^{N_i}\right\}$. □

15.4 An Example of Application

The problem of maintaining the crankshaft speed within a given range was formalized as a hybrid system control problem with safety specifications in [1] (see also [2], where the synthesis of an idle control strategy is based on the assume-guarantee paradigm). We consider here a simplified model of the power train dynamics with the objective of illustrating the procedures presented in Section 15.3. Consider the system:

$$\begin{cases} \dot{p} = a_m np + b_m \alpha \\ \dot{n} = a_n n + b_n T \end{cases} \quad (15.21)$$

where n is the engine speed expressed in rpm (revolutions per minute), p is the manifold pressure expressed in mbar, a_m, b_m, a_n, b_n, are constants, $T = k_1 \eta (AV) p$ is the torque produced by the engine, given the efficiency function $\eta(AV)$, the spark advance angle AV, and the constant k_1. We use the following numerical values:

$$a_n = -1.5308, \quad b_n = 95.4930$$
$$a_m = 0.0262, \quad b_m = 1.8212e3$$

The two control inputs are α, the throttle opening angle, and AV, the spark advance angle.

The idle-speed control problem consists of finding under what conditions it is possible to maintain the engine speed into the desired range 800 ± 30 rpm, with the following constraints on the control inputs:

$$0° \leq \alpha \leq 20°, \quad 0° \leq AV \leq 20°$$

In order to solve the problem, we need to linearize (15.21) and then compute the exponential discretization. The linearization is made around the operating point $n = 800$ rpm, $p = 300$ mBar, $\alpha = 3.45$ degrees, $AV = 5.75$ degrees. We then consider the exponential discretization of the resulting continuous-time linear system with $T = 0.05$.

Using the techniques presented in [3], the maximal controlled invariant set $I(T)$ for the discrete-time system (15.5) is found; the set $I(T)$ is a polytope and its external representation is given by

$$\begin{bmatrix} 1.0000 & 0 \\ -1.0000 & 0 \\ 0 & -1.0000 \\ 0.9920 & 0.0193 \\ 0.9833 & 0.0365 \\ 0.9741 & 0.0519 \\ 0.9644 & 0.0655 \\ 0.9543 & 0.0776 \\ 0.9438 & 0.0882 \\ 0.9329 & 0.0976 \\ 0.9218 & 0.1059 \\ 0.9105 & 0.1131 \\ 0.8991 & 0.1194 \end{bmatrix} \begin{bmatrix} x_1 \\ x_2 \end{bmatrix} \leq \begin{bmatrix} 30 \\ 30 \\ 300 \\ 31.1430 \\ 32.8510 \\ 35.0615 \\ 37.7175 \\ 40.7679 \\ 44.1660 \\ 47.8698 \\ 51.8411 \\ 56.0454 \\ 60.4518 \end{bmatrix}$$

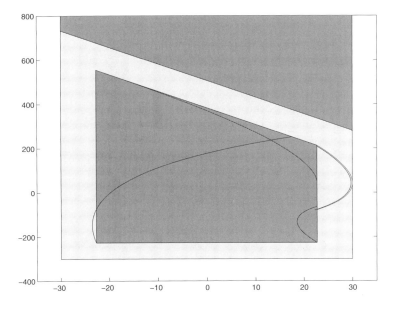

Figure 15.1 The controlled safe-set

where

$$x_1 = n - 800$$
$$x_2 = p - 300$$

Since the state and input constraints are polyhedral and $I(T)$ is a polytope, we can directly apply the techniques presented in Theorem 15.13, obtaining:

$$\mu(I(T)) = 1.3220$$

The controlled safe set is:

$$I(T)/\mu(I(T)) = I(T)/1.3220$$

In Figure 15.1, the outer polytope represents the state constraining set Λ; the inner polytope is $I(T)/\mu(I(T))$ and the polytope in between the other two is $I(T)$. Some trajectories starting from the vertices of $I(T)/\mu(I(T))$ and using some piecewise control law $u \in U(v_i), i = 1, 2, ..., q$ are also shown. It can be seen that such trajectories do not violate the constraints on the state values. In Figure 15.2, the values of $\mu(I(T))$ are plotted for $T \in [0.05, 0.45]$. For the problem at hand, we obtain the smallest expanding factor $\mu(I(T)) = 1.0154$ with a sampling period of $T = 0.003$; smaller sampling times give rise to numerical problems.

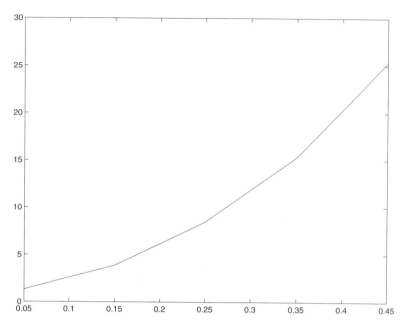

Figure 15.2 convergence to 1

15.5 Conclusions

In this paper, we have proposed approximating the maximal controlled invariant set contained in a set of constraints for a continuous-time system with a T-controlled safe set. We have shown that a T-controlled safe set can be obtained by scaling a controlled invariant set for the exponential discretization system by the expanding factor. Moreover, we have also precisely determined how far from the maximal controlled invariant set our approximating set is. An application to the idle speed control problem has been illustrated.

15.6 References

[1] A. Balluchi, L. Benvenuti, M. D. Di Benedetto, C. Pinello, and A. Sangiovanni-Vincentelli. Automotive engine control and hybrid systems: Challenges and opportunities, *Proceedings of the IEEE*, 88, 888–912, Special Issue on Hybrid Systems, Invited Paper, 2000.

[2] A. Balluchi, L. Benvenuti, M. D. Di Benedetto, and A. Sangiovanni-Vincentelli. Idle speed controller synthesis using an assume-guarantee approach, in This Volume, 2001.

[3] L. Berardi, E. De Santis, and M. D. Di Benedetto. *Controller syntesis for hybrid systems: structural procedures and approximations of robustly controlled invariant sets*, Tech. Rep. Research Report, no. R.00-57, Department of Electrical Engineering, University of L'Aquila, 2001.

[4] L. Berardi, E. De Santis, M. D. Di Benedetto, and G. Pola. Controlled safe sets for continuous time systems, in *Proc. European Control Conference*, 803–808, Porto, Portugal, 2001.

[5] L. Berardi, E. De Santis, and M. D. Di Benedetto. Invariant sets and control synthesis for switching systems with safety specifications, in *Hybrid Systems: Computation and Control*, N. Lynch and B. Krogh, eds., Vol. 1790 of Lecture Notes in Computer Science, Springer–Verlag, 2000.

[6] F. Blanchini. *Constrained control for systems with unknown disturbances*, in *Control and Dynamic Systems*, C. T. Leondes, ed., 51, Academic Press, 1992.

[7] F. Blanchini. Nonquadratic lyapunov functions for robust control, *Automatica*, 31, 451–461, 1995.

[8] E. De Santis, M. D. Di Benedetto, and G. Pola. Inner approximations of domains of attraction for constrained continuous time linear systems, Research Report R.01-63, Department of Electrical Engineering, University of L'Aquila, 2001. (submitted for publication).

[9] C. E. T. Dorea and J. C. Hennet. Computation of maximal admissible sets of constrained linear systems, *Proc. of 4th IEEE Med. Symposium*, 286–291, Krete (Greece), 1996.

[10] F. Blanchini F. and S. Miani. Constrained stabilization for continuous-time systems, *Systems and Control Letters*, 28, 95–102, 1996.

[11] L. Farina and L. Benvenuti. Invariant polytopes of linear systems, *IMA J. Math. Control and Information*, 15, 233–240, 1998.

[12] P. Gutman and M. Cwikel, *Admissible sets and feedback control for discrete-time linear dynamical systems with bounded controls and states*, IEEE Trans. Automatic Control, AC-31, 373–376, 1986.

[13] M. Kimura, Preservation of stabilizability of a continuous time-invariant linear system after discretization, *Int. Journal System Science*, 21, 65–91, 1990.

16

Hamiltonian Formulation of Bond Graphs

G. Golo   A. van der Schaft   P. C. Breedveld   B. M. Maschke

Abstract

This paper deals with the mathematical formulation of bond graphs. It is proven that the power continuous part of bond graphs, the junction structure, can be associated with a Dirac structure and that the equations describing a bond graph model correspond to a port Hamiltonian system. The conditions for well-posedness of the modelled system are given, and representations suitable for numerical simulation are derived. The index of the representations is analysed and sufficient conditions for computational efficiency are given. The results are applied to some models arising in automotive applications.

16.1 Introduction

In most of the current modelling and simulation approaches of physical systems, some sort of *network representation* is used to make it clear that the physical system model under consideration is an interrelated set of elementary concepts. This way of modelling has several advantages. From a physical point of view, it is usually natural to regard the system as composed of functional components, possibly from different domains (mechanical, electrical, and so on). The knowledge about subsystems can be stored in libraries, and is reusable for later occasions. Due to this modularity, the modelling process can be performed in an *iterative* way, gradually refining if necessary the model by adding other subsystems. Furthermore, the approach is suited to general control design, where the overall behaviour of the system is sought to be improved by the addition of the other subsystems or controlling devices. From a systematic-theoretic point of view this modular approach naturally emphasises the need for the models of systems with external variables. In this paper we concentrate on the mathematical description of a network representation of *energy-conserving physical systems* called bond graphs. The bond graph approach is graphically oriented and its outcome, the bond graph model, represents a multiport system based on energy flows [1], [2], [3]. An important feature of this technique is that the cumbersome job of writing down the equations that describe a physical system is replaced by drawing a picture of it, using a finite set of symbols. Once the bond graph is obtained, we can start with its analysis and with the derivation of the equations suitable for numerical simulation. These tasks can be fully automated [4], [5].

The purpose of this paper is to develop a theory of bond graphs in a mathematical language. The mathematical results that have been obtained so far are sporadic. The approaches reported in [6], [7], [8], [9], [10], [11], [12] represent combinatorial theory of bond graphs. Here, a geometric approach is developed, elaborating on previous work on the structure of such dynamical systems [13], [14], [15], [16]. Firstly, we consider the power continuous part of a bond graph model, called the *junction structure*, and prove that every junction structure can be associated to the geometric notion of a Dirac structure [17], [18], [19], which represents a general power conserving interconnection structure of a physical system. Secondly, we prove that equations describing a bond graph model correspond to a port Hamiltonian system. Also, the condition for well-posedness of the modelled system is given and representations suitable for the numerical simulation are derived. The index of representations is analysed and sufficient conditions for the representations to be computationally efficient are given. This theory is applied to some models arising in automotive applications.

16.2 Bond Graph Models

A generic bond graph is shown in Figure 16.1. The bond graph formalism

Hamiltonian Formulation of Bond Graphs

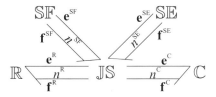

Figure 16.1 Generic bond graph

used here is based on the classification of physical variables rooted in thermodynamics. This leads to the generalised bond graph formalism that admits only one type of storage element [2], [3]. Here, $\mathbb{C}$ stands for the collection of C-type of multiports (energy storage multiport), $\mathbb{R}$ for the collection of R-type of multiports (energy dissipative element), $\mathbb{SE}$ for the collection of SE-type of ports and $\mathbb{SF}$ for the collection of SF-type of ports. These elements are called *power discontinuous elements* [2], [3]. JS stands for the junction structure. JS is connected to power discontinuous elements by multibonds. With every multibond we can associate two power variables: effort $\mathbf{e}^\alpha$ and flow $\mathbf{f}^\alpha$, $\alpha \in \{\text{SE}, \text{SF}, \text{C}, \text{R}\}$. They belong to dual spaces and their duality product $\langle \mathbf{e}^\alpha | \mathbf{f}^\alpha \rangle$ represents the power that goes from (to) the junction structure (depending on the positive orientation indicated by half-arrow) to (from) the α-type of multiport. The integer n^α inside a multibond denotes the dimension of the vectors $\mathbf{e}^\alpha$ and $\mathbf{f}^\alpha$. If $n^\alpha = 1$ then a multibond is simply called bond. The constitutive relations for the $\mathbb{C}$ element are given by

$$\mathbf{x} = \mathbf{f}^C \tag{16.1a}$$

$$\mathbf{e}^C = \frac{\partial H}{\partial \mathbf{x}}(\mathbf{x}) \tag{16.1b}$$

where $\mathbf{x} \in \mathcal{X}$ is the vector of energy variables (such as displacement, momentum, charge, flux, volume, etc.), $H \in C^\infty(\mathcal{X})$ is energy of the system and $\mathcal{X}$ is an n^C-dimensional smooth manifold.

The constitutive relation for the $\mathbb{R}$ element is given by

$$\Phi\left(\mathbf{e}^R, \mathbf{f}^R, \mathbf{x}\right) = 0 \tag{16.2}$$

where (16.2) defines the n^R-dimensional set $\forall \mathbf{x} \in \mathcal{X}$. The set has the properties that every pair belonging to it satisfies the inequality $\left(\mathbf{e}^R\right)^T \mathbf{f}^R \geq 0$. Furthermore, the constitutive relations of $\mathbb{SE}$, $\mathbb{SF}$ elements are described by

$$\mathbf{e}^{SE} = \mathbf{u}^{SE} \in \mathcal{R}^{n^{SE}}, \tag{16.3a}$$

$$\mathbf{e}^{SF} = \mathbf{u}^{SF} \in \mathcal{R}^{n^{SF}} \tag{16.3b}$$

As said before, the junction structure is a power continuous part of a bond graph. It means that it can not accumulate, dissipate, or generate power.

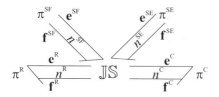

Figure 16.2 Junction structure

It only distributes the power between the power discontinuous elements in a power continuous way. A generic junction structure is shown in Figure 16.2. Here, π^α stands for the α-type of multiport (port if $n^\alpha = 1$). The ports represent the connections of junction structure to the power discontinuous elements. The main feature of the junction structure is power continuity, i.e., a zero power balance at its ports, that is

$$\left\langle \mathbf{e}^{\mathrm{SE}} | -\mathbf{f}^{\mathrm{SE}} \right\rangle + \left\langle -\mathbf{e}^{\mathrm{SF}} | \mathbf{f}^{\mathrm{SF}} \right\rangle + \left\langle \mathbf{e}^{\mathrm{C}} | \mathbf{f}^{\mathrm{C}} \right\rangle + \left\langle \mathbf{e}^{\mathrm{R}} | \mathbf{f}^{\mathrm{R}} \right\rangle = 0.$$

Therefore junction structure relates the flows $\mathbf{f} = (-\mathbf{f}^{\mathrm{SE}}, \mathbf{f}^{\mathrm{SF}}, \mathbf{f}^{\mathrm{C}}, \mathbf{f}^{\mathrm{R}})$ and efforts $\mathbf{e} = (\mathbf{e}^{\mathrm{SE}}, -\mathbf{e}^{\mathrm{SF}}, \mathbf{e}^{\mathrm{C}}, \mathbf{e}^{\mathrm{R}})$ of the ports to each other. The nature of these relations is examined in section 4.

16.3 Dirac Structures

In this section we recall the definition of a Dirac structure as a general representation of a power conserving interconnection structure of a physical system [18], [19].

We start with the space of power variables $(T_\mathbf{x} X \times \mathcal{V}) \times (T_\mathbf{x}^* X \times \mathcal{V}^*)$, for some finite dimensional linear space $\mathcal{V}$ and the smooth finite dimensional manifold X, with the power defined by

$$P = \langle \mathbf{e} | \mathbf{f} \rangle, \quad (\mathbf{e}, \mathbf{f}) \in (T_\mathbf{x} X \times \mathcal{V}) \times (T_\mathbf{x}^* X \times \mathcal{V}^*)$$

where $\langle \mathbf{e} | \mathbf{f} \rangle$ denotes the duality product. We call $T_\mathbf{x} X \times \mathcal{V}$ the space of flows $\mathbf{f}$, and the dual space $T_\mathbf{x}^* X \times \mathcal{V}^*$ the space of efforts. Closely related to the definition of power there exists a canonically defined bilinear form $\langle\!\langle , \rangle\!\rangle$ on the space of power variables $(T_\mathbf{x} X \times \mathcal{V}) \times (T_\mathbf{x}^* X \times \mathcal{V}^*)$, defined as

$$\langle\!\langle (\mathbf{f}^a, \mathbf{e}^a), (\mathbf{f}^b, \mathbf{e}^b) \rangle\!\rangle := \langle \mathbf{e}^a | \mathbf{f}^b \rangle + \langle \mathbf{e}^b | \mathbf{f}^a \rangle.$$

DEFINITION 16.1—DIRAC STRUCTURE [19]

A Dirac structure on a differentiable manifold $X \times \mathcal{V}$ is given by a smooth vector subbundle $\mathcal{D} \subset (TX \times \mathcal{V}) \times (T^*X \times \mathcal{V}^*)$ such that the linear space $\mathcal{D}(\mathbf{x}) \subset (T_\mathbf{x} X \times \mathcal{V}) \times (T_\mathbf{x}^* X \times \mathcal{V}^*)$ satisfies the relation

$$\mathcal{D}^\perp(\mathbf{x}) = \mathcal{D}(\mathbf{x}),$$

Hamiltonian Formulation of Bond Graphs

where $\perp$ denotes orthogonal compliment with respect to the bilinear form $\langle\langle,\rangle\rangle$. □

Locally about every point $(\mathbf{x}, \mathbf{v}) \in X \times V$, we may find $n \times n$ (n is the dimension of the manifold $X \times V$) matrices $\mathbf{E}(\mathbf{x})$ and $\mathbf{F}(\mathbf{x})$ depending smoothly on $\mathbf{x}$, such that locally [19], [18]

$$\mathcal{D}(\mathbf{x}) = \{(\mathbf{f}, \mathbf{e}) \in (T_\mathbf{x} X \times V) \times (T_\mathbf{x}^* X \times V^*)\}$$

Here, the matrices $\mathbf{F}(\mathbf{x}), \mathbf{E}(\mathbf{x})$ satisfy the following two conditions

rank condition: $\text{rank}\,[\mathbf{F}(\mathbf{x})\,\mathbf{E}(\mathbf{x})] = n$;

power-conservation condition: $\mathbf{E}(\mathbf{x})\mathbf{F}^\mathrm{T}(\mathbf{x}) + \mathbf{F}(\mathbf{x})\mathbf{E}^\mathrm{T}(\mathbf{x}) = 0$

This representation is called *kernel representation*. Closely related to the kernel representation there exists an image representation defined by [19]

$$\mathcal{D}(\mathbf{x}) = \mathrm{Im}\left(\begin{bmatrix} \mathbf{E}^\mathrm{T}(\mathbf{x}) \\ \mathbf{F}^\mathrm{T}(\mathbf{x}) \end{bmatrix}\right) = \{(\mathbf{f}, \mathbf{e}) : \mathbf{f} = \mathbf{E}^\mathrm{T}(\mathbf{x})\lambda,\, \mathbf{e} = \mathbf{F}^\mathrm{T}(\mathbf{x})\lambda,\, \lambda \in \mathcal{R}^n\}$$

16.4 Geometric Formulation of a Bond Graphs

As said in the section 2, junction structure relates efforts and flows to each other. Suppose that this relation at the point $\mathbf{x}$ is given by

$$\mathbf{J}(\mathbf{e}, \mathbf{f}, \mathbf{x}) = 0.$$

Define the space $\mathcal{D}(\mathbf{x})$ as

$$\mathcal{D}(\mathbf{x}) = \{(\mathbf{e}, \mathbf{f}) : \mathbf{J}(\mathbf{e}, \mathbf{f}, \mathbf{x}) = 0.\}$$

PROPOSITION 16.1
$\mathcal{D}(\mathbf{x})$ is a Dirac structure on $X \times V$, where X is an n^C-dimensional smooth manifold and $V = \mathcal{R}^{n^\mathrm{SE}+n^\mathrm{SF}+n^\mathrm{R}}$. □

Proof. A junction structure is composed of power continuous elements: junctions (1-junctions, 0-junctions) and transducers (transformers and gyrators). First, we prove that any power continuous element represents a Dirac structure by itself. The constitutive relations of a 1-junction (see Figure 16.3) are given by

$$-\sum_{i=1}^{j} e_i + \sum_{i=j+1}^{k} e_i = 0, \quad f_1 = \cdots = f_j = \cdots = f_k \qquad (16.4\mathrm{a})$$

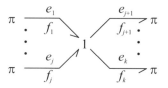

Figure 16.3   1-junction

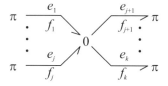

Figure 16.4   0-junction

Multiplying (16.4) by f_1 and taking into account (16.4), one proves that the total incoming power

$$\sum_{i=1}^{j} e_i f_i$$

is equal to the total outgoing power

$$\sum_{i=j+1}^{k} e_i f_i, \quad \text{(power-conservation condition)}$$

Furthermore k power variables, say $e_1, f_2, ..., f_k$, can be expressed as a function of the other k power variables $f_1, e_2, ..., e_k$ (rank condition). Therefore, the set of all power variables satisfying (16.4), (16.4) represents a Dirac structure on $\mathcal{R}^k$.

The constitutive relation of a 0-junction (see Figure 16.4) are given by

$$-\sum_{i=1}^{j} f_i + \sum_{i=j+1}^{k} f_i = 0,, \quad e_1 = \cdots = e_j = \cdots = e_k$$

It can be similarly proved that the set of all power variables satisfying the constitutive relations of the 0-junction represents a Dirac structure on $\mathcal{R}^k$. A transformer, having n_1 incoming ports and n_2 outgoing ports, is shown in Figure 16.5. The constitutive relations of the transformer are given by (kernel representation)

$$\begin{bmatrix} \mathbf{R}_1(\mathbf{x}) & \mathbf{R}_2(\mathbf{x}) \end{bmatrix} \begin{bmatrix} \mathbf{f}_1 \\ \mathbf{f}_2 \end{bmatrix} = 0, \tag{16.5a}$$

$$\left(\begin{bmatrix} \mathbf{R}_1(\mathbf{x}) & \mathbf{R}_2(\mathbf{x}) \end{bmatrix}^\perp \right)^T \begin{bmatrix} -\mathbf{e}_1 \\ \mathbf{e}_2 \end{bmatrix} = 0 \tag{16.5b}$$

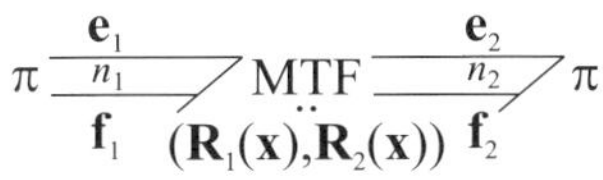

Hamiltonian Formulation of Bond Graphs

$$\pi \xrightarrow[f_1]{e_1}{n_1} \text{MTF} \xrightarrow[f_2]{e_2}{n_2} \pi$$
$$(\mathbf{R}_1(\mathbf{x}), \mathbf{R}_2(\mathbf{x}))$$

Figure 16.5 Transformer

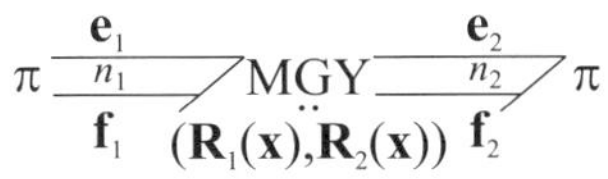

$$\pi \xrightarrow[f_1]{e_1}{n_1} \text{MGY} \xrightarrow[f_2]{e_2}{n_2} \pi$$
$$(\mathbf{R}_1(\mathbf{x}), \mathbf{R}_2(\mathbf{x}))$$

Figure 16.6 Gyrator

where $\mathbf{A}^\perp$ is the maximal rank matrix such that $\mathbf{A}\mathbf{A}^\perp = 0$. The elements of the matrices $\mathbf{R}_1(\mathbf{x}), \mathbf{R}_2(\mathbf{x})$ are smooth functions on the manifold $\mathcal{X}$. The constitutive relations of the transformer can be rewritten as (image representation)

$$\begin{bmatrix} \mathbf{f}_1 \\ \mathbf{f}_2 \end{bmatrix} = \begin{bmatrix} \mathbf{R}_1(\mathbf{x}) & \mathbf{R}_2(\mathbf{x}) \end{bmatrix}^\perp \lambda,$$

$$\begin{bmatrix} -\mathbf{e}_1 \\ \mathbf{e}_2 \end{bmatrix} = \left(\left(\begin{bmatrix} \mathbf{R}_1(\mathbf{x}) & \mathbf{R}_2(\mathbf{x}) \end{bmatrix}^\perp \right)^T \right)^\perp \lambda, \ \lambda \in \mathbf{R}^{n_1+n_2}$$

Now, it is clear that $-\mathbf{f}_1^T \mathbf{e}_1 + \mathbf{f}_2^T \mathbf{e}_2 = 0$ and that the dimension of the space of admissible efforts and flows is $n_1 + n_2$. Therefore, the set of the power variables satisfying (16.5), (16.5) represents a Dirac structure on $\mathcal{X}$. Note that if $\mathbf{R}_1(\mathbf{x}) = -\mathbf{R}(\mathbf{x})$ and $\mathbf{R}_2(\mathbf{x}) = \mathbf{I}_{n_2}$, then the constitutive relations of the transformer are given by

$$\mathbf{e}_1 = \mathbf{R}^T(\mathbf{x}) \mathbf{e}_2$$
$$\mathbf{f}_2 = \mathbf{R}(\mathbf{x}) \mathbf{f}_1$$

which represents the regular form of transformer. Also, if the matrices $\mathbf{R}_1(\mathbf{x}), \mathbf{R}_2(\mathbf{x})$ do not depend on $\mathbf{x}$, then the symbol **MTF** is replaced by **TF**.

Finally, a gyrator having n_1 incoming ports and n_2 outgoing ports is shown in Figure 16.6. The constitutive relations of the gyrator are given by

$$\begin{bmatrix} \mathbf{R}_1(\mathbf{x}) & \mathbf{R}_2(\mathbf{x}) \end{bmatrix} \begin{bmatrix} \mathbf{f}_1 \\ \mathbf{e}_2 \end{bmatrix} = 0,$$

$$\left(\begin{bmatrix} \mathbf{R}_1(\mathbf{x}) & \mathbf{R}_2(\mathbf{x}) \end{bmatrix}^\perp \right)^T \begin{bmatrix} -\mathbf{e}_1 \\ \mathbf{f}_2 \end{bmatrix} = 0$$

Similarly, it can be proved that the set of all power variables satisfying the constitutive relations of the gyrator represents an n_1+n_2 dimensional Dirac structure on $\mathcal{X}$. Note that if the matrices $\mathbf{R}_1(\mathbf{x}), \mathbf{R}_2(\mathbf{x})$ do not depend on

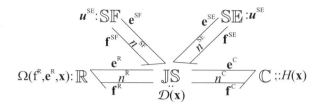

Figure 16.7 Generic bond graph

x then the symbol **MGY** is replaced by **GY**.
Using the fact that the composition of two or more Dirac structures is a Dirac structure [21], one concludes that $\mathcal{D}(\mathbf{x})$ is a Dirac structure on $\mathcal{X} \times \mathcal{V}$. A consequence of Proposition 1 is that the relations describing the junction structure are given by (kernel form)

$$\underbrace{\begin{bmatrix} \left(\mathbf{F}^{SE}(\mathbf{x})\right)^T \\ \left(\mathbf{F}^{SF}(\mathbf{x})\right)^T \\ \left(\mathbf{F}^{C}(\mathbf{x})\right)^T \\ \left(\mathbf{F}^{R}(\mathbf{x})\right)^T \end{bmatrix}^T}_{\mathbf{F}(\mathbf{x})} \begin{bmatrix} -\mathbf{f}^{SE} \\ \mathbf{f}^{SF} \\ \mathbf{f}^{C} \\ \mathbf{f}^{R} \end{bmatrix} + \underbrace{\begin{bmatrix} \left(\mathbf{E}^{SE}(\mathbf{x})\right)^T \\ \left(\mathbf{E}^{SF}(\mathbf{x})\right)^T \\ \left(\mathbf{E}^{C}(\mathbf{x})\right)^T \\ \left(\mathbf{E}^{R}(\mathbf{x})\right)^T \end{bmatrix}^T}_{\mathbf{E}(\mathbf{x})} \begin{bmatrix} \mathbf{e}^{SE} \\ -\mathbf{e}^{SF} \\ \mathbf{e}^{C} \\ \mathbf{e}^{R} \end{bmatrix} = 0 \quad (16.6)$$

The graphical representation of a generic bond graph together with all constitutive relations is shown in Figure 16.7. The equation (16.6) can be rewritten in the following way

$$\left(-\mathbf{f}^{SE}, \mathbf{f}^{SF}, \mathbf{f}^{C}, \mathbf{f}^{R}, \mathbf{e}^{SE}, -\mathbf{e}^{SF}, \mathbf{e}^{C}, \mathbf{e}^{R}\right) \in \mathcal{D}(\mathbf{x})$$

By inserting (16.1), (16.1), (16.2), (16.3), (16.3) into the last relations, the equations describing the dynamics of a system, whose bond graph model is shown in Figure 16.4, are obtained:

$$\left(-\mathbf{f}^{SE}, \mathbf{u}^{SF}, \mathbf{x}, \mathbf{f}^{R}, \mathbf{u}^{SE}, -\mathbf{e}^{SF}, \frac{\partial H(\mathbf{x})}{\partial \mathbf{x}}, \mathbf{e}^{R}\right) \in \mathcal{D}(\mathbf{x}) \quad (16.7a)$$

$$\Omega\left(\mathbf{f}^{R}, \mathbf{e}^{R}, \mathbf{x}\right) = 0 \quad (16.7b)$$

Therefore, the equations describing the system whose bond graph model is shown in Figure 16.7 are in the form of a **port Hamiltonian system (PHS)** [19].

16.5 Well-posedness and Equation Suitable for Numerical Simulation

The dynamical equations given by (16.7), (16.7) are not in form suitable for numerical simulation. In this section we show how (16.7), (16.7) may

Hamiltonian Formulation of Bond Graphs

be transformed into more suitable form. Before that, we introduce some assumptions about the system described by (16.7), (16.7). First we introduce the definitions of well-posedness of PHS.

DEFINITION 16.2—WELL-POSEDNESS OF PHS
PHS is well-posed if

$$\dim\left(\mathcal{D}^{\mathrm{U}}(\mathbf{x})\right) = n^{\mathrm{SE}} + n^{\mathrm{SF}}, \ \forall \mathbf{x} \in \mathbf{X}$$

where

$$\mathcal{D}^{\mathrm{U}}(\mathbf{x}) = \left\{ \left(\mathbf{e}^{\mathrm{SE}}, \mathbf{f}^{\mathrm{SF}}\right) : \exists \left(\mathbf{f}^{\mathrm{SE}}, \mathbf{f}^{\mathrm{C}}, \mathbf{f}^{\mathrm{R}}, \mathbf{e}^{\mathrm{SE}}, \mathbf{e}^{\mathrm{R}}\right) \ \text{s.t.} \ (\mathbf{f}, \mathbf{e}) \in \mathcal{D}(\mathbf{x}) \ \text{and} \ \mathbf{e}^{\mathrm{C}} = 0 \right\}$$

□

PROPOSITION 16.2—WELL-POSED PHS
The system described by (16.7), (16.7) is well-posed if and only if

$$\mathrm{rank}\left(\begin{bmatrix} \left(\mathbf{F}^{\mathrm{SE}}(\mathbf{x})\right)^{\mathrm{T}} \\ \left(\mathbf{E}^{\mathrm{SF}}(\mathbf{x})\right)^{\mathrm{T}} \\ \left(\mathbf{F}^{\mathrm{C}}(\mathbf{x})\right)^{\mathrm{T}} \end{bmatrix}\right) = n^{\mathrm{SE}} + n^{\mathrm{SF}} + \mathrm{rank}\left(\mathbf{F}^{\mathrm{C}}(\mathbf{x})\right), \ \forall \mathbf{x} \in \mathcal{X}$$

□

Proof. An admissible effort $\mathbf{e}^{\mathrm{C}}$ is represented by $\mathbf{e}^{\mathrm{C}} = \left(\mathbf{F}^{\mathrm{C}}(\mathbf{x})\right)^{\mathrm{T}} \lambda$, $\lambda \in \mathcal{R}^n$. Since $\mathbf{e}^{\mathrm{C}} = 0$, then

$$\lambda \in \ker\left(\left(\mathbf{F}^{\mathrm{C}}(\mathbf{x})\right)^{\mathrm{T}}\right)$$

It means that the space $\mathcal{D}^{\mathrm{U}}(\mathbf{x})$ may be rewritten as

$$\mathcal{D}^{\mathrm{U}}(\mathbf{x}) = \begin{bmatrix} \left(\mathbf{F}^{\mathrm{SE}}(\mathbf{x})\right)^{\mathrm{T}} \\ \left(\mathbf{E}^{\mathrm{SF}}(\mathbf{x})\right)^{\mathrm{T}} \end{bmatrix} \ker\left(\left(\mathbf{F}^{\mathrm{C}}(\mathbf{x})\right)^{\mathrm{T}}\right)$$

It is clear that

$$\mathrm{rank}\left(\begin{bmatrix} \left(\mathbf{F}^{\mathrm{SE}}(\mathbf{x})\right)^{\mathrm{T}} \\ \left(\mathbf{E}^{\mathrm{SF}}(\mathbf{x})\right)^{\mathrm{T}} \\ \left(\mathbf{F}^{\mathrm{C}}(\mathbf{x})\right)^{\mathrm{T}} \end{bmatrix}\right) = \mathrm{rank}\left(\begin{bmatrix} \left(\mathbf{F}^{\mathrm{SE}}(\mathbf{x})\right)^{\mathrm{T}} \\ \left(\mathbf{E}^{\mathrm{SF}}(\mathbf{x})\right)^{\mathrm{T}} \\ \left(\mathbf{F}^{\mathrm{C}}(\mathbf{x})\right)^{\mathrm{T}} \end{bmatrix} \left(\left(\mathbf{F}^{\mathrm{C}}(\mathbf{x})\right)^{\mathrm{T}}\right)^{\perp}\right)$$

$$+\mathrm{rank}\left(\mathbf{F}^{\mathrm{C}}(\mathbf{x})\right) = \mathrm{rank}\left(\begin{bmatrix} \left(\mathbf{F}^{\mathrm{SE}}(\mathbf{x})\right)^{\mathrm{T}} \left(\left(\mathbf{F}^{\mathrm{C}}(\mathbf{x})\right)^{\mathrm{T}}\right)^{\perp} \\ \left(\mathbf{E}^{\mathrm{SF}}(\mathbf{x})\right)^{\mathrm{T}} \left(\left(\mathbf{F}^{\mathrm{C}}(\mathbf{x})\right)^{\mathrm{T}}\right)^{\perp} \\ 0 \end{bmatrix}\right) + \mathrm{rank}\left(\mathbf{F}^{\mathrm{C}}(\mathbf{x})\right)$$

$$= \dim\left(\mathcal{D}^{\mathrm{U}}(\mathbf{x})\right) + \mathrm{rank}\left(\mathbf{F}^{\mathrm{C}}(\mathbf{x})\right)$$

Now, the claiming of the proposition is straightforward. For proving this, the following identity is used

$$\mathrm{rank} \begin{bmatrix} \mathbf{A} \\ \mathbf{B} \end{bmatrix} = \mathrm{rank}\,(\mathbf{A}) + \mathrm{rank}\left(\mathbf{B}\mathbf{A}^{\perp}\right),\ \mathbf{A}^{\perp} = \mathrm{ker}\,(\mathbf{A})$$

where $\mathbf{A}^{\perp}$ is the maximal rank matrix such that $\mathbf{A}\mathbf{A}^{\perp} = 0$.

REMARK 16.1—WELL-POSED SYSTEM

The system is not well-posed if either $\mathrm{rank}\left(\left[\mathbf{F}^{\mathrm{SE}}(\mathbf{x})\,\mathbf{E}^{\mathrm{SF}}(\mathbf{x})\right]\right) < n^{\mathrm{SE}} + n^{\mathrm{SF}}$ or $\mathrm{rank}\left(\left[\mathbf{F}^{\mathrm{SE}}(\mathbf{x})\,\mathbf{E}^{\mathrm{SF}}(\mathbf{x})\right]\right) = n^{\mathrm{SE}} + n^{\mathrm{SF}}$ but some columns of $\mathbf{F}^{\mathrm{C}}(\mathbf{x})$ linearly depend on columns of the matrices $\mathbf{F}^{\mathrm{SE}}(\mathbf{x}), \mathbf{E}^{\mathrm{SF}}(\mathbf{x})$. In the first case, the matrix $\left[\mathbf{F}^{\mathrm{SE}}(\mathbf{x})\,\mathbf{E}^{\mathrm{SF}}(\mathbf{x})\right]^{\mathrm{T}}$ is not a full rank matrix and one can find a non-zero matrix $[\mathbf{A}(\mathbf{x})\,\mathbf{B}(\mathbf{x})]$ such that

$$[\mathbf{A}(\mathbf{x})\,\mathbf{B}(\mathbf{x})]\left[\mathbf{F}^{\mathrm{SE}}(\mathbf{x})\,\mathbf{E}^{\mathrm{SF}}(\mathbf{x})\right]^{\mathrm{T}} = 0,\ \forall \mathbf{x} \in \mathcal{X}$$

By post-multiplying the last equation with $\lambda \in \mathcal{R}^n$, one obtains

$$\mathbf{A}(\mathbf{x}) \underbrace{\left(\mathbf{F}^{\mathrm{SE}}(\mathbf{x})\right)^{\mathrm{T}} \lambda}_{\mathbf{e}^{\mathrm{SE}}} + \mathbf{B}(\mathbf{x}) \underbrace{\left(\mathbf{E}^{\mathrm{SF}}(\mathbf{x})\right)^{\mathrm{T}} \lambda}_{\mathbf{f}^{\mathrm{SF}}} = \mathbf{A}(\mathbf{x})\,\mathbf{e}^{\mathrm{SE}} + \mathbf{B}(\mathbf{x})\,\mathbf{f}^{\mathrm{SF}} = 0$$

This means that the junction structure implies a dependency between the efforts of SE-type of ports and the flows of SF-type of ports. In other words, the input signals can not be chosen arbitrarily. In the second case, one can find a full rank matrix $[\mathbf{A}(\mathbf{x})\,\mathbf{B}(\mathbf{x})\,\mathbf{C}(\mathbf{x})]$ such that both $[\mathbf{A}(\mathbf{x})\,\mathbf{B}(\mathbf{x})]$ and $\mathbf{C}(\mathbf{x})$ are non-zero matrices and such that the following relation is satisfied

$$[\mathbf{A}(\mathbf{x})\,\mathbf{B}(\mathbf{x})\,\mathbf{C}(\mathbf{x})]\left[\mathbf{F}^{\mathrm{SE}}(\mathbf{x})\,\mathbf{E}^{\mathrm{SF}}(\mathbf{x})\,\mathbf{F}^{\mathrm{C}}(\mathbf{x})\right]^{\mathrm{T}} = 0,\quad \forall \mathbf{x} \in \mathcal{X}$$

Similarly, the last relation implies $\mathbf{A}(\mathbf{x})\,\mathbf{e}^{\mathrm{SE}} + \mathbf{B}(\mathbf{x})\,\mathbf{f}^{\mathrm{SF}} + \mathbf{C}(\mathbf{x})\,\mathbf{f}^{\mathrm{C}} = 0$. If the power variables $\mathbf{e}^{\mathrm{SF}}, \mathbf{f}^{\mathrm{SF}}$ are discontinuous functions in time, then the energy variables (the states of the system) are also discontinuous functions. Therefore, the sources have to be capable to generate an infinite amount of power, which has no physical justification. □

Suppose that the system described by (16.7), (16.7) is well-posed. The set of admissible efforts of C-type of ports is represented by

$$\mathcal{D}^{\mathrm{C}}(\mathbf{x}) = \left\{ \mathbf{e}^{\mathrm{C}} : \exists \left(\mathbf{e}^{\mathrm{SE}}, -\mathbf{e}^{\mathrm{SF}}, \mathbf{e}^{\mathrm{R}}, -\mathbf{f}^{\mathrm{SE}}, \mathbf{f}^{\mathrm{SF}}, \mathbf{f}^{\mathrm{C}}, \mathbf{f}^{\mathrm{R}}\ \mathrm{s.t.}\ (\mathbf{f}, \mathbf{e}) \in \mathcal{D}(\mathbf{x})\right)\right\}$$

Hamiltonian Formulation of Bond Graphs

ASSUMPTION 16.1—CONSTANT DIMENSIONALITY OF $\mathcal{D}^C(\mathbf{x})$
It is assumed that
$$\dim\left(\mathcal{D}^C(\mathbf{x})\right) = n_1^C, \ \forall \mathbf{x} \in \mathcal{X}$$

□

Now, we show how equations suitable for the numerical simulation can be derived.

PROPOSITION 16.3
Consider a PHS described by (16.7), (16.7). Supposed that the system is well-posed and that assumption 1 holds. Then (16.7), (16.7) has the following representation:

$$\dot{\mathbf{x}} = \mathbf{J}^C(\mathbf{x})\frac{\partial H}{\partial \mathbf{x}}(\mathbf{x}) + \mathbf{G}(\mathbf{x})\lambda + \mathbf{G}^{C,R}(\mathbf{x})\mathbf{f}^R + \mathbf{G}^{C,U}(\mathbf{x})\mathbf{u} \quad (16.8a)$$

$$0 = \mathbf{G}^T(\mathbf{x})\frac{\partial H}{\partial \mathbf{x}}(\mathbf{x}), \quad (16.8b)$$

$$\mathbf{e}^R = -\left(\mathbf{G}^{C,R}(\mathbf{x})\right)^T\frac{\partial H}{\partial \mathbf{x}}(\mathbf{x}) + \mathbf{G}^{R,U}(\mathbf{x})\mathbf{u}, \quad (16.8c)$$

$$0 = \bar{\Omega}\left(\mathbf{f}^R, \mathbf{e}^R, \mathbf{x}\right), \quad (16.8d)$$

$$\mathbf{y} = -\left(\mathbf{G}^{C,U}(\mathbf{x})\right)^T\frac{\partial H}{\partial \mathbf{x}}(\mathbf{x}) - \left(\mathbf{G}^{R,U}(\mathbf{x})\right)^T\mathbf{f}^R + \mathbf{J}^U(\mathbf{x})\mathbf{u} \quad (16.8e)$$

where

$$\mathbf{u} = \begin{bmatrix} \mathbf{e}^{SE} \\ \mathbf{f}^{SF} \end{bmatrix}, \ \mathbf{y} = \begin{bmatrix} -\mathbf{f}^{SE} \\ -\mathbf{e}^{SF} \end{bmatrix}, \ \begin{bmatrix} \mathbf{e}^R \\ \mathbf{f}^R \end{bmatrix} = \mathbf{T}(\mathbf{x})\begin{bmatrix} \mathbf{f}^R \\ \mathbf{e}^R \end{bmatrix}$$

$$\bar{\Omega}\left(\mathbf{f}^R, \mathbf{e}^R, \mathbf{x}\right) = \Omega\left(\mathbf{T}(\mathbf{x})\left(\mathbf{f}^R, \mathbf{e}^R\right), \mathbf{x}\right)$$

Here, $\mathbf{T}(\mathbf{x})$ is a regular and power-conserving transformation of power variables of R-type of ports and $\mathbf{J}^C(\mathbf{x}), \mathbf{J}^U(\mathbf{x})$ are skew-symmetric matrices.

□

Proof. The relation (16.6) can be rewritten as

$$\mathbf{E}^w(\mathbf{x})\mathbf{e}^w + \mathbf{F}^w(\mathbf{x})\mathbf{f}^w + \mathbf{E}^R(\mathbf{x})\mathbf{e}^R + \mathbf{F}^R(\mathbf{x})\mathbf{f}^R = 0 \quad (16.9)$$

where

$$\mathbf{E}^w(\mathbf{x}) = \begin{bmatrix} \left(\mathbf{E}^{SE}(\mathbf{x})\right)^T \\ \left(\mathbf{F}^{SF}(\mathbf{x})\right)^T \\ \left(\mathbf{E}^C(\mathbf{x})\right)^T \end{bmatrix}^T, \ \mathbf{F}^w(\mathbf{x}) = \begin{bmatrix} \left(\mathbf{F}^{SE}(\mathbf{x})\right)^T \\ \left(\mathbf{E}^{SF}(\mathbf{x})\right)^T \\ \left(\mathbf{F}^C(\mathbf{x})\right)^T \end{bmatrix}^T,$$

$$\mathbf{e}^w = \begin{bmatrix} \mathbf{e}^{SE} \\ \mathbf{f}^{SF} \\ \mathbf{e}^C \end{bmatrix}, \ \mathbf{f}^w = \begin{bmatrix} -\mathbf{f}^{SE} \\ -\mathbf{e}^{SF} \\ \mathbf{f}^C \end{bmatrix}.$$

The well-posedness of the system and Assumption 1 guarantee that the matrix $\mathbf{F}^W(\mathbf{x})$ is constant rank matrix. Thus, by performing row-like operation on (16.9), the following can be obtained:

$$\begin{bmatrix} \mathbf{F}_1^W(\mathbf{x}) \\ \mathbf{0}_{(n_2^C + n^R) \times (n - n^R)} \end{bmatrix} \mathbf{f}^W + \begin{bmatrix} \mathbf{E}_1^W(\mathbf{x}) \\ \mathbf{E}_2^W(\mathbf{x}) \end{bmatrix} \mathbf{e}^W + \begin{bmatrix} \mathbf{E}_1^R(\mathbf{x}) \\ \mathbf{E}_2^R(\mathbf{x}) \end{bmatrix} \mathbf{e}^R + \begin{bmatrix} \mathbf{F}_1^R(\mathbf{x}) \\ \mathbf{F}_2^R(\mathbf{x}) \end{bmatrix} \mathbf{f}^R = 0$$
(16.10)

The matrix $\mathbf{F}_1^W(\mathbf{x})$ has the following form:

$$\mathbf{F}_1^W(\mathbf{x}) = \mathrm{diag}\left(\mathbf{I}_{n^{SE}}, \mathbf{I}_{n^{SF}}, \mathbf{F}_1^C(\mathbf{x})\right)$$

and $\mathbf{F}_1^C(\mathbf{x})$ is a full rank matrix $\forall \mathbf{x} \in \mathcal{X}$. Furthermore, $\mathbf{e}^C$ can be expressed as $\mathbf{e}^C = \left(\mathbf{F}^C(\mathbf{x})\right)^T \lambda = \begin{bmatrix} \mathbf{0}_{n^C \times n^{SE}} & \mathbf{0}_{n^C \times n^{SF}} & \left(\mathbf{F}_1^C(\mathbf{x})\right)^T & \mathbf{0}_{n^C \times n^R} \end{bmatrix} \lambda$. It means that

$$\left(\left(\mathbf{F}^C(\mathbf{x})\right)^\perp\right)^T \mathbf{e}^C = 0$$

Therefore, by performing row-like operations on the last $n_2^C + n^R$ rows of (16.10), the following is obtained

$$\begin{bmatrix} \mathbf{F}_1^W(\mathbf{x}) \\ \mathbf{0}_{n_2^C \times (n - n^R)} \\ \mathbf{0}_{n^R \times (n - n^R)} \end{bmatrix} \mathbf{f}^W + \begin{bmatrix} \mathbf{E}_1^W(\mathbf{x}) \\ \mathbf{E}_2^W(\mathbf{x}) \\ \mathbf{E}_3^W(\mathbf{x}) \end{bmatrix} \mathbf{e}^W + \begin{bmatrix} \mathbf{F}_1^R(\mathbf{x}) \\ \mathbf{0}_{n_2^C \times n^R} \\ \mathbf{F}_3^R(\mathbf{x}) \end{bmatrix} \mathbf{f}^R + \begin{bmatrix} \mathbf{E}_1^R(\mathbf{x}) \\ \mathbf{0}_{n_2^C \times n^R} \\ \mathbf{E}_3^R(\mathbf{x}) \end{bmatrix} \mathbf{e}^R = 0$$
(16.11)

and the matrix $\mathbf{E}_2^W(\mathbf{x})$ has the following form:

$$\mathbf{E}_2^W(\mathbf{x}) = \begin{bmatrix} \mathbf{0}_{n_2^C \times n^{SE}} & \mathbf{0}_{n_2^C \times n^{SF}} & \mathbf{E}_2^C(\mathbf{x}) \end{bmatrix}.$$

The matrix $\mathbf{E}_2^C(\mathbf{x})$ is a full rank matrix and $\mathbf{F}_1^C(\mathbf{x})\left(\mathbf{E}_2^C(\mathbf{x})\right)^T = 0$, $\forall \mathbf{x} \in \mathcal{X}$. Since n^R power variables of $\mathbb{R}$-element have to be expressed as a function of other n^R ones then rank $\begin{bmatrix} \mathbf{F}_3^R(\mathbf{x}) & \mathbf{E}_3^R(\mathbf{x}) \end{bmatrix} = n^R$, $\forall \mathbf{x} \in \mathcal{X}$. Suppose that

$$\mathbf{F}_3^R(\mathbf{x})\left(\mathbf{F}_3^R(\mathbf{x})\right)^T + \mathbf{E}_3^R(\mathbf{x})\left(\mathbf{E}_3^R(\mathbf{x})\right)^T = \mathbf{I}_{n^R}, \forall \mathbf{x} \in \mathcal{X}$$

If it is not the case, then, it can be achieved by pre-multiplying the third row in (16.11) by

$$\left(\mathbf{F}_3^R(\mathbf{x})\left(\mathbf{F}_3^R(\mathbf{x})\right)^T + \mathbf{E}_3^R(\mathbf{x})\left(\mathbf{E}_3^R(\mathbf{x})\right)^T\right)^{-\frac{1}{2}}$$

Power conservation yields (block matrices at the position (1,3) and (3,3) of the expression $\mathbf{E}(\mathbf{x})\mathbf{F}^T(\mathbf{x}) + \mathbf{F}(\mathbf{x})\mathbf{E}^T(\mathbf{x}) = 0$)

$$\mathbf{F}_1^w(\mathbf{x})\left(\mathbf{E}_3^w(\mathbf{x})\right)^T + \left[\mathbf{E}_1^R(\mathbf{x})\ \mathbf{F}_1^R(\mathbf{x})\right]\left[\mathbf{F}_3^R(\mathbf{x})\ \mathbf{E}_3^R(\mathbf{x})\right]^T = 0, \quad (16.12a)$$

$$\mathbf{E}_3^R(\mathbf{x})\left(\mathbf{F}_3^R(\mathbf{x})\right)^T + \mathbf{F}_3^R(\mathbf{x})\left(\mathbf{E}_3^R(\mathbf{x})\right)^T = 0 \quad (16.12b)$$

Consider the following transformation of power variables of R-type of ports

$$\begin{bmatrix} \mathbf{f}^R \\ \mathbf{e}^R \end{bmatrix} = \underbrace{\begin{bmatrix} \mathbf{F}_3^R(\mathbf{x}) & \mathbf{E}_3^R(\mathbf{x}) \\ \mathbf{E}_3^R(\mathbf{x}) & \mathbf{F}_3^R(\mathbf{x}) \end{bmatrix}^T}_{\mathbf{T}^T(\mathbf{x})} \begin{bmatrix} \mathbf{e}^R \\ \mathbf{f}^R \end{bmatrix} \quad (16.13)$$

Straightforward computation shows that

$$\mathbf{T}(\mathbf{x})\mathbf{T}^T(\mathbf{x}) = \mathbf{I}_{n^R},$$

$$\mathbf{T}(\mathbf{x})\begin{bmatrix} 0 & \mathbf{I}_{n^R} \\ \mathbf{I}_{n^R} & 0 \end{bmatrix}\mathbf{T}^T(\mathbf{x}) = \begin{bmatrix} 0 & \mathbf{I}_{n^R} \\ \mathbf{I}_{n^R} & 0 \end{bmatrix}$$

The second condition guarantees that the transformation is power conserving, i.e., $\langle \mathbf{e}^R | \mathbf{f}^R \rangle = \langle \mathbf{e}^R | \mathbf{f}^R \rangle$. By inserting (16.13) into (16.11) and taking into account (16.12), (16.12) the following is obtained

$$\begin{bmatrix} \mathbf{F}_1^w(\mathbf{x}) \\ 0_{n_2^C \times (n-n^R)} \\ 0_{n^R \times (n-n^R)} \end{bmatrix}\mathbf{f}^w + \begin{bmatrix} \mathbf{E}_1^w(\mathbf{x}) \\ \mathbf{E}_2^w(\mathbf{x}) \\ \mathbf{E}_3^w(\mathbf{x}) \end{bmatrix}\mathbf{e}^w + \begin{bmatrix} -\mathbf{F}_1^w(\mathbf{x})\left(\mathbf{E}_3^w(\mathbf{x})^T\right) \\ 0_{n_2^C \times n^R} \\ 0_{n^R \times n^R} \end{bmatrix}\mathbf{f}^R + \begin{bmatrix} 0_{(n^{SE}+n^{SF}+n_1^C) \times n^R} \\ 0_{n_2^C \times n^R} \\ \mathbf{I}_{n^R} \end{bmatrix}\mathbf{e}^R = 0 \quad (16.14)$$

where

$$\mathbf{E}_1^w(\mathbf{x}) = \mathbf{E}_1^w(\mathbf{x}) - \left[\mathbf{F}_1^R(\mathbf{x})\ \mathbf{E}_1^R(\mathbf{x})\right]\left[\mathbf{F}_3^R(\mathbf{x})\ \mathbf{E}_3^R(\mathbf{x})\right]^T\mathbf{E}_3^w(\mathbf{x})$$

The power-conservation yields

$$\mathbf{F}_1^w(\mathbf{x})\left(\mathbf{E}_1^w(\mathbf{x})\right)^T + \mathbf{E}_1^w(\mathbf{x})\left(\mathbf{F}_1^w(\mathbf{x})\right)^T = 0 \quad (16.15a)$$

$$\mathbf{F}_1^w(\mathbf{x})\left(\mathbf{E}_2^w(\mathbf{x})\right)^T = 0 \quad (16.15b)$$

Since the matrix $\mathbf{F}_1^w(\mathbf{x})$ is full rank matrix $\forall x \in X$, then the matrix $\mathbf{E}_1^w(\mathbf{x})$ may be represented as follows

$$\mathbf{E}_1^w(\mathbf{x}) = -\mathbf{F}_1^w(\mathbf{x})\mathbf{J}(\mathbf{x}) \quad (16.16)$$

By inserting (16.16) into (16.14), one obtains

$$\mathbf{f}^{\mathrm{w}} = \mathbf{J}(\mathbf{x})\mathbf{e}^{\mathrm{w}} + (\mathbf{E}_3^{\mathrm{w}}(\mathbf{x}))^{\mathrm{T}}\mathbf{f}^{\mathrm{R}} + (\mathbf{F}_1^{\mathrm{w}}(\mathbf{x}))^{\perp}\lambda$$
$$0 = \mathbf{E}_2^{\mathrm{w}}(\mathbf{x})\mathbf{e}^{\mathrm{w}},$$
$$\mathbf{e}^{\mathrm{R}} = -\mathbf{E}_3^{\mathrm{w}}(\mathbf{x})\mathbf{e}^{\mathrm{w}}$$

Here, $(\mathbf{F}_1^{\mathrm{w}}(\mathbf{x}))^{\perp} = \mathrm{diag}\left(0_{n^{\mathrm{SE}} \times n^{\mathrm{SE}}}, 0_{n^{\mathrm{SF}} \times n^{\mathrm{SF}}}, (\mathbf{F}_1^{\mathrm{C}}(\mathbf{x}))^{\perp}\right)$. Inserting (16.16) into (16.15) gives

$$\mathbf{J}(\mathbf{x}) + \mathbf{J}^{\mathrm{T}}(\mathbf{x}) = 0$$

Splitting $\mathbf{J}(\mathbf{x})$ as

$$\mathbf{J}(\mathbf{x}) = \begin{bmatrix} \mathbf{J}^{\mathrm{U}}(\mathbf{x}) & -(\mathbf{G}^{\mathrm{C,U}}(\mathbf{x}))^{\mathrm{T}} \\ \mathbf{G}^{\mathrm{C,U}}(\mathbf{x}) & \mathbf{J}^{\mathrm{C}}(\mathbf{x}) \end{bmatrix}$$

with $\mathbf{E}_3^{\mathrm{w}}(\mathbf{x})$ as

$$\mathbf{E}_3^{\mathrm{w}}(\mathbf{x}) = \begin{bmatrix} \mathbf{G}^{\mathrm{R,U}}(\mathbf{x}) \\ -(\mathbf{G}^{\mathrm{C,R}}(\mathbf{x}))^{\mathrm{T}} \end{bmatrix}$$

and replacing $\mathbf{E}_2^{\mathrm{C}}(\mathbf{x})$ by $\mathbf{G}^{\mathrm{T}}(\mathbf{x})$, $(\mathbf{F}_1^{\mathrm{C}}(\mathbf{x}))^{\perp}$ by $\mathbf{G}(\mathbf{x})$, then Equations (16.8)-(16.8) are obtained.

16.6 Index of System

In this section, the computational issue of the system described by (16.8)-(16.8) is investigated. A measure for expected difficulties in the numerical simulation of the system of equations is the index. The definition of the (differential) index is given.

DEFINITION 16.3—LOCAL DIFFERENTIAL INDEX [20]
Consider a system described by the following implicit equation

$$\mathbf{Q}(\mathbf{z}, \mathbf{z}, \mathbf{u}) = 0 \tag{16.17}$$

Equation (16.17) has a *local differential index* m at the point $(\mathbf{z}_o, \mathbf{u}_o)$ if m is the minimal number such that there exists a neighbourhood of the point $(\mathbf{z}_o, \mathbf{u}_o)$, in which the system of equations

$$\mathbf{Q}(\mathbf{z}, \mathbf{z}, \mathbf{u}) = 0,$$
$$\frac{\mathrm{d}\mathbf{Q}(\mathbf{z}, \mathbf{z}, \mathbf{u})}{\mathrm{d}t} = 0$$
$$\vdots$$
$$\frac{\mathrm{d}^m \mathbf{Q}(\mathbf{z}, \mathbf{z}, \mathbf{u})}{\mathrm{d}t^m} = 0$$

Hamiltonian Formulation of Bond Graphs

can be uniquely solved for $\mathbf{z}$ as a function of $\mathbf{z}, \mathbf{u}$ only. $\square$

Equation (16.8) is an output equation and it can be omitted from the index analysis. By inserting (16.8) into (16.8), the following set of equations are obtained

$$\dot{\mathbf{x}} = \mathbf{J}^C(\mathbf{x}) \frac{\partial H}{\partial \mathbf{x}}(\mathbf{x}) + \mathbf{G}(\mathbf{x})\lambda + \mathbf{G}^{C,R}(\mathbf{x})\mathbf{f}^R + \mathbf{G}^{C,U}(\mathbf{x})\mathbf{u} \quad (16.18a)$$

$$0 = \mathbf{G}^T(\mathbf{x}) \frac{\partial H}{\partial \mathbf{x}}(\mathbf{x}) \quad (16.18b)$$

$$0 = \tilde{\Omega}(\mathbf{f}^R, \mathbf{x}, \mathbf{u}) \quad (16.18c)$$

PROPOSITION 16.4—INDEX TWO SYSTEM
Consider the system described by Eqs. (16.18)-(16.18). Assume that for $\mathbf{x} = \mathbf{x}_0$ and $\mathbf{u} = \mathbf{u}_0$ the equation (16.18) is satisfied and that there exists $\mathbf{f}_0^R$ such that (16.18) is satisfied. If

$$\det\left(\frac{\partial}{\partial \mathbf{x}^T}\left(\mathbf{G}^T(\mathbf{x}) \frac{\partial H}{\partial \mathbf{x}}(\mathbf{x})\right)\mathbf{G}(\mathbf{x})\right)_{\mathbf{x}=\mathbf{x}_0} \neq 0 \quad (16.19a)$$

$$\det\left(\frac{\partial \tilde{\Omega}(\mathbf{f}^R, \mathbf{x}, \mathbf{u})}{\partial (\mathbf{f}^R)^T}\right)_{\substack{\mathbf{x}=\mathbf{x}_0 \\ \mathbf{f}^R=\mathbf{f}_0^R \\ \mathbf{u}=\mathbf{u}_0}} \neq 0 \quad (16.19b)$$

then the system (16.18)-(16.18) has a differential index two at the point $(\mathbf{x}_0, \mathbf{f}_0^R, \lambda_0, \mathbf{u}_0)$. $\square$

Proof. In this case, $\mathbf{z} = (\mathbf{x}, \lambda, \mathbf{f}^R)$. Differentiation of (16.18) gives

$$\frac{\partial \tilde{\Omega}(\mathbf{f}^R, \mathbf{x}, \mathbf{u})}{\partial (\mathbf{f}^R)^T} \dot{\mathbf{f}}^R + \mathbf{L}_1(\mathbf{x}, \mathbf{u}, \lambda, \mathbf{f}^R) = 0.$$

Now, it is clear that if Condition (16.19) is satisfied, then the last equation can be solved for $\dot{\mathbf{f}}^R$. Differentiation of (16.18) gives

$$\frac{\partial}{\partial \mathbf{x}^T}\left(\mathbf{G}^T(\mathbf{x}) \frac{\partial H}{\partial \mathbf{x}}(\mathbf{x})\right)\mathbf{G}(\mathbf{x})\lambda + \mathbf{L}_2(\mathbf{x}, \mathbf{u}, \lambda, \mathbf{f}^R) = 0$$

If Condition (16.19) is satisfied, then the last equation can be uniquely solved for λ. Therefore, for the given value of $\mathbf{x}_0, \mathbf{f}_0^R, \mathbf{u}_0$, the value of λ_0 can be uniquely computed. Differentiation of the last equation gives

$$\frac{\partial}{\partial \mathbf{x}^T}\left(\mathbf{G}^T(\mathbf{x}) \frac{\partial H}{\partial \mathbf{x}}(\mathbf{x})\right)\mathbf{G}(\mathbf{x})\dot{\lambda} + \mathbf{L}_3(\mathbf{x}, \lambda, \mathbf{f}^R, \mathbf{u}, \dot{\mathbf{u}}) = 0$$

It is clear that if Condition (16.19) is satisfied, then the last equation can be uniquely solved for $\dot{\lambda}$. So, the index of (16.18)–(16.18) is two.

REMARK 16.2—INDEX TWO SYSTEM
Since (16.18) does not depend on λ then the index two system (16.18)–(16.18) is computationally efficient [20]. □

By premultiplying (16.18) with $\mathbf{G}^\perp(\mathbf{x})$, the following system of equations is obtained:

$$\mathbf{G}^\perp(\mathbf{x})\mathbf{x} = \mathbf{G}^\perp(\mathbf{x})\left(\mathbf{J}^C(\mathbf{x})\frac{\partial H}{\partial \mathbf{x}}(\mathbf{x}) + \mathbf{G}^{C,R}(\mathbf{x})\mathbf{f}^R + \mathbf{G}^{C,U}(\mathbf{x})\mathbf{u}\right) \quad (16.20a)$$

$$0 = \mathbf{G}^T(\mathbf{x})\frac{\partial H}{\partial \mathbf{x}}(\mathbf{x}) \quad (16.20b)$$

$$0 = \tilde{\Omega}\left(\mathbf{f}^R, \mathbf{x}, \mathbf{u}\right) \quad (16.20c)$$

PROPOSITION 16.5—INDEX ONE SYSTEM
Consider the system described by Equations (16.20 a)-(16.20 c). Assume that for $\mathbf{x} = \mathbf{x}_o$ and $\mathbf{u} = \mathbf{u}_o$ the equation (16.20) is satisfied and that there exists $\mathbf{f}_0^R$ such that (16.20) is satisfied. If the conditions (16.19) and (16.19) are satisfied then the system (16.20 a)-(16.20 c) has a differential index one at the point $(\mathbf{x}_o, \mathbf{f}_0^R, \mathbf{u}_o)$. □

Proof. In this case $\mathbf{z} = (\mathbf{x}, \mathbf{f}^R)$. The part of the proof regarding the calculation of $\mathbf{f}^R$ is the same as in the proof of Proposition 16.4. Now, we concentrate on calculation of $\mathbf{x}$. Differentiating (16.20) and regrouping the derivatives on the left side gives the following

$$\left[\begin{array}{c}\mathbf{G}^\perp(\mathbf{x}) \\ \frac{\partial}{\partial \mathbf{x}^T}\left(\mathbf{G}^T(\mathbf{x})\frac{\partial H}{\partial \mathbf{x}}(\mathbf{x})\right)\end{array}\right]\mathbf{x} = \mathbf{L}_4\left(\mathbf{x}, \mathbf{f}^R, \mathbf{u}\right)$$

If Condition (16.19) is fulfilled, then the rank of the matrix on the left side is full $\forall \mathbf{x} \in \mathcal{X}$.

16.7 Example

A classical simplified half-car model [4], known as *bicycle model*, is shown in Figure 16.8. There is no suspension, nor any bushing to consider. It is assumed that the width of the system is negligible. The front wheel can be steered over the angle δ. The inertial frame is XOY. The velocities v_{xf}, v_{yf} (v_{xr}, v_{yr}) are the velocities of the front wheel (rear wheel) and v_{xb}, v_{yb} are the velocities of the center of mass of the body. The angular velocities of the wheels are denoted by ω^f, ω^r, and the angular velocity of the body is denoted by ω^b, and it is assumed that δ is constant. A model of the wheel-tire system is considered. A schematic of tire is shown in Figure 16.9. Here, F_{x^α} is longitudinal force ($\alpha \in \{f, r\}$), F_{y^α} is the cornering force, and $\lambda_{x^\alpha}, \lambda_{y^\alpha}$

Hamiltonian Formulation of Bond Graphs

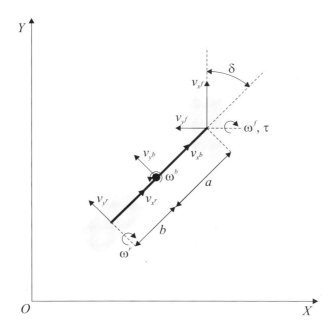

Figure 16.8 Bicycle model: Ideal physical model.

are the constraint forces between the body and wheels. For $\alpha = f$, $\tau^\alpha = \tau$, and for $\alpha = r$, $\tau^\alpha = 0$, thhe forces $F_{x^\alpha}, F_{y^\alpha}$ are given by

$$F_{x^\alpha} = d_{x^\alpha}\left(e^R_{x^\alpha}, p_{x^\alpha}, p_{\omega^\alpha}\right)$$
$$F_{y^\alpha} = d_{y^\alpha}\left(e^R_{y^\alpha}, p_{x^\alpha}, p_{y^\alpha}\right)$$

where $p_{x^\alpha}, p_{y^\alpha}$ are the momenta of the wheel with respect to the body frame, p_{ω^α} is the angular momentum of the wheel with respect to the polar axis of α wheel, $e^R_{x^\alpha} = R\omega^\alpha - v_{x^\alpha}$, and $e^R_{y^\alpha} = v_{y^\alpha}$. The mass of the tire is denoted by m_t, polar inertia by I_t, and radius by R. The bond graph model of wheel-tire system is given in Figure 16.10. A schematic of the vehicle's body is shown in Figure 16.11. The mass of the body is m and its inertia is I. The bond graph model is shown in Figure 16.12. The transformation matrices from the body frame to the wheel frames, $\mathbf{C}^f(\delta)$ and $\mathbf{C}^r$, are given by

$$\mathbf{C}^f(\delta) = \begin{bmatrix} \cos(\delta) & \sin(\delta) & a\sin(\delta) \\ -\sin(\delta) & \cos(\delta) & a\cos(\delta) \end{bmatrix} \quad (16.21a)$$

$$\mathbf{C}^r = \begin{bmatrix} 1 & 0 & 0 \\ 0 & 1 & -b \end{bmatrix} \quad (16.21b)$$

The bond graph model of the whole system is given in Figure 16.13. The equations describing the system (16.8)-(16.8) now become

$$\mathbf{x} = \mathbf{G}(\delta)\lambda + \mathbf{G}^{C,R}\mathbf{f}^R + \mathbf{G}^{C,U}\mathbf{u} \quad (16.22a)$$

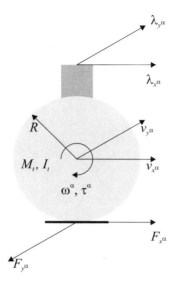

Figure 16.9 Wheel-tire system: Ideal physical model.

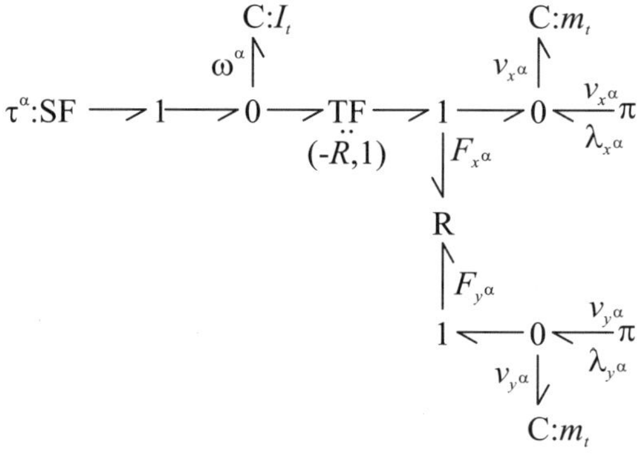

Figure 16.10 Bond graph model of wheel-tire system.

$$0 = \mathbf{G}^{\mathrm{T}}(\delta) \frac{\partial H}{\partial \mathbf{x}}(\mathbf{x}) \tag{16.22b}$$

$$\mathbf{e}^{\mathrm{R}} = -\left(\mathbf{G}^{\mathrm{C,R}}\right)^{\mathrm{T}} \frac{\partial H}{\partial \mathbf{x}}(\mathbf{x}) \tag{16.22c}$$

$$y = -\left(\mathbf{G}^{\mathrm{C,U}}\right)^{\mathrm{T}} \frac{\partial H}{\partial \mathbf{x}}(\mathbf{x}) \tag{16.22d}$$

$$0 = \bar{\Omega}\left(\mathbf{f}^{\mathrm{R}}, \mathbf{e}^{\mathrm{R}}, \mathbf{x}\right) \tag{16.22e}$$

where

$$\mathbf{x}^{\mathrm{T}} = \begin{bmatrix} \mathbf{x}_f^{\mathrm{T}} & \mathbf{x}_r^{\mathrm{T}} & \mathbf{x}_b^{\mathrm{T}} \end{bmatrix} \tag{16.23a}$$

Hamiltonian Formulation of Bond Graphs

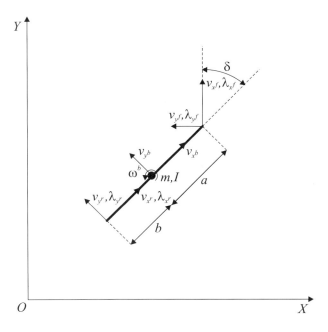

Figure 16.11 Body of the vehicle: The ideal physical model

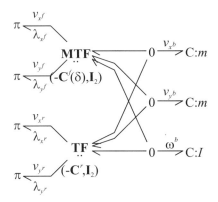

Figure 16.12 Bond graph model of vehicle's body

$$\mathbf{x}_\alpha = \begin{bmatrix} p_{x^\alpha} & p_{y^\alpha} & p_{\omega^\alpha} \end{bmatrix}^T \tag{16.23b}$$

The matrix $\mathbf{G}^T(\delta)$ is

$$\mathbf{G}^T(\delta) = \begin{bmatrix} 1 & 0 & 0 & 0 & 0 & 0 & -\cos(\delta) & -\sin(\delta) & -a\sin(\delta) \\ 0 & 1 & 0 & 0 & 0 & 0 & \sin(\delta) & -\cos(\delta) & -a\cos(\delta) \\ 0 & 0 & 0 & 1 & 0 & 0 & -1 & 0 & 0 \\ 0 & 0 & 0 & 0 & 1 & 0 & 0 & -1 & b \end{bmatrix} \tag{16.24}$$

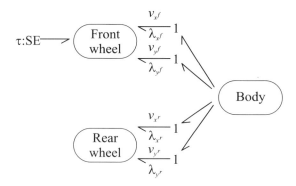

Figure 16.13 Bond graph model of the system

and

$$\left(\mathbf{G}^{C,R}\right)^T = \begin{bmatrix} 1 & 0 & -R & 0 & 0 & 0 & 0 & 0 & 0 \\ 0 & -1 & 0 & 0 & 0 & 0 & 0 & 0 & 0 \\ 0 & 0 & 0 & 1 & 0 & -R & 0 & 0 & 0 \\ 0 & 0 & 0 & 0 & -1 & 0 & 0 & 0 & 0 \end{bmatrix} \quad (16.25a)$$

$$\left(\mathbf{G}^{C,U}\right)^T = \begin{bmatrix} 0 & 0 & 1 & 0 & 0 & 0 & 0 & 0 & 0 \end{bmatrix} \quad (16.25b)$$

$$H(\mathbf{x}) = \frac{1}{2}\mathbf{x}^T \mathbf{M}^{-1} \mathbf{x}, \quad (16.25c)$$

$$\mathbf{M} = \mathrm{diag}\,(m_t, m_t, I_t, m_t, m_t, I_t, m, m, I) \quad (16.25d)$$

$$\mathbf{f}^R = \begin{bmatrix} F_{xf} & F_{yf} & F_{xr} & F_{yr} \end{bmatrix}^T \quad (16.25e)$$

$$\bar{\Omega}\left(\mathbf{f}^R, \mathbf{e}^R, \mathbf{x}\right) = \begin{bmatrix} -F_{xf} + d_{xf}\left(e^R_{xf}, p_{xf}, p_{\omega f}\right) \\ -F_{yf} + d_{yf}\left(e^R_{yf}, p_{xf}, p_{yf}\right) \\ -F_{xr} + d_{xr}\left(e^R_{xr}, p_{xr}, p_{\omega r}\right) \\ -F_{yr} + d_{yr}\left(e^R_{yr}, p_{xr}, p_{yr}\right) \end{bmatrix} \quad (16.25f)$$

Now, the index of the system is analysed. Condition (16.19) is checked.

$$\frac{\partial}{\partial \mathbf{x}^T}\left(\mathbf{G}^T(\delta)\frac{\partial H}{\partial \mathbf{x}}(\mathbf{x})\right)\mathbf{G}(\delta) = \mathbf{G}^T(\delta)\frac{\partial H}{\partial \mathbf{x}\partial \mathbf{x}^T}(\mathbf{x})\mathbf{G}(\delta) = \mathbf{G}^T(\delta)\mathbf{M}^{-1}\mathbf{G}(\delta).$$

The rank of the matrix $\mathbf{G}(\delta)$ is full and $\mathbf{M}$ is a positive definite matrix. Therefore, Conditions (16.19) is satisfied. Now, Condition (16.19) is analysed. Since

$$\frac{\partial \bar{\Omega}\left(\mathbf{f}^R, \mathbf{e}^R, \mathbf{x}\right)}{\partial \mathbf{f}^R} = -\mathbf{I}_4$$

then also Condition (16.19) is satisfied. Therefore, the index of the system is two. The index one system may be obtained by multiplying Equation (16.22a) by $\mathbf{G}^\perp(\delta)$ where

$$\mathbf{G}^\perp(\delta) = \begin{bmatrix} a\sin(\delta) & a\cos(\delta) & 0 & 0 & -b & 0 & 0 & 0 & 1 \\ \sin(\delta) & \cos(\delta) & 0 & 0 & 1 & 0 & 0 & 1 & 0 \\ \cos(\delta) & -\sin(\delta) & 0 & 1 & 0 & 0 & 1 & 0 & 0 \\ 0 & 0 & 0 & 0 & 0 & 1 & 0 & 0 & 0 \\ 0 & 0 & 1 & 0 & 0 & 0 & 0 & 0 & 0 \end{bmatrix}.$$

16.8 Conclusion

In this paper, a mathematical formulation of bond graphs has been developed. It has been proven that every junction structure can be associated with a Dirac structure. Also, it has been proven that equations describing a bond graph model correspond to a port Hamiltonian system. The condition for well-posedness of a modelled system has been given and representations suitable for the numerical simulation have been derived. The index of the representations have been analysed and sufficient conditions for the representations to be computationally efficient have been given. The theory has been applied to some models arising in automotive applications.

16.9 References

[1] H. M. Paynter. *Analysis and Design of Engineering Systems*, The MIT Press, Cambridge, MA, 1961

[2] P. C. Breedveld. Multibond Graph Elements in Physical Systems Theory, *Journal of the Franklin Institute*, 319, 1-36, 1985.

[3] P. C. Breedveld. *Physical Systems Theory in Terms of Bond Graphs*. Ph.D. Thesis, University of Twente, Enschede, NL, 1984.

[4] D. C. Karnopp, D. L. Margolis, and R. C. Rosenberg. *System Dynamics: A Unified Approach*, John Wiley. 1990.

[5] G. Golo, P. C. Breedveld, and B. M. Maschke, A.J. van der Schaft. Geometric formulation of generalised bond graphs models Part I: Generalised junction structure, Technical Report 1555, Faculty of Mathematical Sciences, University of Twente, 2000.

[6] J. D. Lamb, D. R. Woodall, and G. M. Asher. Bond Graphs I: Acausal Equivalence, *Discrete Applied Mathematics* 72, 261-293, 1997

[7] J. D. Lamb, D. R. Woodall, and G. M. Asher. G. M. (1997) Bond Graphs II: Causality and Singularity, *Discrete Applied Mathematics*, 73, 143-173, 1997.

[8] J.D. Lamb, D. R. Woodall, and G. M. Asher. Bond Graphs III: Bond Graph and Electrical Networks, *Discrete Applied Mathematics*, 73, 211-250, 1997.

[9] H. Birkett and P. H. Roe The Mathematical Foundations of Bond Graphs-I. Algebraic Theory, *Journal of the Franklin Institute*, 326, 329-350, 1989.

[10] H. Birkett and P. H. Roe. The Mathematical Foundations of Bond Graphs-II. Duality, *Journal of the Franklin Institute*, 326, 691-708, 1989.

[11] H. Birkett and P. H. Roe. The Mathematical Foundations of Bond Graphs-III. Matroid Theory, *Journal of the Franklin Institute*, 327, 87-108, 1990.

[12] H. Birkett and P. H. Roe. The Mathematical Foundations of Bond Graphs-IV. Matrix Representations and Causality, *Journal of the Franklin Institute*, 327, 109-128, 1990.

[13] A. S. Perelson and G. F. Oster. Chemical Reaction Dynamics Part II: Reaction Networks, *Archive of Rational Mech. Anal.*, 55, 31-98, 1975.

[14] G. F. Oster and A. S. Perelson. Chemical Reaction Dynamics Part I: Geometrical Structure, *Archive of Rational Mech. Anal.*, 57, 230-274, 1975.

[15] B. M. Maschke, A. J. van der Schaft and P. C. Breedveld. An Intrinsic Hamiltonian Formulation of Networks dynamics: Non-Standard Poisson Structures and Gyrators, *Journal of the Franklin Institute*, 329, 923-966, 1992.

[16] B. M. Maschke. Elements on the Modelling of Mechanical Systems in Modelling and Control of Mechanisms and Robots Bertinaro, World Scientific Publishing Ltd, 1-38, 1996.

[17] M. Dalsmo, A. J. van der Schaft, and P. C. Breedveld. On Representation and Integrability of Mathematical Structures in Energy-Conserving Physical Systems, *SIAM Journal on Control and Optimisation*, 37, 54-91, 1999.

[18] A. J. van der Schaft and B. M. Maschke. The Hamiltonian Formulation of Energy Conserving Physical Systems with External Ports, *Archiv fur Elektronik und Ubertragungstechnik*, 49, 362-371, 1995.

[19] A. J. van der Schaft. *L_2-Gain and Passivity Techniques in Nonlinear Control*, Springer-Verlag, 2000.

[20] E. Hairer and G. Wanner. *Numerical Solution of Nonlinear Differential Equations II: Stiff and Differential-Algebraic Problems*, Springer-Verlag, 1991

[21] A. J. van der Schaft. Interconnection and Geometry, in J. W. Polderman and H. L. Trentelman, eds., *The Mathematics of Systems and Control: From Intelligent Control to Behaviour Systems*, 203-218, 1999.

17

Stability Analysis of Hybrid Systems —A Gearbox Application

S. Pettersson   B. Lennartson

Abstract

This paper includes an application consisting of an automatic gearbox and cruise controller which naturally is modelled as a hybrid system including state jumps in the continuous state of the controller. Motivated by this application, we extend existing stability results to include state jumps as well. The proposed stability results are based on Lyapunov techniques. The search for the (piecewise quadratic) Lyapunov functions is formulated as a linear matrix inequality (LMI) problem. It is shown how the proposed stability analysis is applied to the automatic gearbox and cruise controller.

17.1 Introduction

Many physical systems today are modeled by interacting continuous and discrete-event systems. Such *hybrid systems* contain both continuous and discrete states that influence the dynamic behavior. There is a lot of interest in these kinds of systems today, since a large number of systems are neither pure continuous nor discrete but a combination. This is mostly due to the growing use of computers in the control of physical plants, but also as a result of the hybrid nature of many physical processes. There are numerous types of hybrid systems present in the literature, for example, relay systems [21], the management of fishery resources [14], computer disk systems [9], control of continuous plants by discrete event systems [18], motion systems [5], robotics [6], power systems [10], systems in classical mechanics [4], air traffic management systems [20], and automated vehicle systems [13].

This paper includes an application consisting of an automatic gear-box and cruise controller. Both the automatic gear-box (plant) and the cruise controller (controller) are naturally modeled as hybrid systems that interact to control the velocity at a desired value. The automatic gear-box is modeled as simple as possible, with the velocity and the gear position as continuous and discrete state, respectively. The discrete state is changed when the velocity reaches different values, which affects the continuous dynamics. The cruise controller consists of a PI-controller, where the continuous state is the integrator state, implying that the velocity converges to the desired state despite the influence of disturbances. To obtain a comfortable ride, there are restrictions imposed on the derivative of the acceleration. This implies that the gain in the controller must have different values for different gear positions, implying that it acts as a discrete state with different values. Furthermore, the restrictions also imply state jumps in the integrator state at the times when the gear position is changed (bumpless transfer).

Many stability results for hybrid systems using a Lyapunov approach have been proposed in the literature, *e.g*, [15], [7], [3], [24]. However, none of the approaches considers the additional complexity of including also state jumps in the analysis. Therefore, stability results applicable to hybrid systems with state jumps are proposed in this paper. The problem of finding the different local (piecewise quadratic) Lyapunov functions is formulated as a linear matrix inequality (LMI) problem [2], for which there exists numerical software, see, for instance, [8].

The proposed stability results in this paper can be generalized to consider even larger classes of hybrid systems (systems with nonlinear vector fields) than the application model. Such results have, in fact, been carried out in the Ph.D. thesis [16]. However, to keep the paper short and reduce the complexity, possible generalizations will not be discussed herein but we refer to the thesis for interested readers. A similar approach has been carried out by [11].

The outline of this paper is as follows: the application is given in detail in the next section motivating the use of stability results for hybrid systems including state jumps. Section 17.3 proposes conditions for exponential stability. It is shown how the stability result can be formulated as a linear matrix inequality (LMI) problem in Section 17.4. The paper is concluded by showing how the proposed stability result is applied to the gearbox application.

17.2 Application and Hybrid Model

The proposed theory in this paper is motivated by the following application:

Gearbox Application

A motor, together with transmission through a gearbox, is naturally modeled by continuous and discrete states. Nonlinear models describing the dynamics of a vehicle with throttle angle as input are given in [22]. Continuous state variables are the manifold pressure and velocity of the vehicle, and a discrete state variable is the gear position. In this example, a satisfactory model illustrating the hybrid behavior is obtained by assuming that the input signal is the torque T out from the motor (hence, all dynamics in the motor are neglected). The gearbox transforms the torque T and angular velocity ω according to

$$T_1 = pT \text{ and } \omega_1 = \frac{1}{p}\omega \tag{17.1}$$

where T_1 is the torque and ω_1 is the angular velocity of the wheels, and p is the gear position. If the radius of the wheel is r, the force F accelerating the vehicle and the velocity v of the vehicle becomes

$$F = T_1/r \text{ and } v = r\omega_1 \tag{17.2}$$

The vehicle acceleration is according to Newton's law of motion

$$M\dot{v} = F - F_l, \tag{17.3}$$

where M is the weight of the car and F_l is the load force induced from the road. If it is assumed that F_l is proportional to the square of the vehicle velocity v and the road angle is α, this force can be modeled as

$$F_l = kv^2 \text{sign}\, v + Mg\sin\alpha \tag{17.4}$$

By combining (17.1), (17.2) and (17.4) into (17.3), the vehicle dynamics is given by

$$\begin{aligned}\dot{v} &= \frac{p_r T}{M} - \frac{k}{M}v^2 \text{sign}\, v - g\sin\alpha, \\ \omega &= p_r v\end{aligned} \tag{17.5}$$

where $p_r = p/r$ is assumed to take values in the set $\{p_{r_1}, p_{r_2}, p_{r_3}, p_{r_4}\}$, $p_{r_1} > p_{r_2} > p_{r_3} > p_{r_4}$. Hence, there are four possible discrete gear positions, where p_{r_1} corresponds to gear 1, p_{r_2} to gear 2 and so on.

In this illustrative example, the automatic gearbox is designed in such a way that the change of gear occurs if the engine rotational speed exceeds ω_{high}, implying a higher gear (if not already gear 4), or goes below ω_{low}, implying a lower gear (if not already gear 1). Depending on the gear, the values ω_{high} and ω_{low} corresponds to different velocities of the vehicle; see (17.5). The desired behavior is obtained by changing gear position at velocities given by the switch sets

$$\begin{aligned} S_{i,i+1} &= \{v \in \Re \mid v = \tfrac{1}{p_{r_i}}\omega_{high}\}, \\ S_{i+1,i} &= \{v \in \Re \mid v = \tfrac{1}{p_{r_{i+1}}}\omega_{low}\}, \end{aligned} \quad i = 1, 2, 3$$

where $S_{i,i+1}$ denotes gear position changes from i to $i+1$ and *vice versa* for $S_{i+1,i}$.

The cruise controller is designed in the following way. The torque T consists of the terms

$$T = T_P + T_I + \frac{k}{p_r}v^2 \operatorname{sign} v \qquad (17.6)$$

where

$$\begin{aligned} T_P &= K_r(v_{ref} - v), \\ \dot{T}_I &= \frac{K_r}{T_r}(v_{ref} - v), \end{aligned} \qquad (17.7)$$

and v_{ref} is the desired velocity. Hence, the cruise controller (17.6) is essentially a PI-controller that compensates for the nonlinearity due to the load force. If the closed-loop system is (asymptotically) stable, the integrator part of the controller implies that the vehicle velocity v converges to the desired velocity for stationary input values v_{ref} despite the influence of a constant road angle α. Every time a new value of the desired velocity v_{ref} is given by the driver, the integrator state T_I is put to zero.

Besides stabilizing the closed-loop system, the parameters K_r and T_r should be selected in such a way that a desirable performance is obtained. A comfortable ride is maintained if the acceleration is limited to $|\dot{v}| \leq 2m/s^2$; cf. [1]. This condition restricts the gain K_r. In the design of the integration time T_r, there is a trade-off between fast convergence and small overshoot.

Besides conditions on the acceleration, it is desirable also to have restrictions on the derivative of the acceleration, since abrupt changes of this variable can be quite uncomfortable. One reason for possible abrupt changes of $\ddot{v}$ occurs when the gear position is changed. If t_k denotes the time when the change of gear occurs and t_k^- and t_k^+ denote the times just before and after that time, and K_r takes values in the set $\{K_{r_1}, K_{r_2}, K_{r_3}, K_{r_4}\}$, where K_{r_1} corresponds to gear 1, and so on, then (17.5) and (17.7) imply that there are no abrupt changes of $\dot{v}$ due to a change of gear if

$$\begin{aligned} p_{r_i} K_{r_i} &= p_{r_{i+1}} K_{r_{i+1}} \\ p_{r_i} T_I(t_k^-) &= p_{r_{i+1}} T_I(t_k^+) \quad \text{gear } i \text{ to } i+1 \quad i = 1, 2, 3. \\ p_{r_{i+1}} T_I(t_k^-) &= p_{r_i} T_I(t_k^+) \quad \text{gear } i+1 \text{ to } i \end{aligned} \quad (17.8)$$

Hence, by designing the gain parameters $K_{r_1}, \ldots, K_{r_4}$ and abruptly changing the value of the T_I-variable such that (17.8) is satisfied, discontinuities in $\dot{v}$ (and hence $\dot{v}$) due to change of gear position are avoided. The jump in the state variable T_I avoiding jumps in the control signal T is commonly called *bumpless transfer* [12].

Let the numerical values be equal to: $p_{r_1} = 50$, $p_{r_2} = 32$, $p_{r_3} = 20$, $p_{r_4} = 14$, $k = 0.7$, $M = 1500$, $g = 10$, $\omega_{low} = 230$, $\omega_{high} = 500$, $K_{r_1} = 3.75$, $K_{r_2} = 5.86$, $K_{r_3} = 9.37$, $K_{r_4} = 13.39$, $T_r = 40$, $v_{ref} = 30$, $m(0) = m_3$, $v(0) = 14$, and $T_I(0) = 0$. For a specified desired velocity, v_{ref}, the system converges exponentially to v_{ref}, which will be verified by LMIs after the stability theory.

Hybrid Model

The hybrid model in the application has the form:

$$\begin{aligned} \dot{x} &= A(m)x \\ x^+ &= \psi(x, m) \\ m^+ &= \phi(x, m) \end{aligned} \quad (17.9)$$

where $x \in \Re^n$ is the continuous state and $m \in \mathcal{M} = \{m_1, \ldots, m_N\}$ is the discrete state. The hybrid state space H is the Cartesian product $\Re^n \times \mathcal{M}$. The continuous dynamics is given by a linear differential equation, including the possibility of expressing state jumps by the function $\psi : \Re^n \times \mathcal{M} \to \Re^n$, and $\phi : \Re^n \times \mathcal{M} \to \mathcal{M}$ is a function describing the dynamics of the discrete state. The notation x^+ and m^+ means the next state of x and m respectively. The hybrid system described in (17.9) is autonomous, i.e., there are no external inputs affecting the dynamics. This may be the result when external inputs are feedback functions of the continuous and discrete state, which is the case in the application.

The discrete state changes when x and m take certain values can, instead of being described by a function ϕ, be expressed by a number of *switch sets* $S_{i,j}$ (as in the application), which are related to ϕ according to

$$S_{i,j} = \{x \in \Re^n \mid m_j = \phi(x, m_i)\}, \quad i \in I_N, \, j \in I_N$$

where N is the number of elements in $\mathcal{M}$, and $I_N = \{1, 2, \ldots, N\}$; cf. [23],[19]. Hence, the switch sets indicate where in the continuous state space $\Re^n$ the discrete state m_i changes to m_j. It is usual to specify only those switch sets that cause a change of discrete state variable from m_i to m_j where $m_i \neq m_j$. The switch sets are often given as geometrical hypersurfaces or hyperplanes (as in the application). Similarly, the set of states where state jumps occur for the different discrete states can equivalently be described by sets

$$J_i = \{x \in \Re^n \mid x^+ = \psi(x, m_i)\}, \quad i \in I_N$$

It is assumed that $S_{i,j}$ and J_i coincide in this paper and the next continuous state is related to the previous one according to

$$x^+ = G(m_i)x \tag{17.10}$$

at these states.

In the stability analysis given next, it is assumed that there only is a finite number of switches in finite time. Hence, the continuous and discrete dynamicsare well behaved.

17.3 Exponential Stability

We are now prepared to show exponential stability of hybrid systems including state jumps.

Region Partitioning

To show stability, $\Omega \subseteq H$ of the hybrid state space is partitioned into ℓ disjoint regions. If for a given initial point in Ω, t_k, $k = 1, 2 \ldots$ are the consecutive times when the trajectory passes from one region to another, it is assumed that the *partitioning* is made in such a way that t_k is strictly less than t_{k+1}, i.e., $t_k < t_{k+1}$.

Let Ω be a hybrid set. Then, the following projection sets are defined:

$$\begin{aligned}
\Omega^x &= \{x \in \Re^n \mid (x, m) \in \Omega\} \\
\Omega^{x, m_i} &= \{x \in \Re^n \mid (x, m_i) \in \Omega\} \\
\Omega^m &= \{m \in \mathcal{M} \mid (x, m) \in \Omega\}
\end{aligned} \tag{17.11}$$

Hence, Ω^x and Ω^{x,m_i} are sets consisting of continuous states while Ω^m consists of discrete states.

Assume that a trajectory satisfies (17.9) for any initial value in H_0. Let $\varepsilon > 0$ andefine the *neighboring regions* $\Lambda_{q,r}$, $q \in I_\ell$, $r \in I_\ell$, $(q \neq r)$ by

$$\begin{aligned}
\Lambda_{q,r} = \{(x, m) \in \Omega \mid \exists t > 0 \text{ such that } (x(t - \varepsilon), m(t - \varepsilon)) \in \Omega_q \text{ and} \\
(x(t + \varepsilon), m(t + \varepsilon)) \in \Omega_r, \text{ when } \varepsilon \to 0\}
\end{aligned}$$

which are sets where trajectories pass from Ω_q to Ω_r. Let

$$I_\Lambda = \{(q, r) \mid \Lambda_{q,r} \neq \emptyset\}$$

which is the set of tuples indicating that there is at least one point for which the trajectory passes from Ω_q to Ω_r.

Let $V_q = x^T P_q x$, $q \in I_\ell$, be a quadratic function which is used as a measure of the system's (abstract) energy in region Ω_q. Let the overall energy be defined as

$$V(x) = V_q(x) \quad \text{when} \quad (x,m) \in \Omega_q \tag{17.12}$$

which, in general, is a discontinuous function at the neighboring regions $\Lambda_{q,r}$, $(q,r) \in I_\Lambda$. Since it is assumed that the partitioning is made in such a way that $t_k < t_{k+1}$ for every trajectory with initial point in Ω, it is ensured that the overall energy defined in (17.12) is *piecewise continuous* as a function of time. The time derivative of $V_q(x)$ in region Ω_q can be written as:

$$\dot{V}_q(x) = A(m_i)^T P_q + P_q A(m_i), \quad x \in \Omega_q^{x,m_i}, \ m_i \in \Omega_q^m$$

using the projection sets in (17.11).

Exponential Stability Conditions

DEFINITION 17.1
The region of exponential attraction $R(k_1, k_2)$ of a hybrid system (17.9) is the set of initial hybrid states for which the continuous trajectory exponentially converges to the origin according to

$$R(k_1, k_2) = \{(x_0, m_0) \in H_0 \mid \|x(t)\| \leq k_1 e^{-k_2 t} \|x_0\|, \ t \geq 0, \ k_1 > 0, k_2 > 0\}$$

□

Exponential stability in a region can be verified by the following theorem. The proof can be found in [17], [16].

THEOREM 17.1
If there exist P_q, $q \in I_\ell$, and constants $\alpha > 0$ and $\beta > 0$, such that

1. $x \in \Omega_q^x$, $\quad \alpha x^T x \leq x^T P_q x \leq \beta x^T x$, $\quad q \in I_\ell$
2. $x \in \Omega_q^{x,m_i}$, $\quad x^T (A(m_i)^T P_q + P_q A(m_i))x \leq -x^T x$, $\quad m_i \in \Omega_q^m$, $q \in I_\ell$
3. $x \in \Lambda_{q,r}^x$, $\quad x^T G(m_i)^T P_r G(m_i) x \leq x^T P_q x$, $\quad m_i \in \Omega_q^m$, $(q,r) \in I_\Lambda$

then the equilibrium point 0 is exponentially stable in the sense of Lyapunov. If the assumptions hold globally, then the equilibrium point 0 is globally exponentially stable.

If V is defined as in (17.12), then

$$R_c = \{(x, m) \in H \mid V(x) \leq c\} \subseteq R(k_1, k_2)$$

where $R_c \subseteq \Omega$ and

$$k_1 = \left(\frac{\beta}{\alpha}\right)^{\frac{1}{2}}, \quad k_2 = \frac{1}{2\beta} \qquad (17.13)$$

□

The left-hand side of the first condition guarantees that each *local* quadratic Lyapunov function is positive. The right hand side is introduced to calculate the upper bound of the exponential convergence rate. The second and third condition guarantee that the overall energy V (17.12) decreases, both in regions (second condition) and when another region is entered (third condition).

To show exponential stability, the existence of P_qs satisfying the stability conditions has to be verified. This can be done by solving an LMI problem, described next.

17.4 Linear Matrix Inequalities

All conditions of the stability theorem are constrained to be satisfied, not in the entire state space but in a part of the continuous state space. The first condition is restricted to the region Ω_q^x, the second condition is restricted to Ω_q^{x,m_i} and the third condition is restricted to $\Lambda_{q,r}^x$. It is now described how the constrained conditions can be replaced by unconstrained conditions, by first expressing the regions by positive (quadratic) functions and then using a general technique called the S-procedure to obtain an unconstrained condition. This procedure is first explained in general terms and then applied to the constrained conditions in the stability theorem.

From Constrained Conditions to Unconstrained Conditions

Replacement of constraint to regions with constraint by functions.
Assume that $F^0(x) : \Re^n \to \Re$ is a function having unknown variables that are to be decided, satisfying the condition

$$F^0(x) \geq 0 \text{ for all } x \text{ in the region } \mathcal{R} \qquad (17.14)$$

Assume that $F^k(x) : \Re^n \to \Re$, $k \in I_\kappa$, are known functions satisfying

$$F^k(x) \geq 0, \; k \in I_\kappa, \text{ for all } x \text{ in the region } \mathcal{R}$$

Condition (17.14) can then be replaced with the possibly stronger condition

$$F^0(x) \geq 0 \text{ for all } x \text{ satisfying } F^k(x) > 0, \; k \in I_\kappa \qquad (17.15)$$

Hence, the condition $F^0(x) \geq 0$ constrained to the region $\mathcal{R}$ has been replaced by constraints by the functions $F^k(x) \geq 0$, $k \in I_\kappa$. The replacement of $\mathcal{R}$ by functions $F^k(x) \geq 0$ is illustrated in Figure 17.1.

Stability Analysis of Hybrid Systems —A Gearbox Application

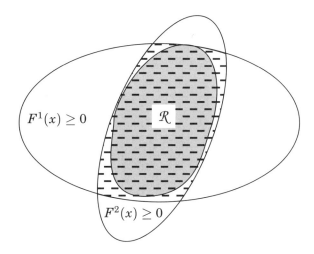

Figure 17.1 The shaded region $\mathcal{R}$ is replaced with a region described by all x satisfying $F^1(x) \geq 0$ and $F^2(x) \geq 0$, which is the dashed region

S-procedure. It is possible to replace the constrained condition (17.15) by a condition without constraints by introducing additional variables $\lambda^k \geq 0$, $k \in I_\kappa$ in the following way:

LEMMA 17.1—[2]
If there exist $\lambda^k \geq 0$, $k \in I_\kappa$, such that

$$\forall x \in \Re^n, \; F^0(x) \geq \sum_{k=1}^{\kappa} \lambda^k F^k(x), \tag{17.16}$$

then (17.15) holds. □

The proof follows directly by noting that the right hand side of (17.16) is greater or equal to zero for all x satisfying $F^k(x) \geq 0$, $k \in I_\kappa$, since $\lambda^k \geq 0$, $k \in I_\kappa$.

The constrained condition (17.14) has been replaced by the unconstrained condition in Lemma 17.1. In the case of quadratic functions

$$F^k(x) = x^T Q^k x, \quad k = 0, \ldots, \kappa \tag{17.17}$$

where $Q^k = (Q^k)^T \in \Re^n \times \Re^n$, Condition (17.16) can be written as an LMI:

$$Q^0 \geq \sum_{k=1}^{\kappa} \lambda^k Q^k, \quad \lambda^k \geq 0, \; k \in I_\kappa \tag{17.18}$$

The finesse of formulating the LMI condition as (17.18) instead of only $Q^0 \geq 0$ is that the condition $x^T Q^0 x \geq 0$ does not have to be fulfilled in the entire state space, implying that Q^0 has to be positive semi-definite, but only in a part of the state space where at least all $x^T Q^k x \geq 0$ are satisfied. The unknown matrix variable Q^0 and the different λ^k can be found by solving the LMI problem (17.18).

Some remarks should be made. First, it is always possible to replace an arbitrary region $\mathcal{R}$ with larger region constraints expressed by quadratic forms $F^k(x) \geq 0$, by simply letting $F^k(x)$ be positive semi-definite functions. This implies that $F^0(x) \geq 0$ has to be satisfied in the entire continuous state space. However, one should avoid replacing a region with quadratic forms $F^k(x) \geq 0$ such that the states satisfying all these inequalities are much larger than $\mathcal{R}$, since this conservatism may imply that the stronger condition (17.15) does not have a solution although (17.14) has. In some cases, the conservatism is no problem since a solution will exist anyway (as in the application). However, in other cases, not being too conservative is crucial for a solution to exist. In Section 17.4, it is explained more thoroughly how to specify the parameters in the quadratic forms (17.17) such that regions $\mathcal{R}$ can be replaced by quadratic forms $F^k(x) \geq 0$ without too much conservatism.

Second, the replacement of (17.15) by Lemma 17.1 may also be conservative. However, it can be shown that the converse is true in the case of a single quadratic form, $\kappa = 1$ [2], provided that there is some x such that $F^1(x) > 0$.

Third, in the case of hypersurfaces defined by $F^k(x) = 0$, $k \in I_\kappa$, it is not necessary to require that the different λ^k, $k \in I_\kappa$, have to be greater or equal to zero in Lemma 17.1, since this lemma holds despite the sign of these constants.

The above procedure is now applied to the constrained conditions in the stability theorem.

Stability conditions. All conditions of the stability theorem are described by $F^0(x) \geq 0$, where $F^0(x)$ is a quadratic function defined as in (17.17). The first and second conditions in the stability theorems are restricted to regions Ω_q^x and Ω_q^{x,m_i} respectively. These conditions can be replaced by unconstrained conditions of the form (17.18). Matrices Q^k corresponding to regions Ω_q^x are denoted Q_q^k, and regions Ω_q^{x,m_i} are denoted Q_{q,m_i}^k.

The third condition is restricted to hypersurfaces $\Lambda_{q,r}^x$. When these are given by $F^k(x) = 0$, $k \in I_\kappa$, where each $F^k(x)$ has the form (17.17), there will be no restrictions on the additional variables λ^k in (17.18). However, if some switch surface cannot exactly be described by $F^k(x) = 0$, $k \in I_\kappa$, then it is possible to include such a region with quadratic functions satisfying $F^k(x) \geq 0$, in which case the additional variables λ^k in (17.18) have to be greater or equal to zero. Matrices Q^k corresponding to hypersurfaces $\Lambda_{q,r}^x$ are denoted $Q_{q,r}^k$.

LMIs for Hybrid Systems with Linear Vector Fields

In the case of verifying exponential stability, it may be desirable not only to find a solution but to search for a solution that gives a better estimate of the convergence rate k_2 in (17.13). This can be achieved by searching for a solution where β is minimized. The LMI problem then becomes as follows:

LMI Problem. If there is a solution to

min β subject to

0. $\alpha > 0$, $\mu_q^k \geq 0$, $v_q^k \geq 0$, $\vartheta_{q,m_i}^k \geq 0$, $\quad q \in I_\ell$
1. $\alpha I + \sum_{k=1}^{K_q} \mu_q^k Q_q^k \leq P_q \leq \beta I - \sum_{k=1}^{K_q} v_q^k Q_q^k$, $\quad q \in I_\ell$
2. $A(m_i)^T P_q + P_q A(m_i) + \sum_{k=1}^{K_{q,m_i}} \vartheta_{q,m_i}^k Q_{q,m_i}^k \leq -I$, $\quad m_i \in \Omega_q^m$, $q \in I_\ell$
3. $G(m_i)^T P_r G(m_i) + \sum_{k=1}^{K_{q,r}} \eta_{q,r}^k Q_{q,r}^k \leq P_q$, $\quad (q,r) \in I_\Lambda$

then the equilibrium point 0 is exponentially stable in the sense of Lyapunov.

The variables α, μ_q, v_q, ϑ_{q,m_i} and matrices P_q are unknowns, while the different Q:s are known matrices corresponding to the different local regions where the conditions have to be valid. The convergence rate is estimated as in Theorem 17.1.

The LMI formulation can be extended to consider affine vector fields by extending the quadratic local Lyapunov functions from $V_q(x) = x^T P_q x$ to $V_q(x) = \pi_q + 2p_q^T x + x^T P_q x$. Nonlinear vector fields may also be handled by slightly modifying the second condition in the LMI problem [16].

Describing Regions by Quadratic Forms

It is now explained how to specify the parameters in the quadratic forms (17.17) such that the set of states satisfying all quadratic forms $F^k(x) \geq 0$ includes the set $\mathcal{R}$. We are focusing on regions partitioned by hyperplanes. More general regions are discussed in [16].

Quadratic forms describing half-planes. If a region $\mathcal{R}$ containing the origin is given by the set of states restricted by two half-planes

$(c^a)^T x \geq 0$ and $(c^b)^T x \geq 0$,

then $\mathcal{R}$ will be described by a quadratic form

$$x^T Q^1 x \geq 0, \text{ where } Q^1 = c^a(c^b)^T + c^b(c^a)^T \tag{17.19}$$

The set of states satisfying a quadratic form (17.19) has the property that if x_1 satisfies the inequality, so does $-x_1$; see Figure 17.2.

If the dimension n is equal to 2, there is no reason for replacing a region $\mathcal{R}$ by a quadratic form (17.19) described by more than two hyperplanes,

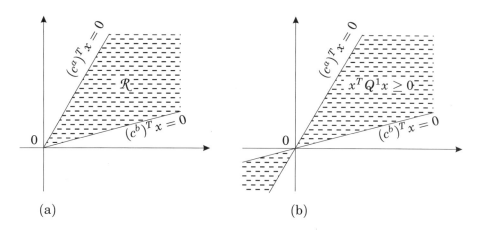

Figure 17.2 (a) Region (dashed) $\mathcal{R}$ restricted by hyperplanes $(c^a)^T x \geq 0$ and $(c^b)^T x \geq 0$. (b) Region (dashed) of states satisfying $x^T Q^1 x \geq 0$, where $Q^1 = c^a(c^b)^T + c^b(c^a)^T$.

since the set of states satisfying several half-planes can equivalently be described by only two half-planes. However, this is reasonable in higher dimensions. In this case, the quadratic forms $x^T Q^k x \geq 0$ are obtained by taking all possible combinations of two different half-planes. This results in $\frac{\rho(\rho-1)}{2}$ different quadratic forms as in (17.19), and hence also variables λ^k in (17.16), where ρ is the number of half-planes. There is no reason to add the combinations of the same half-planes since these quadratic forms are greater or equal to zero for all states.

The reason for specifying a number of quadratic inequalities instead of only one, in case $\mathcal{R}$ is restricted by several half-planes, is that $\mathcal{R}$ cannot exactly be described by the set of states given by a quadratic form greater or equal to zero, even if $\mathcal{R}$ is symmetric around the origin (meaning that if $x \in \mathcal{R}$ then $-x \in \mathcal{R}$; cf. Figure 17.2 b). The set of states satisfying $x^T Q^1 x \geq 0$ for a quadratic form given by any combination of two half-planes describing $\mathcal{R}$ will be strictly larger than $\mathcal{R}$. Since it cannot be said that the set of states given by one quadratic form $x^T Q^1 x \geq 0$ is better than another, all reasonable combinations are specified. The variables λ^k obtained by solving the resulting LMI problem (17.18) then decide the quadratic form such that $\mathcal{R}$ is most suitably replaced by the set of states satisfying the quadratic inequality.

Quadratic forms describing hyperplanes. The quadratic forms equal to zero at a hyperplane can be obtained as follows. Assume that $\mathcal{R}$ is given by the set of states satisfying a hyperplane

$$c^T x = 0 \tag{17.20}$$

where $c = [c^1 \ldots c^n]^T \in \Re^n$. The states satisfying (17.20) also satisfy

$$2(\lambda^T x)^T (c^T x) = 0 \tag{17.21}$$

where $\lambda = [\lambda^1, \ldots, \lambda^n]^T \in \Re^n$ are arbitrary additional variables. The equality in (17.21) can be written as

$$x^T \lambda c^T x + x^T c \lambda^T x = \sum_{k=1}^{n} \lambda^k x^T Q^k x = 0$$

where

$$Q^k = e^k c^T + c(e^k)^T$$

and e^k is a column vector with n elements such that

$$e^k(i) = \left\{ \begin{array}{ll} 1, & i = k, \\ 0, & i \neq k \end{array} \right\}$$

where i means the ith element of e^k.

17.5 Stability of the Gearbox Application

We are now prepared to show stability of the gearbox application. By denoting $\Delta v = v_{ref} - v$ and $\Delta T_I = T_I$, the closed-loop dynamics become

$$\begin{bmatrix} \Delta \dot{v} \\ \Delta \dot{T}_I \end{bmatrix} = \begin{bmatrix} -p_r K_r / M & -p_r / M \\ K_r / T_r & 0 \end{bmatrix} \begin{bmatrix} \Delta v \\ \Delta T_I \end{bmatrix}$$

where $p_r \in \{50, 32, 20, 14\}$, $K_r \in \{3.75, 5.86, 9.37, 13.39\}$, $M = 1500$, $T_r = 40$ and $p_r K_r = 187.5$ for all discrete states. For a specified desired velocity v_{ref} (= 30 m/s), the system converges exponentially to v_{ref}, illustrated in Figure 17.3, and formally proven next.

If it is first assumed that there are no state jumps in T_I, stability can be shown by a single partitioning, implying a single Lyapunov function common for all discrete states. This results in a solution

$$P = \begin{bmatrix} 255.589 & 72.262 \\ 72.262 & 40.822 \end{bmatrix}$$

satisfying the conditions in the LMI problem. Hence, the hybrid system is globally exponentially stable without state jumps. The optimal value of $\beta = 277.6388$.

If the state jumps are included in the dynamics, they occur when the discrete state is changed. Trajectories satisfying the condition $T_I >$

$K_r(v - v_{ref})$ cross the switch set $S_{i,i+1}$ and $S_{i+1,i}$ from left to right (see Figur 17.3) and oppositely for $T_I < K_r(v - v_{ref})$. In the operating region of this cruise controller ($T_I(0)$ is always put to zero when a new desired velocity is given), the gear shiftings will always occur from lower to higher gear when the first condition is satisfied and conversely in the second case. Hence, the third condition of the LMI problem is formulated such that the energy decreases passing from gears i to $i + 1$ satisfying $T_I > K_r(v - v_{ref})$ and gears $i + 1$ to i satisfying $T_I < K_r(v - v_{ref})$.

Consider the case when the trajectories start in the first region. The jump condition (17.8) on the form (17.10) gives

$$G(m_i) = \begin{bmatrix} 1 & 0 \\ 0 & \frac{p_{r_{i+1}}}{p_{r_i}} \end{bmatrix} \quad i = 1, 2, 3$$

According to the LMI problem, the third LMI condition then becomes

$$G(m_i)^T P G(m_i) \leq P$$

which is equivalent to

$$p^{2,2} \left(\frac{p_{r_{i+1}}}{p_{r_i}}\right)^2 \leq p^{2,2}$$

where $p^{2,2}$ is the (2,2) element of P. Since $\left(\frac{p_{r_{i+1}}}{p_{r_i}}\right)^2 < 1$, the energy will decrease due to the state jumps for any quadratic function $x^T P x$. Therefore, the same solution as above verifies stability in this case. However, when the trajectories start in the second region, the jump condition (17.8) is the same as above except that p_{r_i} and $p_{r_{i+1}}$ change position. In this case, there will not exist any solution since $\left(\frac{p_{r_i}}{p_{r_{i+1}}}\right)^2 > 1$.

To overcome this problem, the state space is further partitioned to verify exponential stability. One quadratic Lyapunov-like function candidate is associated with each of the discrete states. The switch surfaces $\Lambda_{i,i+1}$ then coincide with $S_{i,i+1}$ for $i = 1, 2, 3$. Solving the LMI problem leads to a solution

$$P_1 = \begin{bmatrix} 304.082 & 87.089 \\ 87.089 & 376.934 \end{bmatrix}, \quad P_2 = \begin{bmatrix} 248.013 & 79.625 \\ 79.625 & 144.215 \end{bmatrix}$$

$$P_3 = \begin{bmatrix} 212.101 & 59.328 \\ 59.328 & 53.788 \end{bmatrix}, \quad P_4 = \begin{bmatrix} 147.495 & 53.571 \\ 53.571 & 24.112 \end{bmatrix}$$

Hence, the hybrid system is also exponentially stablein the case of state jumps. The optimal value of $\beta = 439.9$. The level curves for the local quadratic Lyapunov functions are shown in Figure 17.3.

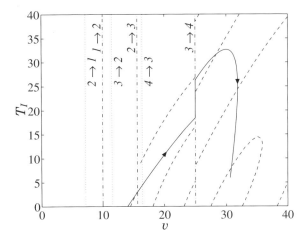

Figure 17.3 The phase plot of v and T_I during a simulation of 80 seconds. The dash-dotted lines (-·-) are the hyperplanes when the gear shifts from lower to higher gears ($i \to i+1$) and oppositely ($i+1 \to i$) at the dotted lines (··). The dashed lines (— —) are the level curves of the local quadratic Lyapunov functions. The continuous trajectories cross these curves such that the energy decreases all the time, verifying that the system is exponentially stable.

17.6 Conclusions

A gearbox application has served as a motivation for investigating stability of hybrid systems including state jumps. None of the results reported in the literature can deal with this additional complexity. Stability results for hybrid systems, including state jumps, are proposed in this paper using Lyapunov techniques. It has been shown how to formulate the search for the (piecewise quadratic) Lyapunov functions as a linear matrix inequality (LMI) problem. The theory has been applied to the gearbox application to formally show stability.

17.7 References

[1] A. Björnberg. Design of control algorithms for intelligent cruise control. Technical Report 184L, Control Engineering Lab, Chalmers University of Technology, 1994.

[2] S. Boyd, L. El Ghaoui, E. Feron, and V. Balakrishnan. *Linear Matrix Inequalities in System and Control Theory*. SIAM, 1994.

[3] M. S. Branicky. Stability of switched and hybrid systems. In *Proc. of the 33rd IEEE Conference on Decision and Control*, 3498–3503, Lake Buena Vista, FL, 1994.

[4] R. W. Brockett. Hybrid systems in classical mechanics. In *Proc. of 13th IFAC*, c:473–476, 1996.

[5] R. W. Brockett. Hybrid models for motion control systems. In H. L. Trentelman and J. C. Willems, editors, *Essays on Control: Perspectives in the Theory and its Applications*, chapter 2, 29–53. Birkhäuser, 1993.

[6] M. Doğruel, S. Drakunov, and Ü. Özguner. Sliding mode control in discrete state systems. In *Proceedings of the 32nd CDC*, 1194–99, 1993.

[7] M. Doğruel and Ü. Özgüner. Stability of hybrid systems. In *IEEE International Symposium on Intelligent Control*, 129–134, 1994.

[8] P. Gahinet, A. Nemirovski, A. J. Laub, and M. Chilali. *LMI Control Toolbox, For use with MATLAB*. The Math Works Inc., 1995.

[9] A. Göllü and P. Varaiya. Hybrid dynamical systems. In *Proc. of the 28th IEEE Conference on Decision and Control*, 2708–2712, Tampa, Florida, Dec. 1989.

[10] I. A. Hiskens. Analysis tools for power systems—contending with nonlinearities. *Proceedings of the IEEE*, 83(11):1573–1587, 1995.

[11] M. Johansson and A. Rantzer. Computation of piecewise quadratic Lyapunov functions for hybrid systems. In *Proc. 4th European Control Conference*, Brussels, Belgium, 1997. ID. 516.

[12] W. S. Levine, Ed. *The Control Handbook*. CRC Press, 1996.

[13] J. Lygeros, D. N. Godbole, and S. Sastry. Verified hybrid controllers for automated vehicles. *IEEE Trans. Automatic Control*, 43(4):522–539, 1998.

[14] P. Peleties and R. DeCarlo. Modeling of interacting continuous time and discrete event systems: An example. In *Twenty-Sixth Annual Allerton Conference on Communication, Control and Computing*, 1150–9, 1988.

[15] P. Peleties and R. DeCarlo. Asymptotic stability of m-switched systems using Lyapunov-like functions. In *Proc. American Control Conference*, 1679–1684, Boston, 1991.

[16] S. Pettersson. *Analysis and Design of Hybrid Systems*. PhD thesis, Control Engineering Laboratory, Chalmers University of Technology, 1999.

[17] S. Pettersson and B. Lennartson. Controller design of hybrid systems. In Oded Maler, editor, *Lecture Notes in Computer Science 1201*, 240–254. Springer-Verlag, 1997.

[18] J. A. Stiver and P. J. Antsaklis. Modeling and analysis of hybrid control systems. In *Proc. of the 31st IEEE Conference on Decision and Control*, 3748–3751, Tucson, Arizona, 1992.

[19] L. Tavernini. Differential automata and their discrete simulators. *Nonlinear Analysis, Theory, Methods & Applications*, 11(6):665–83, 1987.

[20] C. Tomlin, G. J. Pappas, and S. Sastry. Conflict resolution for air traffic management: A study in multiagent hybrid systems. *IEEE Trans. on Automatic Control*, 43(4):509–521, 1998.

[21] Y. Z. Tsypkin. *Relay Control Systems*. Cambridge University Press, 1984.

[22] L. Y. Wang, A. Beudoun, J. Cook, J. Sun, and I. Kolmanovsky. Optimal hybrid control with automotive applications. In A. S. Morse, Ed., *Logic-based Switching Control*. Springer, 1996.

[23] H. S. Witsenhausen. A class of hybrid-state continuous-time dynamic systems. *IEEE Trans. on Automatic Control*, 11(2):161–67, 1966.

[24] H. Ye, A. N. Michel, and L. Hou. Stability analysis of discontinuous dynamical systems with applications. In *Proc. of 13th IFAC*, E:461–466, 1996.

18

On the Existence and Uniqueness of Solution Trajectories to Hybrid Dynamical Systems

W. P. M. H. Heemels   M. K. Çamlıbel   A. J. van der Schaft
J. M. Schumacher

Abstract

In this paper, we study the fundamental system-theoretic property of well-posedness for several classes of hybrid dynamical systems. Hybrid systems are characterized by the presence and interaction of continuous dynamics and discrete actions. Many different description formats have been proposed in recent years for such systems; some proposed forms are quite direct, others lead to rather indirect descriptions. The more indirect a description form is, the harder it becomes to show that solutions are well-defined. This paper intends to provide a survey on the available results on existence and uniqueness of solutions for given initial conditions in the context of various description formats for hybrid systems.

18.1 Introduction

Very broadly speaking, scientific modeling may be defined as the process of finding common descriptions for groups of observed phenomena. Often, several description forms are possible. To take an example from not very recent technology, suppose we want to describe the flight of iron balls fired from a cannon. One description can be obtained by noting that such balls approximately follow parabolas, which may be parameterized in terms of firing angle, cannon ball weight, and amount of gun powder used. Another possible description characterizes the trajectories of the cannon balls as solutions of certain differential equations. The latter description may be viewed as being fairly *indirect*; after all, it represents trajectories only as solutions to some problem, rather than expressing directly what the trajectories are, as the first description form does. On the other hand, the description by means of differential equations is applicable to a wider range of phenomena, and one may therefore feel that it represents a deeper insight. Besides, interconnection (composition) becomes much easier since it is, in general, much easier to write down equations than to determine the solutions of the interconnected system.

There are many examples in science where, as above, an implicit description (that is, a description in terms of a mathematical problem that needs to be solved) is useful and possibly more powerful than explicit descriptions. Whenever an implicit description is used, however, one has to show that the description is a "good" one in the sense that the stated problem has a well-defined solution. This is essentially the issue of well-posedness.

In this article, we are concerned with *hybrid dynamical systems*, that is, systems in which continuous dynamics and discrete actions both occur and influence each other. Many different description formats have been proposed in recent years for such systems; some proposed forms are quite direct, others lead to rather indirect descriptions. The direct forms have advantages from the point of view of *analysis*, but the indirect forms are often preferable from the perspective of *modeling* (specification); examples will be seen below. The more indirect a description form is, the harder it becomes to show that solutions are well-defined. This paper intends to provide a survey on the available results on existence and uniqueness of solutions for given initial conditions in the context of various description formats for hybrid systems.

We consider here systems in which the description of continuous dynamics is based on ordinary differential equations; in particular, we do not consider delayed arguments, partial differential equations, or stochastic differential equations. All of these settings require their own notions of well-posedness. Even in the context of ordinary differential equations, there are situations in which one is naturally led to the consideration of well-posedness problems for systems with mixed boundary conditions, (*i.e.*, partly initial conditions, partly final conditions); see [36, Section 3.5] for an example derived from an optimal control problem. Here, however, we shall concentrate on initial value problems. Furthermore we only consider models that are formulated in continuous time. Discrete-time models are often stated in explicit form so that well-posedness is not much of an issue; that is not to say, of course, that implicit discrete-time models would not be sometimes useful.

18.2 Model Classes

We begin by introducing a number of description formats for hybrid systems.

The Hybrid Automaton Model

Hybrid systems research is sometimes viewed as a merger between dynamical systems/control theory on one side and computer science/automata theory on the other. It is therefore natural to look for description forms that combine elements from both sides. One way is to start with models that are used in computer science and to extend these with elements from continuous system theory.

In computer science, direct description forms appear to dominate. A typical specification of a finite automaton consists of a list of all states together with the transitions that may occur from each of these states and the conditions under which these transitions may take place. In more structured descriptions, such as Petri nets, the collection of states is not listed explicitly, but there is still for each state a simple rule that defines the possible successor states. *Determinism* (in the sense that a uniquely determined trajectory exists for a given initial condition and, if applicable, a given input sequence) is not always required; for instance, if the model is to be used to prove a certain property and it is suspected that the proof will not depend on certain details of the dynamics, it is very convenient to leave these details unspecified. The discrete systems studied by computer scientists are often very large and so a key issue is *compositionality*, that is, the feasibility of putting subsystems together to form a larger system.

The hybrid automaton model as proposed in [2] may be described briefly as follows. The discrete part of the dynamics is modeled by means of a graph whose vertices are called *locations* and whose edges are *transitions*. The continuous state takes values in a vector space X. To each location there is a set of trajectories, which are called *activities* in [2], and which represent the continuous dynamics of the system. Interaction between the discrete dynamics and the continuous dynamics takes place through *invariants* and *transition relations*. Each location has an invariant associated to it, which describes the conditions that the continuous state has to satisfy at this location. Each transition has an associated transition relation, which describes the conditions on the continuous state under which that particular transition may take place and the effect that the transition will have on the continuous state. Invariants and transition relations play supplementary roles: whereas invariants describe when a transition *must* take place (namely, when otherwise the motion of the continuous state as described in the set of activities would lead to violation of the conditions given by the invariant), the transition relations serve as "enabling conditions" that describe when a particular transition *may* take place.

In the model of [2], transitions are further equipped with *synchronization labels*, which express synchronization constraints between different automata. This construct allows the introduction of a notion of *parallel composition* between two automata. The component automata are assumed to have the same continuous state space, and the set of activities at each location of the composition (which is a pair of locations of the component automata) is the intersection of the sets of activities at the corresponding component locations.

The hybrid automaton model provides a particular description format for discrete dynamics, obviously inspired by the finite automaton model. An alternative would, in principle, have been to use the notion of a formal language. As for the continuous part of the dynamics, the model of [2] does opt for a description at a more general level, with no *a priori* selection of a particular specification form. The "sets of activities" of [2] may be compared to the "behaviors" of [41].

Various ramifications of the hybrid automaton model have been proposed in the literature. Sometimes, the notion of a transition relation is split up into two components, namely a *guard*, which specifies the subset of the state space where a certain transition is enabled, and a *jump function* which is a (set-valued) function that specifies which new continuous states may occur given a particular transition and a particular previous continuous state. Often the hybrid automaton model is extended with a description format for continuous dynamics, typically systems of differential equations. Versions of the hybrid automaton model which include external inputs have been proposed for instance in [31], [29], [5].

Explicit State-space Model

Many studies in continuous-variable control theory are based on the model $\dot{x}(t) = f(x(t), u(t))$, where $x(t)$ denotes a continuous state variable and $u(t)$ is a continuous control variable. Often one just writes $\dot{x} = f(x, u)$, suppressing the dependence of all variables on time. A model in the same spirit for hybrid systems may be written down as follows:

$$\dot{x} = f(x, q, u, r) \tag{18.1}$$
$$q^+ = g(x, q, u, r) \tag{18.2}$$

where x and u are continuous state and control variables as before, q and r denote discrete state and control variables, and superscript "+" is used to indicate "next state". The function g expresses updates of the discrete state that depend on the current values of both the continuous and the discrete state, as well as on the continuous and discrete inputs.

We call the above model "explicit" even though the continuous dynamics are actually given in terms of a problem, to wit a differential equation,

Solution Trajectories to Hybrid Dynamical Systems   395

since the model gives the time derivative of the continuous state variable explicitly as a function of all variables in the system. The discrete-state update is given explicitly as well. For such models, the well-posedness issue is rather easy (if not trivial) because of the explicit nature, see, for instance, [6].

Differential Inclusions

During the past decades, extensive studies have been made of *differential equations with discontinuous right hand sides*; see in particular [18] and [38], [39]. For a typical example, consider the following specification:

$$\dot{x} = f_1(x) \quad (h(x) > 0) \tag{18.3a}$$
$$\dot{x} = f_2(x) \quad (h(x) < 0) \tag{18.3b}$$

where h is a real-valued function. A system of this form can be looked at either as a discontinuous dynamical system or as a hybrid system of a particular form. The specification above is obviously incomplete since no statement is made about the situation in which $h(x) = 0$. One way to arrive at a solution concept is to adopt a suitable *relaxation*. Specifically, in a *convex* relaxation one would rewrite the equations (18.3) as

$$\dot{x} \in F(x) \tag{18.4}$$

where the set-valued function $F(x)$ is defined by

$$F(x) = \{f_1(x)\} \quad (h(x) > 0), \quad F(x) = \{f_2(x)\} \quad (h(x) < 0),$$
$$F(x) = \{y \mid \exists a \in [0,1] \text{ s.t. } y = af_1(x) + (1-a)f_2(x)\} \quad (h(x) = 0) \tag{18.5}$$

where it is assumed (for simplicity) that f_1 and f_2 are given as continuous functions defined on $\{x \mid h(x) \geq 0\}$ and $\{x \mid h(x) \leq 0\}$, respectively. The discontinuous dynamical system has now been reformulated as a *differential inclusion*, and so solution concepts and well-posedness results can be applied that have been developed for systems of this type [3]. Other methods to obtain differential inclusions are proposed by Utkin ("control equivalent definition") and Aizerman and Pyatnitskii (see also Section 18.8). In case the vector fields $f_i(x)$ are linear (*i.e.*, of the form $A_i x$ for some matrix A_i) and the switching surface is given by a linear function h, then the system (18.3) is called a *piecewise linear* or multi-modal linear system (see Section 18.6).

Complementarity systems

Systems of the form (18.3) are sometimes known as *variable-structure systems*; they describe a type of mode-switching. A similar mode-switching behavior is obtained from a class of systems known as *complementarity systems* [19], [9], [35], [21], [37]. Equations for a complementarity system may be written in terms of a state variable x and auxiliary variables v and z, which must be vectors of the same length. Typical equations are:

$$\dot{x} = f(x, v) \tag{18.6a}$$
$$z = h(x, v) \tag{18.6b}$$
$$0 \leqslant z \perp v \geqslant 0 \tag{18.6c}$$

where the last line means that the components of the auxiliary variables v and z should be no-nnegative, and that for each index i and for each time t at least one of the two variables $v_i(t)$ and $z_i(t)$ should be equal to 0. Variables that satisfy such relations occur naturally in various problems; think of current/voltage in connection with ideal diodes, flow/pressure in connection with one-sided valves, Lagrange multiplier/slack variable in optimization subject to inequality constraints, and so on. Like (18.3), the system (18.6) consists of a number of different dynamical systems or "modes" that are glued together. The modes can be thought of as discrete states. They correspond to a fixed choice, for each of the indices i, between the two possibilities $v_i \geqslant 0$, $z_i = 0$ and $v_i = 0$, $z_i \geqslant 0$, so that a complementarity system in which the vectors v and z have length m has 2^m different modes. The specification (18.6) is, in general, not complete yet; one has to add a rule that describes possible jumps of the state variable x when a transition from one mode to another takes place.

The description (18.6) is implicit in the discrete variables. Suppose we are at a point where a transition must occur because otherwise an inequality constraint would be violated. There may or may not be a unique mode in which the differential equations of (18.6a), together with the equality constraints in (18.6c) that are implied by the given mode, produce a solution that satisfies the complementary inequality constraints in (18.6c) at least for some positive time interval. If there is indeed a unique solution to this problem, then this mode is taken as the successor state. In case this procedure can be successfully carried out at all points of the continuous state space, the complementarity system can, in principle, be rewritten in the explicit hybrid automaton format, but the representation that is obtained may be very awkward.

18.3 Solution Concepts

A description format for a class of dynamical systems only specifies a collection of trajectories if one provides a notion of solution. Actually, the term "solution" already more or less suggests an implicit description format; in computer science terms, one may also say that a definition should be given of what is understood by a *run* (or an *execution*) of a system description. Formally speaking, description formats are a matter of syntax: they specify what is a well-formed expression. The notion of solution provides semantics: to each well-formed expression it associates a collection of functions of time. In the presentation of description formats above, the syntactic and semantic aspects have not been strictly separated, for reasons of readability. Here, we review in a more formal way solution concepts for several of the description formats that were introduced.

First, consider the hybrid automaton model. To simplify the situation somewhat, we consider models without synchronization labels. The model is then specified by: a finite set Loc; a finite-dimensional real vector space X; a mapping Act from the set Loc to the set $\mathcal{P}(X^{[0,\infty)})$ of collections of functions from $[0,\infty)$ to X; a finite subset Edg of the set $Loc \times 2^{X \times X} \times Loc$; and a mapping Inv from the set Loc to the set 2^X of subsets of X. A *run* of the hybrid automaton is defined to be a finite or infinite sequence $((\delta_0, \ell_0, v_0, f_0), (\delta_1, \ell_1, v_1, f_1), \ldots)$ of elements of $[0,\infty) \times Loc \times X \times X^{[0,\infty)}$ such that the following conditions are satisfied for each $i = 0, 1, 2, \ldots$:

$f_i \in Act(\ell_i)$

$f_i(0) = v_i$

for all $0 \leqslant t \leqslant \delta_i$, $f(t) \in Inv(\ell_i)$

there exists $(\ell, \mu, \ell') \in Edg$ such that $\ell = \ell_i$, $\ell' = \ell_{i+1}$, and $(f_i(\delta_i), v_{i+1}) \in \mu$.

Note that the numbers $\delta_i \geqslant 0$ denote differences between event times (durations or dwell times) rather than event times themselves. In the terminology of [2], a run is said to *diverge* if $\sum \delta_i$ is infinite. A hybrid automaton is said to be *non-zeno* if all of its runs can be extended to divergent runs. This terminology is focused, in particular, on the obstruction that may arise when the sequence of δ_is is infinite but the sum $\sum \delta_i$ is finite; in this case the system exhibits "live-lock" (an infinite number of events at one time instant) or a right accumulation of event times. The solution concept of [2] does not allow for left accumulations of event times[1].

As we have seen above, some hybrid systems can alternatively be viewed as differential inclusions. The standard solution concept for differential inclusions is the following. A vector function $x(t)$ defined on an interval $[a, b]$ is said to be a *solution* of the differential inclusion $\dot{x} \in F(x)$, where $F(\cdot)$ is a set-valued function, if $x(\cdot)$ is absolutely continuous and satisfies $\dot{x}(t) \in F(x(t))$ for almost all $t \in [a, b]$. The requirement of absolute continuity guarantees the existence of the derivative almost everywhere. One may note that the solution concept for differential inclusions does not have a preferred direction of time, as opposed to the notion of an execution for hybrid automata.

For complementarity systems, one may develop several solution concepts that may be similar to the notion of a run for hybrid automata, or to the solution concept for differential inclusions as discussed above. A solution concept of the first type can for instance be formulated as follows. A triple

---

[1] An element t of a set $\mathcal{E}$ is said to be a *left (right) accumulation point* if for all $t' > t$ ($t' < t$) $(t, t') \cap \mathcal{E}$ $((t', t) \cap \mathcal{E})$ is not empty.

$(v, x, z) : [a, b] \mapsto \mathbb{R}^{m+n+m}$ is said to be a *forward solution* of the system (18.6) on the interval $[a, b]$, if x is continuous on $[a, b]$, there exists a sequence of time points $(t_0, t_1, \ldots)$ with $t_0 = a$, $t_{j+1} > t_j$ for all j, and either $t_N = b$ or $\lim_{j \to \infty} t_j = b$, as well as for each $j = 0, 1, \ldots$ an index set I_j, such that for all j the restrictions of $x(\cdot)$, $v(\cdot)$, and $z(\cdot)$ to (t_j, t_{j+1}) are real-analytic, and for all $t \in (t_j, t_{j+1})$ the following holds:

$$\dot{x}(t) = f(x(t), v(t)), \; z(t) = h(x(t), v(t))$$

$$z_i(t) = 0 \text{ for } i \in I_j, \; v_i(t) = 0 \text{ for } i \notin I_j$$

$$z_i(t) \geq 0 \text{ for } i \notin I_j, \; v_i(t) \geq 0 \text{ for } i \in I_j$$

The definition requires that the x-part of the solutions are continuous across events. For so-called "high-index" systems, this requirement is too strong and one has to add jump rules that connect continuous states before and after an event has taken place. Under suitable conditions (specifically, in the case of linear complementarity systems and in the case of Hamiltonian complementarity systems), a general jump rule may be given; see [21, 35]. Another possibly restrictive aspect of the definition lies in the fact that it assumes that the set of event times is well-ordered[2] by the usual order of the reals, but not necessarily by the reverse order; in other words, event times may accumulate to the right, but not to the left. This lack of symmetry with respect to time can be removed by allowing the set of event times $\mathcal{E}$ to be of a more general type. For instance, one may require that $\mathcal{E}$ is closed and nowhere dense[3]; this guarantees that the complement of $\mathcal{E}$ is open and that for each event time τ one can construct sequences of non-event times converging to τ, both of which may be useful properties for other parts of the definition. In particular, if solutions are assumed to be continuous across events, then the requirements listed below can be made applicable as such to maximal intervals between events. Solutions that are obtained in this way are called *hybrid solutions*, because the corresponding solution concept is still based on explicit reference to event times.

An alternative concept that foregoes explicit mention of events is the following one, which turns out to be convenient for complementarity systems that satisfy a certain passivity condition. A triple $(x, v, z) \in L_2^{n+2m}$ is said to be an L_2-*solution* of (18.6) on the interval $[0, T]$ with initial condition x_0 if for almost all $t \in [a, b]$ the following conditions hold:

$$x(t) = x_0 + \int_0^t f(x(s), v(s)) \, ds$$

$$z(t) = h(x(t), v(t))$$

$$0 \leq z(t) \perp v(t) \geq 0.$$

---

[2] An ordered set S is said to be well-ordered if each non-empty subset of S has a least element.

[3] A closed subset of a topological space is nowhere dense if and only if its interior is empty.

This definition is in the spirit of the definition given above for differential inclusions.

Some other solution concepts have been proposed in which solutions are defined as limits of approximate solutions defined by some approximation scheme ("sampling solutions" [12], "Euler solutions" [13]).

18.4 Well-posedness Notions

In the context of systems of differential equations, the term well-posedness roughly means that there is a nice relation between trajectories and initial conditions (or, more generally, boundary conditions). There are various ways in which this idea can be made more precise, so the meaning of the term may, in fact, be adapted to the particular problem class at hand. Typically, it is required that solutions exist and are unique for any given initial condition. Both for the existence and for the uniqueness statement, one has to specify a function class in which solutions are considered. The function class used for existence may be the same as the one used for uniqueness, or they may be different; for instance, one might prove that solutions exist in some function class and that uniqueness holds in a larger function class. In the latter situation one is able to show specific properties (the ones satisfied by the smaller function class) of solution trajectories in the larger class. In the case where one is dealing with a system description that includes equality and/or inequality constraints, it may be reasonable to limit the set of initial conditions to a suitably chosen set of "feasible" or "consistent" initial conditions.

If solutions exist and are unique, a given system description defines a mapping from initial conditions to trajectories. In the theory of smooth dynamical systems, it is usually taken as part of the definition of well-posedness that this mapping is continuous with respect to suitably chosen topologies. In the case of non-smooth and hybrid dynamical systems, it frequently happens that there are certain boundaries in the continuous state space separating regions of initial conditions that generate widely different trajectories. Therefore, continuous dependence of solutions on initial conditions (at least in the sense of the topologies that are commonly used for smooth dynamical systems) may be too strong a requirement for hybrid systems.

One may also distinguish between various notions of well-posedness on the basis of the time interval that is involved. For instance, in the context of hybrid automata, one may say that a given automaton is *non-blocking* [25] if for each initial condition either at least one transition is enabled or an activity during an interval of positive length is possible. If the continuation is unique (the automaton is *deterministic* [25]), one may then say that the automaton is *initially well-posed*. This definition allows a situation in which a transition from location 1 to location 2 is immediately followed by a transition back to location 1 and so on in an infinite loop, so that all δ_is in the definition of a run of the hybrid automaton are equal to zero (live-lock). A stronger notion is obtained by requiring that a solution exists at least on an interval $[0, \varepsilon)$ with $\varepsilon > 0$; system descriptions for which such solutions exist and are unique are called *locally well-posed*. In computer science terminology, such systems "allow time to progress". Finally, if solutions exist and are unique on the whole half-line $[0, \infty)$, then one speaks of *global well-posedness*.

18.5 Well-posedness of Hybrid Automata

As already mentioned above, a useful framework to describe hybrid dynamical systems is that of a *hybrid automaton*, see [2],[36],[29],[31], [5]. Here, we adapt the description of Section 18.2 using the set-up as in [25], [30] in which the set of activities is defined by ordinary differential equations. Basically, a hybrid automaton *merges* the standard concepts of automata and continuous-time dynamics, by associating to every discrete state or *location* $\ell \in Loc$ of the automaton a continuous-time dynamics[4] $\dot{x} = f_\ell(x)$ generating the set $Act(\ell)$ for the continuous state x. Furthermore, the continuous-time dynamics may induce discrete transitions in the locations by specifying for every location ℓ a so-called *location invariant* $Inv(\ell)$, which is a subset of the continuous state space X (taken to be $\mathbb{R}^n$ for simplicity), specifying the feasible set of continuous states for the location ℓ, in the sense that if exit of the continuous state from the location invariant is imminent, then a transition to another location ℓ' and/or a reset of the continuous state x has to take place (or the system is in a deadlock). The discrete transitions are given by a collection of edges $E \subset Loc \times Loc$. For every discrete transition $(\ell, \ell') \in E$ a guard $G(\ell, \ell') \subset X$ is specified, defining *enabling* conditions on the continuous state in order that the transition to ℓ' may take place. Another interplay between discrete and continuous dynamics is provided by the reset relations $R(\ell, \ell') \subset X \times X$, specifying for every discrete transition $(\ell, \ell') \in E$ the continuous state reset from $x \in G(\ell, \ell')$ to $x' \in X$ such that $(x, x') \in R(\ell, \ell')$. In the terminology of Section 18.3, this means that

$$Edg = \{(\ell, v, v', \ell') \in Loc \times X \times X \times Loc \mid (\ell, \ell') \in E, \, v \in G(\ell, \ell'), \, (v, v') \in R(\ell, \ell')\}$$

Sometimes, a set of initial (hybrid) states $Init \subseteq Loc \times X$ is given that restricts the possible starting points of the executions.

Necessary and sufficient conditions for well-posedness of hybrid automata have been stated in [30], see also [25, 27]. Basically, these conditions mean that transitions with non-trivial reset relations are enabled whenever continuous evolution is impossible, this property is called *non-blocking*, and that discrete transitions must be forced by the continuous flow exiting the invariant set, no two discrete transitions can be enabled simultaneously, and no point x can be mapped onto two different points $x' \neq x''$ by the

---

[4]A more general setting would allow differential and algebraic equations $F_\ell(x, \dot{x}) = 0$ instead of ordinary differential equations only.

Solution Trajectories to Hybrid Dynamical Systems

reset relation $R(\ell, \ell')$, this property is called *determinism*. We will formally state the results of [30], [25] after introducing some necessary concepts and definitions.

DEFINITION 18.1—[25]
A hybrid time trajectory $\tau = \{I_i\}_{i=0}^{N}$ is a finite ($N < \infty$) or infinite ($N = \infty$) sequence of intervals of the real line, such that:

- $I_i = [\tau_i, \tau_i']$ with $\tau_i \leqslant \tau_i' = \tau_{i+1}$ for $0 \leqslant i < N$;
- if $N < \infty$, either $I_N = [\tau_N, \tau_N']$ with $\tau_N \leqslant \tau_N' \leqslant \infty$.

□

Note that a hybrid time trajectory does not allow left accumulation points. The event set $\mathcal{E} := \{0\} \cup \{\frac{1}{n} \mid n \in \mathbb{N}\}$ and the corresponding sequence of intervals cannot be rewritten in terms of a hybrid time trajectory. Hence, the above definition excludes implicitly specific Zeno behaviour.

We say that the hybrid time trajectory $\tau = \{I_i\}_{i=0}^{N}$ is a prefix of $\tau' = \{J_i\}_{i=0}^{M}$, and write $\tau \leqslant \tau'$ if they are identical or τ is finite, $M \geqslant N$, $I_i = J_i$ for $i = 0, 1, \ldots, N-1$, and $I_N \subseteq J_N$. In case τ is a prefix of τ' and they are not identical, τ is a *strict* prefix of τ'.

DEFINITION 18.2
An execution χ of a hybrid automaton is a collection $\chi = (\tau, \lambda, \xi)$ with τ a hybrid time trajectory, $\lambda : \tau \to Loc$ and $x : \tau \to X$, satisfying:

- $(\lambda(\tau_0), \xi(\tau_0)) \in Init$ (initial condition);
- for all i such that $\tau_i < \tau_i'$, ξ is continuously differentiable and λ is constant for $t \in [\tau_i, \tau_i']$, and $\xi(t) \in Inv(\lambda(t))$ and $\dot{\xi}(t) = f_{\lambda(t)}(\xi(t))$ for all $t \in [\tau_i, \tau_i')$ (continuous evolution); and
- for all i, $e = (\lambda(\tau_i'), \lambda(\tau_{i+1})) \in E$, $\xi(\tau_i') \in G(e)$ and $(x(\tau_i'), x(\tau_{i+1})) \in R(e)$ (discrete evolution).

□

An execution $\chi = (\tau, \lambda, \xi)$ is called *finite* if τ is a finite sequence ending with a closed interval, *infinite* if τ is an infinite sequence or if $\sum_i (\tau_i' - \tau_i) = \infty$, and *maximal* if it is not a strict prefix of any other execution of the hybrid automaton. We denote the set of all maximal and infinite executions of the automaton with initial state $(\ell_0, x_0) \in Init$ by $\mathcal{H}_{(\ell_0, x_0)}^{M}$ and $\mathcal{H}_{(\ell_0, x_0)}^{\infty}$, respectively.

DEFINITION 18.3
A hybrid automaton is called *non-blocking*, if $\mathcal{H}_{(\ell_0, x_0)}^{\infty}$ is non-empty for all $(\ell_0, x_0) \in Init$. It is called *deterministic* if $\mathcal{H}_{(\ell_0, x_0)}^{M}$ contains, at most, one element for all $(\ell_0, x_0) \in Init$.

□

These well-posedness concepts are similar to what we called *initial* well-posedness as they do not say anything about live-lock or the continuation beyond accumulation points of event times.

To simplify the characterization of non-blocking and deterministic automata, the following assumption has been introduced in [30], [25].

ASSUMPTION 18.1
The vector field $f_\ell(\cdot)$ is globally Lipschitz continuous for all $\ell \in Loc$. The edge (ℓ, ℓ') is contained in E if and only if $G(\ell, \ell') \neq \varnothing$ and $x \in G(\ell, \ell')$ if and only if there is an $x' \in X$ such that $(x, x') \in R(\ell, \ell')$. □

The first part of the assumption is standard to guarantee global existence and uniqueness of solutions within each location given a continuous initial state. The latter part is without loss of generality as can easily be seen [30].

A state $(\hat{\ell}, \hat{x})$ is called *reachable*, if there exists a finite execution (τ, λ, ξ) with $\tau = \{[\tau_i, \tau_i']\}_{i=0}^N$ and $(\lambda(\tau_N'), \xi(\tau_N')) = (\hat{\ell}, \hat{x})$. The set $Reach \subseteq Loc \times X$ denotes the collection of reachable states of the automaton.

The set of states from which continuous evolution is impossible is defined as

$$Out = \{(\ell_0, x_0) \in Loc \times X \mid \forall \varepsilon > 0 \exists t \in [0, \varepsilon) \ x_{\ell_0, x_0}(t) \notin Inv(\ell_0)\}$$

in which $x_{\ell_0, x_0}(\cdot)$ denotes the unique solution to $\dot{x} = f_{\ell_0}(x)$ with $x(0) = x_0$.

THEOREM 18.1—[30], [25]
A hybrid automaton is non-blocking, if for all $(\ell, x) \in Reach \cup Out$, there exists $(\ell, \ell') \in E$ with $x \in G(\ell, \ell')$. In case the automaton is deterministic, this condition is also necessary. □

THEOREM 18.2
A hybrid automaton is deterministic, if and only if for all $(\ell, x) \in Reach$

- if $x \in G(\ell, \ell')$ for some $(\ell, \ell') \in E$, then $(\ell, x) \in Out$;
- if $(\ell, \ell') \in E$ and $(\ell, \ell'') \in E$ with $\ell' \neq \ell''$, then $x \notin G(\ell, \ell') \cap G(\ell, \ell'')$; and
- if $(\ell, \ell') \in E$ and $x \in G(\ell, \ell')$, then there is, at most, one $x' \in X$ with $(x, x') \in R(\ell, \ell')$.

□

As a consequence of the broad class of systems covered by the results in this section, the conditions are rather implicit in the sense that for a particular example the conditions cannot be verified by direct calculations (*i.e.*, are not in an algorithmic form). Especially, if the model description itself is implicit (*e.g.*, variable structure systems or complementarity models) these results are only a start of the well-posedness analysis as the hybrid automaton model and the corresponding sets *Reach* and *Out* have to be determined first. However, some explicit characterizations of the set Out as can be found in [30], [25] might be convenient in this respect. In the next sections, we will present results that can be checked by direct computations.

The extension of the initial well-posedness results for hybrid automata to local or global existence of executions are awkward as Zeno behaviour is hard to characterize or exclude, and continuation beyond Zeno times is not easy to show. Relaxations play a crucial role in this respect [25]. In case the location can be described as a function of the continuous state (like for complementarity systems or differential equations with discontinuous right hand sides), it is possible to define an evolution beyond the Zeno time by proving that the (left-)limit of the continuous state exists at the Zeno point. Continuation from this limit follows then again by initial or local existence.

18.6 Well-posedness of Multi-modal Linear Systems

A problem of considerable importance is to find necessary and sufficient conditions for well-posedness of multi-modal linear systems

$$\begin{aligned}
\dot{x} &= A_1 x, \text{ if } x \in C_1 \\
\dot{x} &= A_2 x, \text{ if } x \in C_2 \quad x \in \mathbb{R}^n \\
&\vdots \\
\dot{x} &= A_r x, \text{ if } x \in C_r
\end{aligned} \quad (18.7)$$

where C_i are certain subsets of $\mathbb{R}^n$ having the property that

$$C_1 \cup C_2 \cup \cdots \cup C_r = \mathbb{R}^n$$

$$\text{int } C_i \cap \text{int } C_j = \varnothing, \quad i \neq j \quad (18.8)$$

This situation may naturally arise from modeling, as well as from the application of a switching linear feedback scheme (with different feedback laws corresponding to the subsets C_i). Of course, even more general cases may be considered, or, instead, extra conditions may be imposed on the subsets C_i. Note that the first condition in (18.8) is a necessary (but not sufficient) condition for existence of solutions for all initial conditions and the second one is necessary (but again not sufficient) for uniqueness (unless the vector fields are equal on the overlapping parts of the regions C_i).

A particular case of the above problem, which has been investigated in depth is the *bimodal* linear case

$$\dot{x} = A_1 x, \quad Cx \geq 0$$
$$\dot{x} = A_2 x, \quad Cx \leq 0$$
$$x \in \mathbb{R}^n \quad (18.9)$$

under the additional assumption that both pairs (C, A_1) and (C, A_2) are *observable*.

The solution concept that is employed is the *extended Carathéodory solution*, that is, a function $x : [t_0, t_1] \mapsto \mathbb{R}^n$, which is absolutely continuous on $[t_0, t_1]$, satisfies

$$x(t) = x(t_0) + \int_{t_0}^{t} f(x(\tau)) d\tau, \quad (18.10)$$

where $f(x)$ is the (discontinuous) vector field given by the right-hand side of (18.9), and there are no left-accumulation points of event times on $[t_0, t_1]$.

Notice that *Filippov solutions involving sliding modes* are not extended Carathéodory solutions. Moreover, note that if $f(x)$ is *continuous*, then necessarily there exists a K such that $A_1 = A_2 + KC$, and f is automatically Lipschitz continuous, implying local uniqueness of solutions.

Before stating the main result, we introduce some notation. First we define the $n \times n$ observability matrices corresponding to (C, A_1), respectively (C, A_2) :

$$W_1 := \begin{bmatrix} C \\ CA_1 \\ \vdots \\ CA_1^{n-1} \end{bmatrix}, \quad W_2 := \begin{bmatrix} C \\ CA_2 \\ \vdots \\ CA_2^{n-1} \end{bmatrix} \quad (18.11)$$

(by assumption they both have rank n). Furthermore, we define the following subsets of the state space $\mathbb{R}^n$:

$$S_i^+ = \{x \in \mathbb{R}^n | W_i x \succeq 0\}$$
$$S_i^- = \{x \in \mathbb{R}^n | W_i x \preceq 0\}$$
$$i = 1, 2 \quad (18.12)$$

where $\succeq$ denotes *lexicographic* ordering, that is, $x = 0$ or $x \succeq 0$ if the first component of x that is non-zero is positive. Furthermore, $x \preceq 0$ iff $-x \succeq 0$. Then the following result from [24] can be stated:

THEOREM 18.3
The bimodal linear system (18.9) is well-posed if and only if one of the following equivalent conditions are satisfied:

(a) $S_1^+ \cup S_2^- = \mathbb{R}^n$

(b) $S_1^+ \cap S_2^- = \{0\}$

(c) $W_2 W_1^{-1}$ is a lower-triangular matrix with positive diagonal elements. □

REMARK 18.1
Clearly, we may also interchange the indices 1 and 2 in the conditions (a), (b), (c). □

Possible extensions to non-invertible observability matrices, the multimodal situation, as well as to modification of the sets $Cx \geq 0$, $Cx \leq 0$, are discussed in [24], [23]. An interesting application of Theorem 18.3 to a switching control scheme is the following:

PROPOSITION 18.1—[24]
Consider the linear system $\dot{x} = Ax + Bu$, $x \in \mathbb{R}^n, u \in \mathbb{R}$, with switching feedback

$$\begin{aligned} u &= F_1 x, \quad Cx \geq 0 \\ u &= F_2 x, \quad Cx \leq 0 \end{aligned} \quad (18.13)$$

Let ρ be the *relative degree* of the system defined by the triple $(C, A + BF_1, B)$. Then, the controlled system is well-posed if and only if

$$\begin{aligned} F_2 - F_1 &= \alpha_1 C + \alpha_2 C(A + BF_1) + \cdots + \alpha_\rho C(A + BF_1)^{\rho-1} \\ &\quad + \gamma C(A + BF_1)^\rho \end{aligned}$$

for certain constants $\alpha_1, \alpha_2, \cdots, \alpha_\rho, \gamma$, with γ such that

$$\gamma C(A + BF_1)^{\rho-1} B > -1$$

□

18.7 Complementarity Systems

Within specific application domains of complementarity systems, the question of well-posedness has already received ample attention. For instance, in the context of unilaterally constrained mechanical systems (see, *e.g.*, [7], [4], [28], [32]) and projected dynamical systems [15],[33], [22] several results are available.

Linear Complementarity Systems

As the interconnection of a continuous, time-invariant, linear system and complementarity conditions, a *linear complementarity system* (LCS) can be given by

$$\dot{x}(t) = Ax(t) + Bu(t) \qquad (18.14\text{a})$$
$$y(t) = Cx(t) + Du(t) \qquad (18.14\text{b})$$
$$0 \leqslant u(t) \perp y(t) \geqslant 0 \qquad (18.14\text{c})$$

where $x(t) \in \mathbb{R}^n$, $u(t) \in \mathbb{R}^m$, $y(t) \in \mathbb{R}^m$, and A, B, C and D are matrices with appropriate sizes. We denote (18.14a)–(18.14b) by $\Sigma(A, B, C, D)$, and (18.14) by LCS(A, B, C, D).

One may look at LCS as a dynamical extension of the linear complementarity problem (LCP) of mathematical programming. See [14] for an excellent survey on the LCP.

PROBLEM 18.1
LCP(q, M): Given an m-vector q and $m \times m$ matrix M, find an m-vector z such that

$$z \geqslant 0 \qquad (18.15\text{a})$$
$$w := q + Mz \geqslant 0 \qquad (18.15\text{b})$$
$$z^T w = 0 \qquad (18.15\text{c})$$

□

We say z *solves* (or is a *solution* of) LCP(q, M), if z satisfies (18.15). The set of solutions of LCP(q, M) is denoted by SOL(q, M). Some definitions are introduced next.

DEFINITION 18.4
A matrix $M \in \mathbb{R}^{m \times m}$ is called

- *non-degenerate* if its principal minors detM_{II} for $I \subseteq \{1, \ldots, m\}$ are nonzero;

- a *P-matrix* if all its principal minors are positive;

- *positive (nonnegative) definite*[5] if $x^T M x > 0$ ($\geqslant 0$) for all $0 \neq x \in \mathbb{R}^m$.

□

Note that every positive definite matrix is a P-matrix, but the converse is not true. However, every symmetric P-matrix is also positive definite.

---

[5]Note that the matrix is not assumed to be symmetric.

Definition 18.5
The *dual cone* of a given non-empty set $S \subset \mathbb{R}^m$, denoted by S^*, is given by $\{v \in \mathbb{R}^m \mid v^T w \geq 0 \text{ for all } w \in S\}$. □

The final ingredient of our preparation is the *"index"* of a rational matrix.

Definition 18.6
A rational matrix $H(s) \in \mathbb{R}^{l \times l}(s)$ is said to be *of index k* if it is invertible as a rational matrix and $s^{-k} H^{-1}(s)$ is proper. It is said to be *totally of index k* if all its principal sub-matrices are of index k. □

With a slight abuse of terminology, we say that a linear system $\Sigma(A, B, C, D)$ is (totally) of index k, if its transfer function is (totally) of index k.

Linear complementarity systems with index 1. We will start investigating well-posedness of LCS for which the underlying linear system is totally of index 1. First, we define a solution concept for such LCS. First, the definition of event times set is in order.

Definition 18.7
A set $\mathcal{E} \subset \mathbb{R}_+$ is called an *admissible event times set* if it is closed and countable, and $0 \in \mathcal{E}$. To each admissible event times set $\mathcal{E}$, we associate a collection of intervals between events $\tau_\mathcal{E} = \{(t_1, t_2) \subset \mathbb{R}_+ \mid t_1, t_2 \in \mathcal{E} \cup \{\infty\} \text{ and } (t_1, t_2) \cap \mathcal{E} = \varnothing\}$. □

Note that both left and right accumulations[6] of event times are allowed by the above definition. Next, we define the *hybrid* solution concept. Later on, we will compare it with the solution concepts mentioned in Section 18.3.

Definition 18.8
A quadruple $(\mathcal{E}, u, x, y)$, where $\mathcal{E}$ is an admissible event times set, and $(u, x, y) : \mathbb{R}_+ \mapsto \mathbb{R}^{m+n+m}$, is said to be a *hybrid solution* of LCS(A, B, C, D) with the initial state x_0, if $x(0) = x_0$, x is continuous on $\mathbb{R}_+$, and the following conditions hold for each $\tau \in \tau_\mathcal{E}$:

1. The triple $(u, x, y)|_\tau$ is analytic.
2. For all $t \in \tau$, it holds that
$$\dot{x}(t) = Ax(t) + Bu(t)$$
$$y(t) = Cx(t) + Du(t)$$
$$0 \leq u(t) \perp y(t) \geq 0$$
□

Moreover, we say that a hybrid solution $(\mathcal{E}, u, x, y)$ is *non-redundant* if there does not exist a $t \in \mathcal{E}$ and t', t'' with $t' < t < t''$ such that (u, x, y) is analytic on (t', t''). Without loss of generality, we will only consider nonredundant solutions from now on.

---

[6] An element t of an admissible set $\mathcal{E}$ is said to be a *left (right) accumulation point* if for all $t' > t$ ($t' < t$) $(t, t') \cap \mathcal{E}$ (($t', t) \cap \mathcal{E}$) is not empty.

DEFINITION 18.9
An admissible event times set $\mathcal{E}$ is said to be *left (right) Zeno free* if it does not contain any left (right) accumulation points. A hybrid solution is said to be *left (right) Zeno free* if the corresponding event times set is left (right) Zeno free. It is said to be *left (right) Zeno* if it is not left (right) Zeno free, and *non-Zeno* if it is both left and right Zeno free. □

The following proposition summarizes the relations between forward, L_2 and hybrid solutions.

PROPOSITION 18.2—[9]
Consider an LCS(A, B, C, D). The following statements hold.

1. If $(\mathcal{E}, u, x, y)$ is a left Zeno free hybrid solution of LCS(A, B, C, D) for some initial state, then (u, x, y) is a forward solution with the same initial state of LCS(A, B, C, D).

2. Suppose that $D + C(\sigma I - A)^{-1}B$ is a non-degenerate matrix for all sufficiently large σ. If $(\mathcal{E}, u, x, y)$ is a hybrid solution of LCS(A, B, C, D) for some initial state, then (u, x, y) is an L_2 solution of LCS(A, B, C, D) with the same initial state.

□

Note that a left Zeno free hybrid solution is not necessarily a forward solution as a forward solution can in principle not be defined beyond a right-accumulation point (although an extension may be formulated including this possibility).

The following theorem provides sufficient conditions for well-posedness in the sense of existence and uniqueness of LCS with index 1.

THEOREM 18.4—[9]
Consider an LCS(A, B, C, D) with $\Sigma(A, B, C, D)$ is totally of index 1. Suppose that $D + C(\sigma I - A)^{-1}B$ is a P-matrix for all sufficiently large σ. There exists a left Zeno free hybrid solution of LCS(A, B, C, D) with the initial state x_0 if and only if LCP(Cx_0, D) is solvable. Moreover, if such a solution exists, it is left Zeno free unique, *i.e.*, there is no other left Zeno free solution. □

Linear passive complementarity systems. When the underlying system $\Sigma(A, B, C, D)$ is passive (in the sense of [42]) we call the overall system (18.14) a *linear passive complementarity system* (LPCS). For a detailed study on LPCS, the reader may refer to [8]. As shown in [9, Lemma 3.8.5], the passivity of the system (under some extra assumptions) implies that it is of index 1. Hence, Theorem 18.4 is applicable to the LPCS. Additionally, it can be shown that there are no left Zeno solutions for the LPCS as formulated in the following theorem.

THEOREM 18.5—[9]
Consider an LCS(A, B, C, D) with $\Sigma(A, B, C, D)$ being passive, (A, B, C) being minimal, and

$$\mathrm{col}(B, D + D^T) := \begin{pmatrix} B \\ D + D^T \end{pmatrix}$$

of full column rank. Let $Q_D = \{z \mid z \text{ solves LCP}(0, D)\}$. There exists a hybrid solution of LCS(A, B, C, D) with the initial state x_0 if and only if $Cx_0 \in Q_D^*$. Moreover, if a solution exists it is unique[7] and left Zeno free. □

Observe that if $(\mathcal{E}, \mathcal{S}, u, x, y)$ is a solution of LCS(A, B, C, D), then $(\mathcal{E}, \mathcal{S}, t \mapsto e^{\rho t}u(t), t \mapsto e^{\rho t}x(t), t \mapsto e^{\rho t}y(t))$ is a solution of LCS$(A + \rho I, B, C, D)$. This correspondence makes it possible to apply the above theorem to a class of non-passive systems. Indeed, even if $\Sigma(A, B, C, D)$ is not passive $\Sigma(A + \rho I, B, C, D)$ may be passive for some ρ. In this case, we say that $\Sigma(A, B, C, D)$ is *passifiable by pole shifting* (PPS). Necessary and sufficient conditions for PPS property have been given in [9, Theorem 3.4.3]. By using those conditions, we can state the following extension of Theorem 18.5.

THEOREM 18.6—[9]
Consider an LCS(A, B, C, D) with (A, B, C) minimal and $\mathrm{col}(B, D + D^T)$ full column rank. Let E be such that $\ker E = \{0\}$ and $\mathrm{im}\, E = \ker(D + D^T)$. Suppose that D is non-negative definite and $E^T CBE$ is symmetric positive definite. There exists a hybrid solution of LCS(A, B, C, D) with the initial state x_0 if and only if $Cx_0 \in Q_D^*$. Moreover, if a solution exists it is unique[7] and left Zeno free. □

The PPS property can be employed to rule out right Zeno solutions as well. Indeed, the systems for which PPS property is an invariant under time-reversion do not exhibit Zeno behavior at all. Necessarily, such a system has a positive definite feedthrough term. This very particular case is worth stating separately.

THEOREM 18.7—[9]
Consider an LCS(A, B, C, D) with (A, B, C) is minimal and D is positive definite. There exists a unique non-Zeno hybrid solution of LCS(A, B, C, D) for all initial states. □

Note that *Zeno states* (*i.e.*, the states at the accumulation points) are well-defined due to the fact that the x-part of a hybrid solution is uniformly continuous under the condition $\Sigma(A, B, C, D)$ being totally of index 1. Intuitively, the most natural candidates for Zeno states are equilibrium states, in particular the zero state, of the system. The following theorem indicates that the zero state cannot be a Zeno state for linear complementarity systems of index 1.

---
[7]It can also be shown that this solution is unique in L_2.

THEOREM 18.8—[11]
Consider an LCS(A, B, C, D) with $\Sigma(A, B, C, D)$ being totally of index 1. Then, the zero state is not a right Zeno state. □

Piecewise Linear Systems

As is wellknown (see, for instance, [16]), piecewise linear relations may be described in terms of the linear complementarity problem. In the circuits and systems community (see e.g., [26], [40]) the complementarity formulation has already been used for *static* piecewise linear systems; this subsection may be viewed as an extension of the cited work in the sense that we consider *dynamic* systems. For the sake of simplicity, we will focus on a specific type of piecewise linear systems, namely linear saturation systems, i.e., linear systems coupled to saturation characteristics. They are of the form

$$\dot{x}(t) = Ax(t) + Bu(t) \tag{18.16a}$$
$$y(t) = Cx(t) + Du(t) \tag{18.16b}$$
$$(u(t), y(t)) \in \text{saturation}_i \tag{18.16c}$$

where $x(t) \in \mathbb{R}^n$, $u(t) \in \mathbb{R}^m$, $y(t) \in \mathbb{R}^m$, A, B, C and D are matrices of appropriate sizes, and saturation$_i$ is the curve depicted in Figure 18.1 with $e_2^i - e_1^i > 0$ and $f_1^i \geqslant f_2^i$. This curve is formally described by the set

$$\{(v, z) \in \mathbb{R}^2 \mid (v = e_2^i \text{ and } z \leqslant f_2^i) \text{ or } (v = e_1^i \text{ and } z \geqslant f_1^i) \text{ or }$$
$$(e_1^i \leqslant v \leqslant e_2^i \text{ and } (f_1^i - f_2^i)e_2^1 + (e_1^i - e_2^i)(z - f_2^i))\}. \tag{18.17}$$

We denote the overall system (18.16) by SAT(A, B, C, D). Note that relay characteristics can be obtained from saturation characteristics by setting $f_1^i = f_2^i$. We adopt the solution concept defined for LCS to saturation systems as follows.

DEFINITION 18.10
A quadruple $(\mathcal{E}, u, x, y)$ where $\mathcal{E}$ is an admissible event times set, and $(u, x, y) : \mathbb{R}_+ \mapsto \mathbb{R}^{m+n+m}$ is said to be a *hybrid solution* of SAT(A, B, C, D) with the initial state x_0 if $x(0) = x_0$ and the following conditions hold for each $\tau \in \tau_{\mathcal{E}}$:

1. The triple $(u, x, y)|_\tau$ is analytic.

2. For all $t \in \tau$ and $i \in \overline{m}$, it holds that

$$\dot{x}(t) = Ax(t) + Bu(t)$$
$$y(t) = Cx(t) + Du(t)$$
$$(u_i(t), y_i(t)) \in \text{saturation}_i$$

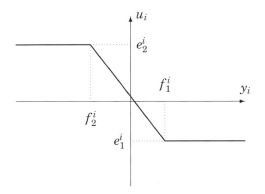

Figure 18.1 Saturation characteristic

Note that if in Definition 18.10 the PL curve saturation$_i$ is replaced by the complementarity conditions (18.14c), we obtain Definition 18.8.

One may argue that the saturation characteristic is a Lipschitz continuous function (provided that $f_1^i - f_2^i > 0$), and hence existence and uniqueness of solutions follow from the theory of ordinary differential equations. The following example shows that this is not correct in general if the feedthrough term D is non-zero.

EXAMPLE 18.1
Consider the single-input single-output system

$$\dot{x} = u \qquad (18.18)$$
$$y = x - 2u \qquad (18.19)$$

where u and y are restricted by a saturation characteristic with $e_1 = -f_1 = -e_2 = f_2 = \frac{1}{2}$ as shown in Figure 18.1. Let the periodic function $\tilde{u} : \mathbb{R}_+ \to \mathbb{R}$ be defined by

$$\tilde{u}(t) = \begin{cases} 1/2 & \text{if } 0 \leqslant t < 1 \\ -1/2 & \text{if } 1 \leqslant t < 3 \\ 1/2 & \text{if } 3 \leqslant t < 4 \end{cases}$$

and $\tilde{u}(t-4) = \tilde{u}(t)$ whenever $t \geqslant 4$. By using this function define $\tilde{x} : \mathbb{R}_+ \to \mathbb{R}$ as

$$\tilde{x}(t) = \int_0^t \tilde{u}(s)\, ds,$$

and $\tilde{y} : \mathbb{R}_+ \to \mathbb{R}$ as

$$\tilde{y} = \tilde{x} - 2\tilde{u}.$$

It can be verified that $(-\tilde{u}, -\tilde{x}, -\tilde{y})$, $(0, 0, 0)$ and $(\tilde{u}, \tilde{x}, \tilde{y})$ are all solutions of SAT$(0, 1, 1, -2)$ with the zero initial state. □

As illustrated in the example, the Lipschitz continuity argument does not work in general when $f_1^i > f_2^i$. Also in the case, where $f_1^i = f_2^i$, this reasoning does not apply. The following theorem gives a sufficient condition for the well-posedness of linear systems with saturation characteristics.

THEOREM 18.9—[9]
Consider SAT(A, B, C, D). Let $R = \text{diag}(e_2^i - e_1^i)$, and $S = \text{diag}(f_2^i - f_1^i)$. Suppose that $G(\sigma)R - S$ is a P-matrix for all sufficiently large σ. Then, there exists a unique left Zeno free hybrid solution of SAT(A, B, C, D) for all initial states. $\square$

Variations and Generalizations

Up to this point, we have presented results on global well-posedness of complementarity systems in which the x-part of the solutions is continuous. In this subsection, the available variations and/or generalizations of those results will be mentioned briefly.

Earlier work on complementarity systems mainly focused on *initial* and *local* well-posedness issues (in the sense described in Section 18.4) including the possibility of re-initializations (state jumps). In these studies, the issue of irregular initial states had to be tackled, *i.e.*, the states for which there is no solution in the senses defined so far for complementarity systems. A distributional framework was used to obtain a new solution concept (see [21] for details). Sufficient conditions for *local* well-posedness have been provided for an LCS [37], [21], Hamiltonian complementarity systems with one complementarity constraint [37], and a class of nonlinear complementarity systems [35], [19]. The result in [21] for LCS with multiple constraints is presented next. Consider the LCS(A, B, C, D) with Markov parameters $H^0 = D$ and $H^i = CA^{i-1}B$, $i = 1, 2, \ldots$ and define the leading row and column indices by

$$\eta_j := \inf\{i \in \mathbb{N} \mid H_{\bullet j}^i \neq 0\}, \quad \rho_j := \inf\{i \in \mathbb{N} \mid H_{j\bullet}^i \neq 0\},$$

where $j \in \{1, \ldots, k\}$ and $\inf \varnothing := \infty$. The *leading row coefficient matrix* $\mathcal{M}$ and *leading column coefficient matrix* $\mathcal{N}$ are then given for *finite* leading row and column indices by

$$\mathcal{M} := \begin{pmatrix} H_{1\bullet}^{\rho_1} \\ \vdots \\ H_{k\bullet}^{\rho_k} \end{pmatrix} \quad \text{and} \quad \mathcal{N} := (H_{\bullet 1}^{\eta_1} \ldots H_{\bullet k}^{\eta_k})$$

THEOREM 18.10—[21]
If the leading column coefficient matrix $\mathcal{N}$ and the leading row coefficient matrix $\mathcal{M}$ are both defined and P-matrices, then LCS(A, B, C, D) has a unique *local* left Zeno free solution on an interval of the form $[0, \varepsilon)$ for some

$\varepsilon > 0$. Moreover, live-lock (an infinite number of events at one time instant) does not occur. □

In another related paper [20], it has been shown that the *initial* well-posedness problem comes down to checking the existence and uniqueness of a family of linear complementarity problems.

The first steps in the direction of getting global well-posedness results for an LCS *with external inputs* are due to [8] for LPCS and [10], where the underlying linear system is of index 1.

18.8 Differential Equations with Discontinuous Right Hand Sides

Differential equations of the form

$$\dot{x}(t) = f(t, x(t)) \qquad (18.20)$$

with f being piecewise continuous in a domain G and with the set M of discontinuity points having measure zero, have received quite some attention in the literature. Major roles have been played in this context by Filippov [18], [17] and Utkin [39]. As mentioned in Section 18.2, solution concepts have been defined by replacing the basic differential equation (18.20) by a differential inclusion of the form

$$\dot{x}(t) \in F(t, x(t)), \qquad (18.21)$$

where F is constructed from f. The solution concept is then inherited from the realm of differential inclusions [3].

DEFINITION 18.11
The function $x : \Omega \to \mathbb{R}^n$ is called a *solution of the differential inclusion* (18.21), if x is absolutely continuous on the time interval Ω and satisfies $\dot{x}(t) \in F(t, x(t))$ for almost all $t \in \Omega$. □

There are several ways to transform f into F and we will restrict ourselves to the two most famous ones and briefly discuss an alternative transformation proposed by Aizerman and Pyatnitskii [1]. For further details, see [18].

In the *convex definition* [18], [17], the set $F_a(t, x)$ is taken to be the smallest convex closed set containing all the limit values of the function $f(\bar{t}, \bar{x})$ for $\bar{x} \longrightarrow x$, $\bar{t} = t$, and $(\bar{t}, \bar{x}) \notin M$.

The *control equivalent definition* proposed by Utkin [39] (see also [18, p. 54]) applies to equations of the form

$$\dot{x}(t) = f(t, x(t), u_1(t, x), \ldots, u_r(t, x)) \qquad (18.22)$$

where f is continuous in its arguments, but $u_i(t, x)$ is a scalar-valued function being discontinuous only on a smooth surface S_i given by $\phi_i(x) = 0$.

We define the sets $U_i(t,x)$ as $\{u_i(t,x)\}$ when $x \notin S_i$ and in the case $x \in S_i$ by the closed interval with end-points $u_i^-(t,x)$ and $u_i^+(t,x)$. The values $u_i^-(t,x)$ and $u_i^+(t,x)$ are the limiting values of the function u_i on both sides of the surface S_i which we assume to exist. The differential equation (18.22) is replaced by (18.21) with $F_b(t,x) = f(t,x,U_1(t,x),\ldots,U_r(t,x))$.

REMARK 18.2
In the case where $F_c(t,x)$ is chosen as the smallest convex closed set containing $F_b(t,x)$, then the general definition of Aizerman and Pyatnitskii [1] is obtained. In the case where f is linear in $u_1,\ldots,u_r$ and the surfaces $S_1,\ldots,S_r$ are all different and at the point of intersection the normal vectors are linearly independent, all the before-mentioned definitions coincide, *i.e.*, $F_a = F_b = F_c$. □

The well-posedness results of the differential equation (18.20) or (18.22) can now be based on the theory available for differential inclusions (see [3], [18] and the references therein).

Let A, B be two non-empty closed sets in a metric space with metric d. The distance between A and B may be characterized by the following quantities

$$\beta(A,B) = \sup_{\alpha \in A} \inf_{\beta \in B} d(\alpha,\beta)$$
$$\alpha(A,B) = \max(\beta(A,B), \beta(B,A))$$

A set-valued function F is called *upper semi-continuous* at p_0, if

$$\beta(F(p), F(p_0)) \to 0 \text{ if } p \to p_0,$$

or stated differently, if for all $\varepsilon > 0$ there is a $\delta > 0$ such that $\|p - p_0\| \leq \delta$ implies $F(p) \subseteq F(p_0) + \varepsilon \mathbb{B}$, where $\mathbb{B}$ denotes the unit ball. F is called *continuous* at the point p_0, if $\alpha(F(p), F(p_0)) \to 0$ if $p \to p_0$. F is called (upper semi)continuous on a set D, if F is (upper semi)continuous in each point of the set D.

DEFINITION 18.12
We say that the set-valued map $F(t,x)$ satisfies the *basic conditions*, if

- for all $(t,x) \in G$ the set $F(t,x)$ is nonempty, bounded, closed and convex;

- F is upper semi-continuous in t, x.

□

The following result is described on page 77 of the monograph [18].

THEOREM 18.11—[18, THEOREM 2.7.1 AND 2.7.2]
If $F(t, x)$ satisfies the basic conditions in the domain G, then for any point $(t_0, x_0) \in G$ there exists a solution of the problem

$$\dot{x}(t) \in F(t, x(t)), \qquad x(t_0) = x_0 \qquad (18.23)$$

If the basic conditions are satisfied in a closed and bounded domain G, then each solution can be continued on both sides up to the boundary of the domain G. □

In combination with the following result, Theorem 18.11 proves the existence of solutions for the differential inclusions related to F_a, F_b and F_c.

THEOREM 18.12—[18, P. 67]
The sets $F_a(t, x)$, $F_b(t, x)$ and $F_c(t, x)$ are nonempty, bounded and closed. $F_a(t, x)$, and $F_c(t, x)$ are also convex. F_a is upper semicontinuous in x, and F_b and F_c are upper semicontinuous in t, x. □

Together Theorem 18.11 and 18.12 now show the existence of solutions when Filippov's convex definition is used under the condition that f is time-invariant. In the case where f is not time-invariant, additional assumptions are needed to arrive at F being upper semi-continuous in t as well (see page 68 in [18]). For the definition of Aizerman and Pyatnitskii (i.e., using F_c) existence of solutions is guaranteed. In the case where $F_b(t, x)$ is convex for all relevant (t, x) (e.g., if the conditions mentioned in Remark 18.2 are satisfied), then existence follows as well. If the convexity assumption is not satisfied, the existence result still holds if upper semi-continuity is replaced by continuity [18, p. 79]. In fact, the two major cases studied in [3, Ch. 3] are related to these two situations: i) the values of F are compact and convex and F is upper semicontinuous; and ii) the values of F are compact, but not necessarily convex, and F is continuous.

Now, we will discuss the issue of uniqueness. Right uniqueness (in Filippov sense) holds for the differential equation (18.20) at the point (t_0, x_0), if there exists $t_1 > t_0$ such that each two solutions of this equation satisfying the initial condition $x(t_0) = x_0$ coincide on the interval $[t_0, t_1]$, or on the interval on which they are both defined. Right uniqueness holds for a domain D, if from each point $(t_0, x_0) \in D$ right uniqueness holds.

Not too many uniqueness results are available in the literature. The most useful result given in [18] is related to the following situation. Let the domain $G \subset \mathbb{R}^n$ be separated by a smooth surface S into domains G^- and G^+. Let f and $\partial f / \partial x_i$ be continuous in the domains G^- and G^+ up to the boundary such that $f^-(t, x)$ and $f^+(t, x)$ denote the limit values of the function f at (t, x), $x \in S$ from the regions G^- and G^+, respectively. We define $h(t, x) = f^+(t, x) - f^-(t, x)$ as the discontinuity vector over the surface S. Moreover, let n be the normal vector to S directed from G^- to G^+.

THEOREM 18.13
Consider the differential equation (18.20) with f as above. Let S be a twice continuously differentiable surface and suppose that the function h is continuously differentiable. If for each $t \in (a,b)$ and each point $x \in S$ at least one of the inequalities $n^T f^-(t,x) > 0$ or $n^T f^+(t,x) < 0$ (possibly different inequalities for different x and t) is fulfilled, then right uniqueness holds for (18.20) in the domain G for $t \in (a,b)$ in the sense of Filippov. □

As mentioned in [36], the criterion above clearly holds for general nonlinear systems, but needs to be verified on a point-by-point basis. Alternatively, the result in Section 18.7 is more straightforward to check as it requires the computation of the determinants of all principal minors of the transfer function of the underlying linear system, and determines the signs of the leading Markov parameters. However, that theory is restricted to piecewise linear systems and uses a *different* solution concept. Hence, uniqueness is not proven in the Filippov sense, but in the forward (or left Zeno free) sense.

The difference between Filippov, forward (or left-Zeno hybrid solutions) and extended Carathéodory solutions will be discussed in the context of the class of systems for which all these concepts apply. In particular, we will study

$$\dot{x}(t) = Ax(t) + Bu(t); \quad y(t) = Cx(t) \qquad (18.24)$$

in a closed loop with the relay feedback

$$u(t) = -\text{sgn}(y(t)) \qquad (18.25)$$

Note that in the context of Theorem 18.9, we are dealing with the situation in which $R = 2I$ and $S = 0$. Note also that $F_a = F_b = F_c$ for such linear relay systems and the corresponding solution concepts coincide and will be referred to as "Filippov solutions" from now on.

The difference between the forward solutions and the Filippov solution is related to Zeno behavior, and is nicely demonstrated by an example constructed by Filippov [18, p. 116], which is given by

$$\dot{x}_1 = -u_1 + 2u_2 \qquad (18.26a)$$
$$\dot{x}_2 = -2u_1 - u_2 \qquad (18.26b)$$
$$y_1 = x_1 \qquad (18.26c)$$
$$y_2 = x_2 \qquad (18.26d)$$
$$u_1 = -\text{sgn}(y_1) \qquad (18.26e)$$
$$u_2 = -\text{sgn}(y_2) \qquad (18.26f)$$

This system has, besides the zero solution (which is both a Filippov and a forward solution), an infinite number of other trajectories (being Filippov, but not forward solutions) starting from the origin. The non-zero solutions leave the origin due to left accumulations of the relay switching times and are Filippov solutions, but are not forward solutions. However, Filippov's example does not satisfy the conditions for uniqueness given in Section 18.7. Hence, it is not clear if the conditions in Section 18.7 are sufficient for Filippov uniqueness as well. This problem is studied in [34] for the case

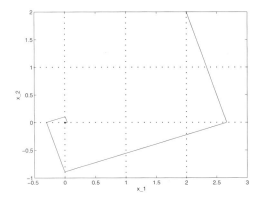

Figure 18.2 Trajectory in the phase plane of (18.26).

where (18.24) is a single-input-single-output (SISO) system. The theory of Section 18.7 states that the positivity of the leading Markov parameter H^ρ with $H^i = CA^{i-1}B$, $i = 1, 2, \ldots$ and $\rho = \min\{i \mid H^i \neq 0\}$ implies uniqueness in forward sense.

THEOREM 18.14—[34]
Consider the system (18.24)–(18.25). The following statements hold for the relative degree ρ being 1 or 2.

$\rho = 1$ The system (18.24)–(18.25) has a unique Filippov solution for all initial conditions if and only if the leading Markov parameter H^ρ is positive;

$\rho = 2$ The system (18.24)–(18.25) has a unique Filippov solution for initial condition $x(0) = 0$ if and only if the leading Markov parameter H^ρ is positive.

Moreover, in the case where $H^1 = CB > 0$, Filippov solutions do not have left accumulations of relay switching times. □

Interestingly, the above theorem presents conditions that exclude particular types of Zeno behaviour.

Up to this point, one might hope that the positivity of the leading Markov parameter is also sufficient for Filippov uniqueness for higher relative degrees. However, in [34] a counter-example is presented of the form (18.24)–(18.25) with (18.24) being a triple integrator. This relay system has one forward solution (being identically zero) starting in the origin (as expected, as the leading Markov parameter is positive), but has at least two

Filippov solutions, of which one is the zero solution and the other starts with a left-accumulation point of relay switching times. This example can also be considered in the light of Section 18.6 in the form

$$\begin{cases} \text{mode 1}: \dot{x} = A_1 x, & \text{if } y = Cx \geq 0 \\ \text{mode 2}: \dot{x} = A_2 x, & \text{if } y = Cx \leq 0 \end{cases} \tag{18.27}$$

with

$$A_1 = \begin{pmatrix} 0 & 1 & 0 & 0 \\ 0 & 0 & 1 & 0 \\ 0 & 0 & 0 & -1 \\ 0 & 0 & 0 & 0 \end{pmatrix}, \quad A_2 = \begin{pmatrix} 0 & 1 & 0 & 0 \\ 0 & 0 & 1 & 0 \\ 0 & 0 & 0 & 1 \\ 0 & 0 & 0 & 0 \end{pmatrix}, \quad C = (1\ 0\ 0\ 0) \tag{18.28}$$

Imura and Van der Schaft [24] use an extended Carathéodory solution concept for this type of systems and present necessary and sufficient conditions for existence and uniqueness (see also Section 18.6). As this solution concept does not allow for sliding modes and left-accumulation points of event times, the above system does not have any extended Carathéodory solution starting from the initial state $(0, 0, 0, 1)^T$ as can easily be seen (see also Theorem 18.3).

In summary, the triple integrator connected to a (negative) relay forms a nice comparison between the three mentioned solution concepts; for the system (18.27) with (18.28) and $x_0 = (0, 0, 0, 1)^T$, there exist [34]:

- *no* extended Carathéodory solution;

- *one* forward solution; and

- *infinitely many* Filippov solutions.

For specific applications in discontinuous feedback control, the Filippov solution concept allows trajectories, which are not practically relevant for the stabilization problem at hand. So-called Euler (or sampling) solutions seem to be more appropriate in this context [12], [13]. Also in this case the discontinuous dynamical system is replaced by a differential inclusion with the difference that a particular choice of the controller is made at the switching surface. This choice determines which trajectories are actually Euler solutions by forming the limits of certain numerical integration routines (see [12], [13] for more details).

In Section 2.10 of [18], some further results can be found on uniqueness. The most general result in [18] for uniqueness in the setting of Filippov's convex definition uses the exclusion of left-accumulation points as one of the conditions to prove uniqueness. Unfortunately, it is not clear how such assumptions should be verified. As a consequence, Theorem 18.14 is quite useful. In [18, Section 2.8], one can also find some results on continuous dependence of solutions on initial data.

18.9 Summary

Well-posedness problems arise in hybrid systems theory as a consequence of the use of implicit descriptions and of solution concepts that are based on relaxations. Examples show that the well-posedness issue is considerably more complex in hybrid systems than in continuous systems, as a result of a number of factors including the possible presence of sliding modes, the interaction of guards and invariants, and the occurrence of left or right accumulations of event times. Description formats that are based on implicit or relaxed specifications are typically connected to particular subclasses of hybrid systems, and so there is no general theory of well-posedness of hybrid systems; however, the questions that need to be answered are similar in each case. This article has surveyed several description formats and solution concepts that are used for hybrid systems. We have concentrated on well-posedness in the sense of existence and uniqueness of solutions, without requiring continuous dependence on initial conditions. A selection of results available in the literature has been presented for the subclasses of multi-modal linear systems, complementarity systems, and differential equations with discontinuous right hand sides.

18.10 References

[1] M. A. Aizerman and E. S. Pyatnitskii. Fundamentals of the theory of discontinuous dynamical systems. I, II. *Automatika i telemekhanika*, 7: 33–47 and 8: 39–61, 1974 (in Russian).

[2] R. Alur, C. Courcoubetis, N. Halbwachs, T. A. Henzinger, P.-H. Ho, X. Nicollin, A. Olivero, J. Sifakis, and S. Yovine. The algorithmic analysis of hybrid systems. *Theoretical Computer Science*, 138:3–34, 1995.

[3] J.-P. Aubin and A. Cellina. *Differential Inclusions: Set-Valued Maps and Viability Theory*. Springer-Verlag, Berlin, 1984.

[4] P. Ballard. The dynamics of discrete mechanical systems with perfect unilateral constraints. Submitted to *Archive for Rational Mechanics and Analysis*.

[5] M. S. Branicky, V. S. Borkar, and S. K. Mitter. A unified framework for hybrid control: model and optimal control theory. *IEEE Transactions on Automatic Control*, 43(1):31–45, 1998.

[6] R. W. Brockett. Hybrid models for motion control systems. In H. L. Trentelman and J.C. Willems, editors, *Essays on Control. Perspectives in the Theory and its Applications* (lectures and the mini-courses of the European control conference (ECC'93), Groningen, the Netherlands, June/July 1993), 29–53. Birkhäuser, 1993.

[7] B. Brogliato. *Nonsmooth Impact Mechanics. Models, Dynamics and Control*, volume 220 of *Lecture Notes in Control and Information Sciences*. Springer-Verlag, London, 1996.

[8] M. K. Çamlıbel, W. P. M. H. Heemels, and J. M. Schumacher. On linear passive complementarity systems. to appear in *European Journal of Control's special issue "Dissipation and Control"*, 2002.

[9] M. K. Çamlıbel. *Complementarity Methods in the Analysis of Piecewise Linear Dynamical Systems*. P.hD. thesis, Tilburg University, 2001.

[10] M. K. Çamlıbel, W. P. M. H. Heemels, and J. M. Schumacher. Well-posedness of a class of linear network with ideal diodes. In *Proc. of the 14th International Symposium of Mathematical Theory of Networks and Systems*, Perpignan, France, 2000.

[11] M. K. Çamlıbel and J. M. Schumacher. Do the complementarity systems exhibit Zeno behavior? 2001, submitted for presentation at CDC'01.

[12] F. H. Clarke, Yu. S. Ledyaev, E. D. Sontag, and A. I. Subbotin. Asymptotic controllability implies feedback stabilization. *IEEE Trans. Automat. Contr.*, 42:1394–1407, 1997.

[13] F. H. Clarke, Yu. S. Ledyaev, R. J. Stern, and P. R. Wolenski. *Nonsmooth Analysis and Control Theory*. Springer-Verlag, Berlin, 1998.

[14] R. W. Cottle, J.-S. Pang, and R. E. Stone. *The Linear Complementarity Problem*. Academic Press, Boston, 1992.

[15] P. Dupuis and A. Nagurney. Dynamical systems and variational inequalities. *Annals of Operations Research*, 44:9–42, 1993.

[16] B. C. Eaves and C. E. Lemke. Equivalence of LCP and PLS. *Mathematics of Operations Research*, 6:475–484, 1981.

[17] A. F. Filippov. Differential equations with discontinuous right-hand side. *Matemat. Sbornik.*, 51:99–128, 1960. In Russian. English translation: *Am. Math. Soc. Transl.* 62 (1964).

[18] A. F. Filippov. *Differential Equations with Discontinuous Righthand Sides*. Mathematics and Its Applications. Kluwer, Dordrecht, The Netherlands, 1988.

[19] W. P. M. H. Heemels. *Linear Complementarity Systems: A Study in Hybrid Dynamics*. Ph.D. dissertation, Eindhoven University of Technology, Dept. of Electrical Engineering, Eindhoven, The Netherlands, 1999.

[20] W. P. M. H. Heemels, J. M. Schumacher, and S. Weiland. The rational complementarity problem. *Linear Algebra and its Applications*, 294:93–135, 1999.

[21] W. P. M. H. Heemels, J. M. Schumacher, and S. Weiland. Linear complementarity systems. *SIAM J. Appl. Math.*, 60:1234–1269, 2000.

[22] W. P. M. H. Heemels, J. M. Schumacher, and S. Weiland. Projected dynamical systems in a complementarity formalism. *Operations Research Letters*, 27(2):83–91, 2000.

[23] J.-I. Imura and A. J. van der Schaft. Well-posedness of a class of dynamically interconnected systems. In *Proc. 38th IEEE Conf. on Decision and Control*, 3031–3036, 1999.

[24] J.-I. Imura and A. J. van der Schaft. Characterization of well-posedness of piecewise linear systems. *IEEE Transactions on Automatic Control*, 45(9):1600–1619, 2000.

[25] K. J. Johansson, M. Egerstedt, J. Lygeros, and S. Sastry. On the regularization of zeno hybrid automata. *Systems and Control Letters*, 38:141–150, 1999.

[26] D. M. W. Leenaerts and W. M. G. van Bokhoven. *Piecewise Linear Modelling and Analysis*. Kluwer Academic Publishers, Dordrecht, The Netherlands, 1998.

[27] M. Lemmon. On the existence of solutions to controlled hybrid automata. In N. Lynch and B. Krogh, editors, *Hybrid Systems: Computation and Control*, volume 1790 of *Lecture Notes in Computer Science*, 229–242, Springer-Verlag, New York, 2000.

[28] P. Lötstedt. Mechanical systems of rigid bodies subject to unilateral constraints. *SIAM Journal on Applied Mathematics*, 42(2):281–296, 1982.

[29] J. Lygeros, D. N. Godbole, and S. Sastry. Verified hybrid controllers for automated vehicles. *IEEE Trans. Automat. Contr.*, 43:522–539, 1998.

[30] J. Lygeros, K. H. Johansson, S. Sastry, and M. Egerstedt. On the existence and uniqueness of executions of hybrid automata. In *Proc. 38th IEEE Conf. Decision and Control*, Phoenix, AZ, 2249–2254, 1999.

[31] N. Lynch, R. Segala, F. Vaandrager, and H. B. Weinberger. Hybrid I/O automata. In *Hybrid Systems III* (Proc. Workshop on Verification and Control of Hybrid Systems, New Brunswick, NJ, Oct. 1995). Springer-Verlag, Berlin, 1996. Lect. Notes Comp. Sci., volume 1066.

[32] M. D. P. Monteiro Marques. *Differential Inclusions in Nonsmooth Mechanical Problems. Shocks and Dry Friction*. Progress in Nonlinear Differential Equations and their Applications. Birkhäuser, Basel, 1993.

[33] A. Nagurney and D. Zhang. *Projected Dynamical Systems and Variational Inequalities with Applications*. Kluwer, Boston, 1996.

[34] A. Yu. Pogromsky, W. P. M. H. Heemels, and H. Nijmeijer. On well-posedness of relay systems. In *Proceedings of NOLCOS 2001*, St. Petersburg (Russia), 2001.

[35] A. J. van der Schaft and J. M. Schumacher. Complementarity modeling of hybrid systems. *IEEE Trans. Automat. Contr.*, AC-43:483–490, 1998.

[36] A. J. van der Schaft and J. M. Schumacher. *An Introduction to Hybrid Dynamical Systems*. Springer-Verlag, London, 2000.

[37] A. J. van der Schaft and J. M. Schumacher. The complementary-slackness class of hybrid systems. *Mathematics of Control, Signals and Systems*, 9:266–301, 1996.

[38] V. I. Utkin. Variable structure systems with sliding modes. *IEEE Transactions on Automatic Control*, 22(1):31–45, 1977.

[39] V. I. Utkin. *Sliding Regimes in Optimization and Control Problems*. Nauka, Moscow, 1981.

[40] L. Vandenberghe, B. L. De Moor, and J. Vandewalle. The generalized linear complementarity problem applied to the complete analysis of resistive piecewise-linear circuits. *IEEE Trans. Circuits Syst.*, CAS-36:1382–1391, 1989.

[41] J. C. Willems. Paradigms and puzzles in the theory of dynamical systems. *IEEE Transactions on Automatic Control*, 36:259–294, 1991.

[42] J. C. Willems. Dissipative dynamical systems. *Archive for Rational Mechanics and Analysis*, 45:321–393, 1972.

Author List

Implementation of an Active Suspension, Preview Controller for Improved Ride Comfort

M. D. Donahue
mdonahue@vehicle.me.berkeley.edu

J. K. Hedrick
khedrick@newton.berkeley.edu

University of California at Berkeley,
Department of Mechanical Engineering,
Berkeley, California, 94720, USA

Active and Passive Suspension Control for Vehicle Dive and Squat

Fu-Cheng Wang
fcw22@eng.cam.ac.uk

M. C. Smith
mcs@eng.cam.ac.uk

Department of Engineering, University of Cambridge,
Cambridge, CB2 1PZ, UK

Modeling of Drivers' Longitudinal Behavior

Johan Bengtsson
Johan.Bengtsson@control.lth.se

Rolf Johansson
Rolf.Johansson@control.lth.se

Department of Automatic Control, Lund Institute of Technology,
Lund University, PO Box 118,
SE-221 00, Lund, Sweden

Agneta Sjögren

Volvo Technological Development Corporation, Chalmers Science Park
SE-412 88, Gothenburg, Sweden

Nonlinear Adaptive Backstepping with Estimator Resetting using Multiple Observers

Jens Kalkkuhl
Jens.Kalkkuhl@daimlerchrysler.com

Jens Lüdemann
Jens.Ludemann@daimlerchrysler.com
DaimlerChrysler,
Research and Technology, Alt-Moabit 96a.
D-10559, Berlin, Germany

Tor Arne Johansen
tor.arne.johansen@itk.ntnu.no
Department of Engineering Cybernetics,
Norwegian University of Science and Technology,
Trondheim, Norway

ABS Control—A Design Model and Control Structure

Stefan Solyom
Stefan.Solyom@control.lth.se

Anders Rantzer
Anders.Rantzer@control.lth.se

Department of Automatic Control, Lund Institute of Technology,
Lund University, PO Box 118,
SE-221 00, Lund, Sweden

Controller Design for Hybrid Systems using Simultaneous D-stabilisation and its Application to Anti-lock Braking Systems

Kenneth J. Hunt
k.hunt@mech.gla.ac.uk

Yongji Wang
ywang@mech.gla.ac.uk

Michael Schinkel
m.schinkel@mech.gla.ac.uk

Tilmann Schmitt-Hartmann
tschmitt@mech.gla.ac.uk

Centre for Systems and Control,
University of Glasgow,
G128QQ, Glasgow, UK

Wheel slip control in ABS Brakes using Gain Scheduled Constrained LQR

Idar Petersen

SINTEF Electronics and Cybernetics,
N-7465 Trondheim, Norway.

Tor Arne Johansen   *see above*

Jens Kalkkuhl   *see above*

Jens Lüdemann   *see above*

Friction Tire/Road Modeling, Estimation and Optimal Braking Control

Carlos Canudas de Wit
canudas@lag.ensieg.inpg.fr

X. Claeys
claeys@lag.ensieg.inpg.fr

Laboratoire d'Automatique de Grenoble, UMR CNRS 5528,
ENSIEG-INPG, B.P. 46,
38 402 St. Martin d'Hères, France

P. Tsiotras
p.tsiotras@ae.gatech.edu

School of Aerospace Engineering,
Georgia Institute of Technolog,y
Atlanta, GA 30332-0150, USA

J. Yi
jgyi@me.berkeley.edu

R. Horowitz
horowitz@me.berkeley.edu

Department of Mechanical Engineering,
University of California,
Berkeley, CA 94720-1740, USA.

Nonlinear Observer Control of Internal Combustion Engines with EGR

Elbert Hendricks
eh@iau.dtu.dk

Institute for Automation,
Technical University of Denmark,
DK-2800, Kongens Lyngby, Denmark

Idle Speed Controller Synthesis Using an Assume-Guarantee Approach

Andrea Balluchi
balluchi@parades.rm.cnr.it

Alberto L. Sangiovanni–Vincentelli
alberto@parades.rm.cnr.it

PARADES
Via San Pantaleo, 66,
00186, Roma, Italy

Luca Benvenuti
luca.benvenuti@uniroma1.it

Dip. di Informatica e Sistemistica

Università di Roma "La Sapienza",
Via Eudossiana, 18
00184, Roma, Italy

Maria Domenica Di Benedetto
dibenede@ing.univaq.it

Dip. di Ingegneria Elettrica,
Università dell'Aquila,
Poggio di Roio,
67040 L'Aquila, Italy

Fault Diagnosis of Switched Nonlinear Dynamical Systems with Application to a Diesel Injection System

Dirk Förstner

Jan Lunze
Lunze@tu-harburg.de

Technical University Hamburg-Harburg,
Institute of Control Engineering,
Eissendorfer Strasse 40,
D-21071, Hamburg, Germany

Modelling the Dynamic Behavior of Three-Way Catalytic Converters during the Warm-up Phase

Giovanni Fiengo
gifiengo@unina.it

Stefania Santini
stsantin@unina.it

Dipartimento di Informatica e Sistemistica
Università di Napoli Federico II,
Via Claudio 21,
80125 Napoli, Italy

Luigi Glielmo
glielmo@unisannio.it

Università del Sannio, Facoltà di Ingegneria,
Corso Garibaldi 107,
82100 Benevento, Italy

Control of Gasoline Direct Injection Engines using Torque Feedback

Magnus Gäfvert
magnus@control.lth.se

Karl-Erik Årzén
karlerik@control.lth.se

Bo Bernhardsson
Department of Automatic Control, Lund Institute of Technology,
Lund University, P.O. Box 118,
SE-221 00, LUND Sweden

Lars Malcolm Pedersen,
United Technologies Research Center,
411 Silver Lane, MS 129-15 East Hartford,
Connecticut, 06108, USA

Closed-loop Combustion Control of HCCI Engines

Per Tunestål
Per.Tunestal@vok.lth.se

Jan-Ola Olsson
Jan-Ola.Olsson@vok.lth.se

Bengt Johansson
Bengt.Johansson@vok.lth.se

Div. Combustion Engines, Dept. Heat and Power Engineering,
Lund Institute of Technology,
Lund University, P.O. Box 118,
SE-221 00, LUND Sweden

Approximations of Maximal Controlled Safe Sets for Hybrid Systems

L. Berardi
berardi@eecs.berkeley.edu

Department of Electrical Engineering and Computer Sciences
University of California
Berkeley CA 94720, USA

E. De Santis
desantis@ing.univaq.it

Maria Domenica Di Benedetto   *see above*

G. Pola
pola@ing.univaq.it

Dipartimento di Ingegneria Elettrica,
Monteluco di Roio,
67040 L'Aquila, Italy

Hamiltonian Formulation of Bond Graphs

Goran Golo
G.Golo@math.utwente.nl

Peter C. Breedveld
Cornelis J. Drebbel Institute for Mechatronics,
Enschede, The Netherlands

Arjan van der Schaft
University of Twente,
Faculty of Mathematical Sciences, SSC group,
P.O. Box 217 7500 AE,
Enschede, The Netherlands

Bernhard M. Maschke
Université Claude Bernard Lyon-1,
Laboratoire d'Automatique et de Génie des Procédés,
France

Stability Analysis of Hybrid Systems—A Gear-box Application

S. Pettersson
stp@s2.chalmers.se

B. Lennartson
bl@s2.chalmers.se

Control Engineering Lab,
Chalmers University of Technology,
SE-412 96, Göteborg, Sweden

On the Existence and Uniqueness of Solution Trajectories to Hybrid Dynamical Systems

W. P. M. H. Heemels
w.p.m.h.heemels@tue.nl

Department of Electrical Engineering,
Eindhoven University of Technology, P.O. Box 513,
5600 MB Eindhoven, The Netherlands.

M. K. Çamlıbel
k.camlibel@math.rug.nl

Department of Mathematics
University of Groningen
P.O. Box 800,
9700 AV Groningen, The Netherlands.

Arjan van der Schaft   see above

J. M. Schumacher
j.m.schumacher@kub.nl

Department of Econometrics and Operations Research,
Tilburg Universit,y
P.O. Box 90153,
5000 LE Tilburg, The Netherlands.

Author Index

Çamlıbel, 392, 419, 428
Özgüner, 95, 144
Årzén, 426
Åström, 82, 95, 206

Abate, 243, 286
Achleitner, 217
Adamczyk, 286
Addison, 42, 53
Ahmed, 53
Ai-Poh Loh, 82
Aimard, 286
Aizerman, 419
Alleyne, 2, 19
Altpeter, 206
Alur, 419
Alvarez, 144, 206
Amin, 18
Anderson, 38
Andersson, 220
Annaswamy, 82
Aquino, 223
Arcak, 82
Asher, 372
Ashrafi, 95, 144
Aubin, 348, 419
Ayres, 38

Baba, 286
Bakker, 38, 95, 152, 206
Balakrishnan, 82
Ballard, 419
Balluchi, 230, 243, 348, 426
Barabanov, 206
Baroni, 260
Bass, 206
Basseville, 286
Bemporad, 144
Bender, 286
Benekohal, 53
Bengtsson, vi, 42, 423

Benninger, 215
Benvenuti, 230, 348, 426
Berardi, 336, 348, 427
Bernard, 206, 286
Bernhardsson, 427
Berzeri, 206
Birkett, 352, 372
Blanchini, 348
Bleile, 42, 53
Bliman, 153, 206
Blondel, 99
Bodenheimer, 243
Boer, 53
Bollerslev, 53
Bonald, 206
Borkar, 419
Botling, 53
Bradley, 286
Brandt, 286
Branicky, 419
Breedveld, 352, 372, 428
Brockett, 419
Brogliato, 419
Browne, 206
Buhlmann, 206
Burckhardt, 82, 144, 152, 206
Butts, 243
Byrnes, 99

Campbell, 38
Canudas-de-Wit, 82, 144, 147, 206, 425
Carathéodory, 418
Carnevale, 243
Cellina, 419
Chang, 82
Chaplin, 19
Chester, 286
Chevalier, 221
Cho, 206
Chow, 206

Claeys, 147, 206, 425
Clarke, 419
Clover, 206
Cohn, 286
Cole, 286
Console, 260
Cottle, 419
Courcoubetis, 419
Curtis, 223
Cussenot, 286
Cwikel, 348

Dahl, 206
Daiß, 82, 152, 206
Dalsmo, 352
Daniel, 111
Davis, 19
De Moor, 419
De Santis, 336, 348, 427
Delebecque, 206
Deur, 206
Dewilde, 53
de Kleer, 260
Di Benedetto, 230, 243, 336, 348
Di Nunzio, 243
Dix, 95, 144
Dixon, 38
Di Benedetto, 426
Donahue, v, 2
Dorea, 348
Doyle, 38
Drakunov, 95, 144
Dua, 144
Dubien, 286
Dupuis, 419

Eaves, 419
Egerstedt, 419
El Ghaoui, 206
Engleman, 19

Förstner, 426
Faria, 206
Farina, 348
Fekete, 224
Fiengo, 286, 426

Filippov, 404, 413, 415, 418, 419
Fons, 216
Frank, 260
Freeman, 144
Förstner, 246, 260

Garbow, 206
Garg, 53
Geering, 243
Gerdes, 53
Gerhardt, 215
Gertler, 260
Ghorbel, 206
Ghosh, 99
Gim, 206
Glielmo, 286, 426
Glover, 38
Godbole, 419
Golo, 352, 428
Goodrich, 42, 53
Goodwin, 82
Gopalasamy, 2, 19
Grizzle, 286
Gutman, 348
Gäfvert, 426

Hac, 19
Hairer, 372
Halbwachs, 419
Hamilton, 53
Hamscher, 260
Harned, 206
Hassan, 19, 38
Hattwig, 95
Haverkamp, 53
Hawthorn, 286
Hedrick, v, 2, 19, 42, 53
Heemels, 392, 419, 428
Heiming, 260
Heintz, 216
Hendricks, 215, 221, 243, 425
Hennet, 348
Henzinger, 419
Herz, 286
Hespanha, 82
Hesslow, 53

Hill, 82
Hillstrom, 206
Hitz, 42, 53
Ho, 260, 419
Horne, 206
Horowitz, 82, 144, 147, 206, 425
Hrovat, 19, 243
Hunt, 424
Hägglund, 95

Iijima, 42, 53
Imura, 418, 419
Ingimundarson, 95
Ioannou, 42, 53
Isermann, 260

Jackson, 286
Jacob, 286
Janković, 82
Jeffrey, 286
Jing Sun, 95
Jingang Yi, 144
Johansen, 60, 82, 95, 126, 144, 424
Johansson, vi, 42, 53, 321, 419, 423, 427
John, 286
Johnston, 206
Jones, 286
Jonner, 95

Kalkkuhl, 60, 82, 95, 126, 144, 424
Kaminer, 144
Kandylas, 286
Kanellakopoulos, 82
Karnopp, 372
Kelley, 206
Kevorkian, 286
Khalil, 144, 286
Khargonekar, 243
Kharitonov, 98
Kiela, 286
Kiencke, 82, 95, 152, 206
Kimura, 348
Kjergaard, 243

Kokotović, 82, 206, 286
Koltsakis, 286
Kopp, 206
Kostantinidis, 286
Koutsopoulos, 53
Koziol, 53
Krstić, 82
Kuge, 42, 53

Lötstedt, 419
Lafortune, 260
Lam, 53
Lamb, 352, 372
Lamperti, 260
Ledyaev, 419
Lee, 224
Leenaerts, 419
Lemke, 419
Lemmon, 419
Lennartson, 428
Leutzbach, 53
Levenspiel, 286
Li, 19, 286
Lidner, 95
Liederman, 286
Lim, 260
Lin, 286
Lischinsky, 206
Liu, 19, 53, 206
Longchamp, 206
Lunze, 260, 426
Lygeros, 419
Lynch, 419
Lüdemann, 60, 82, 95, 126, 144, 424

Maciua, 53
Maier, 95
Maisch, 95
Maschke, 352, 372, 428
Maurice, 206
Mayne, 82
Meyer, 206
Miconi, 243
Middleton, 82
Milano, 286

Minyue Fun, 82
Mitter, 419
Monteiro Marques, 419
Montreuil, 286
Moore, 206
Morari, 144
Morgan, 286
Morse, 82
Moschetti, 243
Moyer, 206
Muller, 219
Müller, 95

Nagurney, 419
Narendra, 82, 286
Netto, 82
Newcomb, 38
Nicollin, 419
Nielsen, 95, 243
Nijmeijer, 419
Nikoukhah, 206
Nikravesh, 206
Nixdorf, 260
Nunan, 286
Nyborg, 38, 206

O'Malley, 206
O'Reilly, 286
Ohsawa, 286
Olin, 219
Olivero, 419
Olsson, 206, 321, 427
Onder, 243
Orszag, 286
Ortega, 82, 206
Osgood, 206
Osorio, 2, 19

Pacejka, 38, 82, 95, 151, 206
Pakkala, 224
Pang, 419
Park, 243
Parthasarathy, 286
Pasterkamp, 206
Paynter, 372
Pearce, 19

Perelson, 352, 372
Persson, 53
Petersen, 126, 144, 424
Pettersson, 428
Pettit, 95
Philips, 260
Pinello, 243, 348
Pistikopoulos, 144
Ploemen, 286
Pogliano, 260
Pogromsky, 419
Pola, 336, 348
Polyak, 111
Pottinger, 206
Pozzi, 243
Preisig, 260
Prestl, 42, 53
Pshenichnyi, 111
Pyatnitskii, 419

Rajamani, 19, 206
Ramberg, 206
Ramkumar, 260
Ranney, 42, 53
Rantzer, vii, 86, 424
Rascoal, 144
Ribbens, 243
Rinaldi, 286
Rizzo, 332
Rizzoni, 19
Roberta, 286
Roberts, 286
Robinson, 111
Rohde, 206
Rothschild, 286

Saito, 223
Sampath, 260
Sangiovanni-Vincentelli, 230, 243, 348, 426
Sannuti, 206
Santini, 286, 426
Sargin, 206
Sastry, 419
Sauer, 53
Scharpf, 206

Author Index

Schiesser, 286
Schiller, 260
Schinkel, 424
Schmitt-Hartmann, 424
Schröder, 260
Schumacher, 82, 392, 419, 428
Schwarz, 144
Schweich, 286
Segala, 419
Segel, 286
Sell, 286
Sengupta, 260
Sepulchre, 82
Serra, 286
Sharp, 19, 38, 82, 151, 206
Shim, 243
Shimoyama, 53
Sifakis, 419
Sigi, 95
Simpson, 243
Sinkevitch, 286
Sinnamohideen, 260
Sivashankar, 243
Sjögren, vi, 42, 423
Skantze, 82
Slotine, 19
Slupphaug, 144
Smith, 24, 38, 423
Solyom, vii, 86, 95, 424
Sontag, 82, 419
Sorine, 206, 286
Stamatelos, 286
Steinle, 53
Stern, 419
Stone, 419
Su, 286
Subbotin, 419
Sugiura, 286
Sun, 206, 243
Swaroop, 53
Szymanski, 206

Taylor, 286
Teneketzis, 260
Thompson, 19
Tomizuka, 2, 19

Treiterer, 53
Tschernoster, 53
Tsiotras, 82, 144, 147, 206, 425
Tunestål, 321, 427

Utkin, 419

Vaandrager, 419
van Bokhoven, 419
van de Molengraft, 286
van der Schaft, 82, 372, 392, 418, 428
van Zanten, 157
Vandenberghe, 419
Vandewalle, 419
Verhaegen, 53
Vesterholm, 243
Villa, 243
Voltz, 286
Vongpanitlerd, 38

Walter, 206
Wang, 24, 38, 286, 423, 424
Weiland, 419
Weinberger, 419
Wellstead, 95
Whicker, 206
Willems, 53, 419
Williams, 38, 286
Wittenmark, 82
Wolenski, 419
Wong, 206
Wong-Toi, 243
Wright, 38

Xiang, 82

Yager, 206
Yamamura, 53
Yao, 286
Yavari, 206
Yi, 147, 206, 425
Yong Liu, 82, 95
Youla, 99
Yovine, 419
Yurkovich, 243

Zanella, 260
Zhang, 419
Zhivoglyadov, 82
Zhou, 38

Subject Index

T-controlled safe set, 338
S-procedure, 381
λ-sensor, 264
$\mathbb{R}$ element, 353
$\mathbb{C}$ element, 353
$\mathbb{SE}$ element, 353
$\mathbb{SF}$ element, 353
NO_X, 322, 324
1-junction, 355

ABS, anti-lock brake system, 126, 148
ABS, anti-lock braking system, 86, 98
ACC, adaptive cruise control, vi, 42
accumulation point, 417, 418
active area/volume ratio, 286
active suspension, 148
activities, 394
adaptive cruise control (ACC), vi, 42
adiabatic MVEM (AMVEM), 217
AFR, air-fuel ratio, 278, 279
air-fuel ratio (AFR), 278, 279
anti-dive, 27
anti-lock brake system (ABS), 126, 148
anti-lock braking system (ABS), vii, 86, 98
anti-squat, 27
approximation
    outer, 341
arousal behavior, 51
automaton nonzeno hybrid, 397

backstepping, 62
BDC, bottom dead center, 322
bond graph, 352
bond graph
    1-junction, 355
    generalised, 353
    geometric formulation, 355
    gyrator, 355
    junction structure, 352, 353
    transformer, 355
bond graph models, 352
bottom dead center (BDC), 322

CA50, 325–332
CA50, crank angle of 50 % heat release, 325
California PATH program, v, 42, 206
CAP, cumulative absorbed power, 14
Carathéodory solution, 416, 418
Carathéodory solution extended, 404
catalytic converter, 264
catalytic converter
    afterburner catalyst, 265
    catalytic bed (matrix), 265
    monolithic type, 265
    three-way, 264
charge stratification, 323
CI, 322, 324
CI, compression ignition, 322
CLF, control Lyapunov function, 62
cold start, 264
complementarity system, 395
compositionality, 394
compression ratio, 322–324, 332
compression stroke, 322, 323, 325
consistency-based diagnosis, 252
control
    engine speed, 235
    idle-speed, 234, 348
    manifold, 235
    model predictive, 9

435

suspension, 24
torque, 235
control Lyapunov function (CLF), 62
controlled invariant set, 337
controlled safe set, 337
controller
    force tracking, 3
    passive suspension, v, 2
convective heat transfer coefficient, 286
Coulomb friction, 156
crank angle, 322–326, 332
cumulative absorbed power (CAP), 14
cylinder pressure, 324–327, 332

Dahl model, 156
determinism, 393, 401
Diesel injection, 245
diesel injection, ix
differential inclusion, 395
differential index, 364
Dirac structure, 354
direct injection, spark ignition (DISI), 291
DISI, direct injection, spark ignition, 291
dive and squat, 24

ECU, engine control units, viii, 211
effort, 354
EGR, 323
EGR, exhaust gas recirculation, 214, 290
electronic stability program (ESP), 126
emission regulations, 264
EMS, engine management system, 292
energy-conserving physical systems, 352
engine control units (ECU), viii, 211

engine management system (EMS), 292
equivalence ratio, 323
ESP, electronic stability program, 126
Euler solution, 399
event
    conditions, 396
Exhaust Gas Recirculation, 323
exhaust gas recirculation (EGR), 214, 290
exhaust stroke, 322, 323
expanding factor $\mu(S(T))$, 338
expansion stroke, 322

fault diagnosis, ix, 245
feasibility problem (FP), 111
Filippov solution, 416, 418
flow, 354
force tracking controller (FTC), 3, 14
forward solution, 416
four-stroke, 322
FP, feasibility problem, 111
friction
    Coulomb, 156
    LuGre model, 148, 155
    observer, 175, 177
FTC, force tracking controller, 3, 14

gain scheduling, 134
GARCH, generalized autoregressive conditional heteroskedasticity, vi, 42, 51
gas direct injection (GDI), 230
gasoline direct injection (GDI), 290
GDI, gas direct injection, 230
GDI, gasoline direct injection, 290
gearbox control, 385
generalized autoregressive conditional heteroskedasticity (GARCH), vi, 42, 51
guard, 394
gyrator, 355

Subject Index 437

half-car model, 25
Hamiltonian complementarity
    system, 398
HCCI, 321–325, 327–329, 332
heat release, 324–327, 332
heave, pitch, roll (HPR), 12
high mobility multipurpose
    wheeled vehicle (HMMWV),
    v
high-mobility multi-purpose
    wheeled vehicle (HMMWV),
    3
HMMWV, 2
HMMWV, high mobility multipur-
    pose wheeled vehicle, v
HMMWV, high-mobility multi-
    purpose wheeled vehicle, 3
HPR, heave, pitch and roll, 11
HRF, human response filter, 14
human response filter (HRF), 14
hybrid automaton
    activities, 393
    compositionality, 393
    determinism, 393
    enabling conditions, 393
    execution, 397
    forward solution, 398
    generalized, 403
    guard, 394
    invariants, 393
    jump function, 394
    live-lock, 413
    location invariance, 400
    locations, 393
    model, 393
    non-blocking, 399
    nonzeno, 397
    parallel composition, 393
    run, 397
    synchronization labels, 394
    transition relations, 393
    transitions, 393
    well-posedness, 400
    well-posedness problem, 413
hydrogen, 323, 328, 330
hydroplaning

dynamic, 172
limit, 172
viscous, 172

idle speed control, 234
idle-speed control, 291, 348
IGA, ignition-advance angle, 293
ignition, 322–325, 329, 332
ignition-advance angle (IGA), 293
image representation of Dirac
    structures, 355
IMVEM observer, 224
IMVEM, isothermal manifold
    filling MVEM, 214
inclusion
    differential, 395
index
    differential, 364
    of system, 364
index
    local differential, 364
index one system, 366
index two system, 365
instability, 329
intake manifold, 231
intake manifold
    pressure, 232
intake stroke, 322
iso-octane, 323, 327, 328, 331
isothermal manifold filling
    MVEM (IMVEM), 214

jump function, 394
junction structure, 352–354

kernel representation of Dirac
    structures, 355

lambda-sensor, 264
LCS, linear complementarity
    system, 406
LFT, linear fractional transforma-
    tion model, 33
linear complementarity system
    (LCS), 406
linear fractional transformation
    (LFT), 33

linear matrix inequalities (LMI), 380
linear matrix inequality, LMI, xi, 374
linear passive complementarity systems (LPCS), 408
linear quadratic regulator (LQR), 9
linear quadratic regulator (LQR), 126
live-lock, 413
LMI, linear matrix inequalities, 380
LMI, linear matrix inequality, xi, 374
location, 393, 394
location
    invariance, 400
LPCS, linear passive complementarity systems, 408
LQR, linear quadratic regulator, 9, 126
LuGre model, viii, 148, 155
Lunze, 246

MAF, mass air flow, 216
manifold control, 235
mass air flow (MAF), 216
mean value engine models (MVEM), viii, 211
Minkowski functional, 342
MMO, multiple model observer, 61, 71
model
    half-car, 25, 35
    quarter-car, 4, 126
model depth, 248
model predictive control (MPC), v, 2, 8, 9, 16, 314
MPC preview controller, 14
MPC, model predictive control, 2, 14, 16, 314
MPC, model predictive control , 8
MTC, fresh air throttle control signal, 293

MTC, fresh air throttle position (MTCPOS), 293
MTCPOS, fresh air throttle position, 293
multiple model observer (MMO), 61, 71
MVEM, mean value engine models, viii, 211

n-heptane, 323, 327, 328, 331
natural gas, 323, 328, 330
NCLOP, nonlinear constrained optimisation problem, 112
network representation, 352
nonlinear constrained optimisation problem (NLCOP), 112
nonzeno hybrid automaton, 397

observer design, 177

Pacejka model, 151, 190
parallel composition, 394
passive suspension controller, v, 2
passivity, 24
passivity
    analysis, 31
PATH program, v, 42, 201
PFI, port fuel injection, 290
PHS, port Hamiltonian system, 358
pneumatic state estimator (PSE), 218
pneumatic thermal state estimator (PTSE), 218
port Hamiltonian system (PHS), 358
port fuel injection (PFI), 290
power continuity, 354
power discontinuous elements, 353
PSE, pneumatic state estimator, 218
PTSE, pneumatic thermal state estimator, 218

quadratic form, 384

Subject Index

quarter-car model, 4, 77, 87, 126, 195

residual gas, 323, 329, 330
RFS, Richmond Field Station, 21
Richmond Field Station (RFS), 21

sampling solution, 399
set
    T-controlled safe, 338
    controlled invariant, 337
    controlled safe, 337
SI, 322, 324
SI, spark ignition, 322
simultaneous stabilisation problem (SSP), 98
skyhook, 9
skyhook
    damping, 9
skyhook damping, 2
slip, 77, 127, 132, 134
SLP, successive linear programming, 113
SOI, start of injection, 293
solution
    Carathéodory, 416
    Filippov, 416
    forward, 416
    Zeno hybrid, 416
spark
    advance, 236
    advance angle, 346
specific heat capacity, 286
specific reaction rate, 286
speed control
    engine, 235
    idle, 234
SQP, successive quadratic programming, 113
SSP, simultaneous stabilisation problem, 98
SSSP, strong simultaneous stabilisation problem, 98
stability, 329
start of injection (SOI), 293
stop-and-go controller, 43

Stribeck effect, 156, 171
strong simultaneous stabilisation problem (SSP), 98
successive linear programming (SLP), 113
successive quadratic programming (SQP), 113
suspension
    active, 32
    control, 24
suspension control, 24
synchronization labels, 394
system
    complementarity, 395
    Hamiltonian complementarity, 398
    variable-structure, 395

TDC, top dead center, 322–324
THC, total hydrocarbons, 280
thermal state estimator (TSE), 218
three-way catalytic converter (TWC), 264
top dead center (TDC), 322
torque control, 235
total hydrocarbons (THC), 280
traction control system, 148
trailing-arm model, 36
transformer, 355, 356
transition relations, 394
TSE, thermal state estimator, 218
TWC, three-way catalytic converter, 264
two-stroke, 322

uniqueness
    in Filippov sense, 415

variable-structure system, 395, 398
void fraction, 286

well-posedness
    global, 400
    notions, 399
    of PHS, 358

problem, 400, 413
wheel-slip control, 77, 132, 134
wheel-slip dynamics, 127

Zeno
    behavior, 401, 403
    free, 408
    hybrid solution, 416
    time point, 403
    times, 403